Climate Change, Forests and REDD

A search for new methods for dealing with climate change led to the identification of forest maintenance as a potential policy option that could cost-effectively reduce greenhouse gas emissions, with the development of measures for Reducing Emissions from Deforestation and Forest Degradation (REDD). This book explores how an analysis of past forest governance patterns, from the global through to the local level, can help us build institutions which more effectively deal with forests within the climate change regime. The book assesses the options for reducing emissions from deforestation in developing countries under the international climate regime, as well as the incentives flowing from them at the national and sub-national level, and examines how these policy levers change human behaviour and interface with the drivers and pressures of land use change in tropical forests. The book considers the trade-offs between certain forestry-related policies within the current climate regime and the larger goal of sustainable forestry.

Based on an assessment of existing multi-level institutional forestry arrangements, the book questions how policy frameworks can be better designed in order to effectively and equitably govern the challenges of deforestation and land degradation under the global climate change regime. This book will be of particular interest to students and scholars of Law and Environmental Studies.

Joyeeta Gupta is professor of climate change policy and law at the VU University Amsterdam and of water and environmental law and policy at the UNESCO-IHE Institute for Water Education in Delft. She is editor-in-chief of *International Environmental Agreements: Politics, Law and Economics*. She was and continues to be lead author in the Intergovernmental Panel on Climate Change which shared the 2007 Nobel Peace Prize with Al Gore and of the Millennium Ecosystem Assessment which won the Zayed Second Prize. She has published extensively and is on the scientific steering committees of many different international programmes including the Global Water Systems Project and Earth System Governance.

Nicolien van der Grijp is a senior researcher at the Institute for Environmental Studies (IVM) at the VU University Amsterdam. Her research interests are related to environmental law and policy at the international, EU and national level. She specialises in applying social scientific research methods to investigate how legal instruments are implemented in practice. She has extensive experience in doing contract research for international organisations, the European Commission and the Dutch government. She is currently working on an ESPON project about governance arrangements for regional seas (ESaTDOR) and an EU project on marine litter (CLEANSEA).

Onno Kuik is senior researcher in environmental economics at the Institute for Environmental Studies (IVM) at VU University Amsterdam. His research interests are in climate change, energy, agriculture and CGE modelling. He has extensive experience as project leader of smaller and larger research projects for a wide variety of clients, including the European Commission, FAO, OECD, Dutch Government, European Parliament and the European Environment Agency. He has written more than 40 peer-reviewed journal articles and book chapters and produced over 100 research reports.

Routledge Research in Environmental Law
Available titles in this series:

International Environmental Law and the Conservation of Coral Reefs
Edward J. Goodwin

Environmental Governance in Europe and Asia
A Comparative Study of Institutional and Legislative Frameworks
Jona Razzaque

Climate Change, Forests and REDD
Lessons for Institutional Design
Joyeeta Gupta, Nicolien van der Grijp, Onno Kuik

Climate Change, Forests and REDD

Lessons for Institutional Design

Edited by
Joyeeta Gupta,
Nicolien van der Grijp and Onno Kuik

LONDON AND NEW YORK

First published 2013
by Routledge
2 Park Square, Milton Park, Abingdon, Oxon OX14 4RN

Simultaneously published in the USA and Canada
by Routledge
711 Third Avenue, New York, NY 10017

Routledge is an imprint of the Taylor & Francis Group, an informa business

© 2013 editorial matter and selection, Joyeeta Gupta, Nicolien van der Grijp and Onno Kuik; individual chapters, the contributors.

The right of Joyeeta Gupta, Nicolien van der Grijp and Onno Kuik to be identified as the editors of this work has been asserted by them in accordance with sections 77 and 78 of the Copyright, Designs and Patents Act 1988.

All rights reserved. No part of this book may be reprinted or reproduced or utilised in any form or by any electronic, mechanical, or other means, now known or hereafter invented, including photocopying and recording, or in any information storage or retrieval system, without permission in writing from the publishers.

Trademark notice: Product or corporate names may be trademarks or registered trademarks, and are used only for identification and explanation without intent to infringe.

British Library Cataloguing in Publication Data
A catalogue record for this book is available from the British Library

Library of Congress Cataloging-in-Publication Data
Climate change, forests, and REDD : lessons for institutional design / edited by Joyeeta Gupta, Nicolien van der Grijp, and Onno Kuik.
  p. cm.
Includes bibliographical references.
ISBN 978-0-415-52699-9 (hardback) – ISBN 978-0-203-07722-1 (e-book)
1. Carbon sequestration. 2. Climate change mitigation–International cooperation. 3. Deforestation–Environmental aspects. 4. Forest degradation. I. Gupta, Joyeeta, 1964- II. Grijp, Nicolien van der. III. Kuik, Onno, 1955- IV. Title: Climate change, forests, and reducing emissions from deforestation and forest degradation.
SD387.C37C585 2013
577.27–dc23   2012023857

ISBN 978-0-415-52699-9 (hbk)
ISBN 978-0-203-07722-1 (ebk)

Typeset in Garamond
by Cenveo Publisher Services

Printed and bound in the United States of America by Publishers Graphics, LLC on sustainably sourced paper.

Contents

List of tables	*x*
List of figures	*x*
List of boxes	*xii*
Notes on contributors	*xiii*
Acknowledgements	*xviii*
Abbreviations	*xxi*

1. **Climate change and forests: from the Noordwijk Declaration to REDD**   1

    1.1 *Introduction* 1
    1.2 *Climate change* 3
        1.2.1 The physical problem 3
        1.2.2 The governance process 3
        1.2.3 The key political challenges 4
    1.3 *Forests* 7
        1.3.1 Current status 7
        1.3.2 Deforestation and degradation 9
        1.3.3 The governance process 9
        1.3.4 The key political challenges 12
    1.4 *Climate change and forests* 13
    1.5 *The research questions and the analytical framework* 16
    1.6 *Inferences* 20

2. **The forest transition, the drivers of deforestation and governance approaches**   25

    2.1 *Introduction* 25
    2.2 *The forest transition* 25
    2.3 *The drivers of deforestation* 29

2.3.1 Generic drivers and forest transitions 29
2.3.2 Drivers of deforestation in different regions 30
2.4 *Instruments of forest governance 33*
2.4.1 Introduction 33
2.4.2 Forest transitions, drivers and forest policy 33
2.4.3 Classifying governance instruments 34
2.4.4 Regulatory instruments 34
2.4.5 Economic and market instruments 38
2.4.6 Suasive, information and research instruments 40
2.4.7 Management measures 40
2.4.8 Forest instruments and drivers 41
2.5 *Inferences 44*

3. Global forest governance   52

3.1 *Introduction 52*
3.2 *Institutions 52*
3.2.1 A brief history 52
3.2.2 Global institutions with a forest-focused mandate 54
3.2.3 Global institutions with a forest-related mandate and/or an indirect impact on forest services 57
3.2.4 Global governance institutions and ecosystem services 60
3.2.5 Regional institutions with a forest-focused mandate 61
3.2.6 Regional institutions with a forest-related mandate and/or an indirect impact on forest services 62
3.2.7 Extraterritorial impacts of national governance 63
3.3 *Key principles and concepts in international forest governance 63*
3.3.1 Key principles 64
3.3.2 Key concepts 65
3.4 *Instruments of international forest governance 65*
3.4.1 Regulatory instruments 66
3.4.2 Economic instruments 66
3.4.3 Suasive instruments 68
3.5 *Inferences 70*

4. The emergence of REDD on the global policy agenda   77

4.1 *Introduction 77*
4.2 *Forests under the UN Climate Convention and the Kyoto Protocol 77*
4.2.1 Early days: the forest-climate pre-Kyoto debate 77
4.2.2 Forests in the Kyoto Protocol – 'integrating forest commitments into the climate regime' 78

Contents   vii

- 4.3   *The emergence of REDD* 80
    - 4.3.1   REDD in the UNFCCC negotiations   80
    - 4.3.2   REDD developments outside the UNFCCC   81
- 4.4   *Key challenges for REDD at the international level: designing an effective, robust mechanism*   84
    - 4.4.1   The right scale for REDD   84
    - 4.4.2   Reference levels   84
    - 4.4.3   Financing REDD   85
    - 4.4.4   Monitoring, reporting and verification   86
    - 4.4.5   Permanence, additionality and leakage   86
    - 4.4.6   Safeguards   87
- 4.5   *Key challenges for REDD at the domestic level: implementation and benefit-sharing*   88
    - 4.5.1   The impact of REDD to date   88
    - 4.5.2   Challenges of good governance, tenure and internal benefit-sharing   89
    - 4.5.3   Risks and implications of commodifying forest carbon   91
- 4.6   *Inferences*   92

5. Case study: Vietnam   99

- 5.1   *Introduction*   99
- 5.2   *Driving forces of deforestation and forest degradation*   100
- 5.3   *The forest policy context*   101
    - 5.3.1   The organizational framework   101
    - 5.3.2   The evolution of forest policy   103
    - 5.3.3   The influence of international treaties and bodies   105
- 5.4   *Key forest policy instruments and their analysis*   105
- 5.5   *Implications for REDD*   109
- 5.6   *Inferences*   112

6. Case study: Indonesia   121

- 6.1   *Introduction*   121
- 6.2   *Driving forces of deforestation and forest degradation*   122
- 6.3   *The forest policy context*   124
    - 6.3.1   The organizational framework   124
    - 6.3.2   The evolution of forest policy   125
    - 6.3.3   The influence of international treaties and bodies   125
- 6.4   *Key forest policy instruments and their analysis*   126
- 6.5   *Implications for REDD*   133

6.6 Inferences 135

7 Case study: Cameroon 143

7.1 Introduction 143
7.2 Driving forces of deforestation and forest degradation 144
7.3 Key policies and instruments 146
   7.3.1 The organizational framework 146
   7.3.2 The evolution of forest policy 147
   7.3.3 The influence of international treaties and bodies 147
7.4 Key forest policy instruments and their analysis 148
7.5 Implications for REDD 154
7.6 Inferences 157

8. Case study: Peru 163

8.1 Introduction 163
8.2 Driving forces of deforestation and forest degradation 164
8.3 The forest policy context 167
   8.3.1 The organizational framework 167
   8.3.2 The evolution of forest policy 168
   8.3.3 The influence of international treaties and bodies 169
8.4 Key forest policy instruments and their analysis 169
8.5 Implications for REDD 176
8.6 Inferences 179

9. Comparative analysis of Vietnam, Indonesia, Cameroon and Peru 185

9.1 Introduction 185
9.2 Driving forces of deforestation and forest degradation 185
   9.2.1 Direct drivers of deforestation and forest degradation 185
   9.2.2 Underlying drivers of deforestation and forest degradation 187
9.3 Forest policy instruments assessed 187
9.4 Equity issues: impact on access and allocation 194
9.5 Implications for REDD 198
9.6 Conclusion 201

10. REDD policies, global food, fibre and timber markets, and 'leakage' 207

10.1 Introduction 207

    10.2 *Methods and data 209*
    10.3 *A scenario of future deforestation 210*
        10.3.1 Introduction 210
        10.3.2 Growth of population and income 211
        10.3.3 Demand for food and timber 211
        10.3.4 Demand for biofuels 211
        10.3.5 Future demand and supply of land 212
        10.3.6 Baseline scenario of land-use change and deforestation 213
    10.4 *Economic effects of REDD-induced forest conservation 214*
        10.4.1 Introduction 214
        10.4.2 A forest conservation policy scenario 215
        10.4.3 Global food and timber markets 215
        10.4.4 Economy-wide effects and environmental benefits 216
        10.4.5 Inferences 219
    10.5 *Leakage 219*
        10.5.1 Introduction 219
        10.5.2 Leakage simulations 221
        10.5.3 Inferences 224
    10.6 *Inferences 224*

11. The future of forests   229

    11.1 *Global forest governance: a twenty-first-century myth of Sisyphus? 229*
    11.2 *'Glocal' forest governance 230*
        11.2.1 Evolutionary phases in forest governance 230
        11.2.2 The politics of scale: Should there
                be 'glocal' forest governance? 231
        11.2.3 Current global forest governance 238
    11.3 *National forest governance 241*
        11.3.1 The forest transition 241
        11.3.2 Forest transitions, drivers and policies 243
        11.3.3 National forest policies 244
    11.4 *REDD revisited 245*
        11.4.1 Practical options for implementing REDD in countries 245
        11.4.2 Buying time or a REDD herring 247
        11.4.3 A North–South analysis 250
        11.4.4 Going beyond REDD: the challenge of
                mainstreaming forests 252
    11.5 *Conclusions 253*

Index   259

Tables, figures and boxes

Tables

1.1	The ecosystem services of forests	8
2.1	Proximate and underlying causes of deforestation and forest degradation	30
2.2	Classifying policy instruments on forest management	35
2.3	Forest policy instruments, measures and drivers	42
3.1	Instruments in forest-related agreements	67
5.1	Drivers of deforestation and forest degradation in Vietnam and Dak Lak province	101
6.1	Drivers of deforestation and forest degradation in Indonesia and Jambi Province	123
7.1	Drivers of deforestation and forest degradation in Cameroon and the South-West region	145
8.1	Drivers of deforestation and forest degradation in Peru and the Ucayali region	165
9.1	Proximate drivers of deforestation and degradation in the case study countries	186
9.2	Some of the instruments used in different countries	191
10.1	Estimates of the costs of avoiding emissions through deforestation in 2030 (USD/tCO_2)	219
10.2	Alternative displacement metrics for unilateral forest conservation policies	223
10.3	'People-based' leakage: employment changes of unskilled labour in different sectors (%)	224
11.1	Reasons for scaling up forests	235
11.2	Reasons for scaling down forests	236

Figures

1.1	Dependence of the 'deforestation rate' for Indonesia in three time periods on the operational forest definition used; negative deforestation rates indicate increase in forest	10

1.2	The analytical framework	18
1.3	The case study countries and regions	19
2.1	Stylized representations of the demographic, economic and forest transitions	26
2.2	Environmental Kuznets Curves for deforestation and forest cover	28
2.3	Drivers and the forest transition: an ideal typical representation	31
2.4	Halting deforestation (a) or moving the forest transition curve upwards (b)	34
3.1	Institutions in international forest governance	53
3.2	Global forest governance arrangements and ecosystem services	61
3.3	Global forest governance and its instruments	70
4.1	Pledged, approved and disbursed finance of key multilateral and bilateral REDD funds	86
5.1	Map of Vietnam and the case study area of Dak Lak province	100
5.2	Applying the analytical framework to forest and REDD policy in Vietnam	111
6.1	Map of Indonesia and case study area of Jambi province	122
6.2	Areas under different land-use-right concessions from the Ministry of Forestry (in hectares)	130
6.3	Applying the analytical framework to forest and REDD policy in Indonesia	136
7.1	Map of Cameroon and the case study area of the South-West region	144
7.2	Applying the analytical framework to forest and REDD policy in Cameroon	156
8.1	Map of Peru and the case study area of Ucayali region	164
8.2	Applying the analytical framework to forest and REDD policy in Peru	178
10.1	The Von Thünen model of spatial allocation of land (adapted from Angelsen 2007)	208
10.2	Deforestation and degradation in four tropical forest regions over the period 2010–30	215
10.3	Change in Gross Domestic Product (%) with respect to baseline due to forest conservation policy	218
11.1	The timeline of the phases of forest governance through human history	231
11.2	Scaling up and down forests: a stylized representation	237
11.3	Case study countries in different stages of the forest transition curve: a stylistic representation	242
11.4	Policies compatible with different stages of the forest transition curve	243

Boxes

1.1	Defining deforestation	15
2.1	An Environmental Kuznets Curve for forest cover	27
2.2	Protected Areas	36
2.3	Payment for ecosystem services (PES)	37
2.4	Debt-for-nature swaps	39
2.5	Forest certification	40
2.6	Community-based forest management	41
3.1	Why is there no global forest convention?	55
3.2	Certification bodies	56
3.3	Policies on indigenous peoples	59
4.1	The World Bank and carbon financing	82
4.2	Indigenous peoples and REDD	89
9.1	Land grabbing worldwide	195

Contributors

Editors

Joyeeta Gupta is professor of environment and development in the global south at the University of Amsterdam and professor at the UNESCO-IHE Institute for Water Education in Delft. Previously, she was professor of climate change policy and law at the VU University of Amsterdam. She is editor-in-chief of *International Environmental Agreements: Politics, Law and Economics* (IF 1.128) and is on the editorial board of journals like *Carbon and Law Review*, *International Journal on Sustainable Development*, *Environmental Science and Policy*, *Current Opinion in Environmental Sustainability*, *Catalan Environmental Law Journal*, *Review of European Community and International Environmental Law* and the new *International Journal of Water Governance*. She was and continues to be lead author in the Intergovernmental Panel on Climate Change which shared the 2007 Nobel Peace Prize with Al Gore and of the Millennium Ecosystem Assessment which won the Zaved Second Prize. She has published extensively. She is on the scientific steering committees of many different international programmes including the Global Water Systems Project and Earth System Governance.

Dr Nicolien van der Grijp is a senior researcher at the Institute for Environmental Studies (IVM) at the VU University Amsterdam. Her research interests are related to environmental law and policy at the international, EU and national level. She is specialised in applying social scientific research methods to investigate how legal instruments are implemented in practice. She has extensive experience in doing contract research for international organisations, the European Commission and the Dutch Government. In 2008, she finalised her dissertation on the regulation of pesticide risks, in which she focused on state and non-state actor approaches to risk management from the perspective of legal pluralism. In 2010, she published a book, together with Professor Joyeeta Gupta, at Cambridge University Press, titled *Mainstreaming Climate Change in Development Cooperation: Theory, Practice and Implications for the European Union*. She is currently working on an ESPON project about governance arrangements for regional seas (ESaTDOR) and an EU project on marine litter (CLEANSEA).

Dr Onno Kuik is senior researcher in environmental economics at the Institute for Environmental Studies (IVM) at VU University Amsterdam. His research interests are in climate change, energy, agriculture and CGE modeling. He has extensive experience as project leader of smaller and larger research projects for a wide variety of clients, including the European Commission, FAO, OECD, Dutch Government, European Parliament, and the European Environment Agency. He has written more than 40 peer-reviewed journal articles and book chapters and produced over 100 research reports.

Authors

Dr Fahmuddin Agus is a researcher at the Soil Conservation Division of the Indonesian Soil Research Institute, Ministry of Agriculture. His current research is mainly on land-based greenhouse gas emissions.

Mairon Bastos Lima is a PhD researcher at the Department of Environmental Policy Analysis, Institute for Environmental Studies, VU University Amsterdam. He specializes in the political and institutional dimensions of land-use policies, agri-energy governance and North–South issues.

Channah Deborah Betgen graduated from the Environment and Resource Management MSc programme at the Institute for Environmental Studies at VU University Amsterdam in 2011 and from the Earth Sciences MSc programme at the Faculty of Earth and Life Sciences at VU University in 2012. She specialized in Water and Climate in both MSc programmes, and has been an intern at the Netherlands Environmental Assessment Agency and the Institute for Environmental Studies.

Léa Bigot graduated from the Institute of Political Studies of Bordeaux in 2007 and from the Environment and Resource Management MSc Programme at the Institute for Environmental Studies at VU University Amsterdam in 2011. She worked and did research in the field of natural disaster management, international cooperation and water politics in Mauritania, Vietnam and Germany. She is presently working at MicroEnergy International in Berlin.

Felix von Blücher graduated from the Environment and Resource Management MSc Programme at the Institute for Environmental Studies at VU University Amsterdam in 2011 and holds a BA in International Relations at the Technical University of Dresden. In 2010, he co-authored the study 'Welche Energiezukunft ist möglich?' comparing four low energy scenarios for the German energy sector until 2050. He is currently working as a consultant at the Division on Investment and Enterprise at United Nations Conference on Trade and Development to support the establishment of regulatory frameworks conducive to attract foreign direct investments into low carbon energy technologies in economies in transition and developing countries.

Constanze Haug is a PhD researcher at the Department of Environmental Policy Analysis at the Institute for Environmental Studies (IVM) at VU University Amsterdam. She specializes in international and European climate and forest policy as well as stakeholder learning.

Jonathan Y. Kuiper holds a Master's in Environment and Resource Management from the Institute for Environmental Studies at VU University Amsterdam and a BA Honours in International Studies from York University in Toronto. Jonathan's interests revolve around the politics and economics of conventional and renewable energy, system transitions, community-based management and private governance. He recently co-founded and works for the Ethical Coffee Chain, a Toronto-based company that works directly with coffee farmers.

Dr Robin Matthews is currently head of the interdisciplinary Vibrant and Low Carbon Communities Theme at the James Hutton Institute, and has a strong national and international reputation in the field of mitigation and adaptation to climate change by the natural resources sector ranging over a period of 30 years. His current research interests are in developing and evaluating ways of storing more carbon in landscapes without adversely affecting livelihoods and other ecosystem services, which includes leading the EU-FP7 REDD-ALERT project on tropical deforestation. He is a member of the Scottish Government Agriculture and Climate Change Stakeholder Group (ACCSG), and also a member of the Global Steering Group of the CGIAR's ASB Partnership for the Tropical Forest Margins.

Dr Patrick Meyfroidt holds a PhD in geography (2009) and a degree in sociology from Université Catholique de Louvain in Belgium. His main research interests are land use and forest transitions, and feedbacks between environmental perceptions and land changes. After working for the REDD-ALERT project funded by EU FP 7 in 2009–2011, since October 2011 he has been postdoctoral researcher at the FRS-FNRS (the Belgian Research Funds) and at Université Catholique de Louvain.

Dr Peter A. Minang is a Senior Climate Policy Scientist and Global Coordinator of the ASB Partnership for the Tropical Forest Margins at The World Agroforestry Centre, Nairobi, Kenya. He is a geographer with over 18 years of diverse experience working on conservation, community forestry and climate change forestry. His current research interests are on reducing emissions from all land use and REDD+ at the tropical forest margins and the dynamics of enabling joint planning and implementation of mitigation and adaptation to climate change in Africa. He has served in various advisory roles on climate change for African Governments, Regional Bodies and organizations.

Dr Meine van Noordwijk is Chief Science Advisor/Leader Global Research Project – Environmental Services. In addition to being Chief Science Advisor for the Centre, Meine van Noordwijk leads the organization-wide environmental services research area. From 2002 to 2008 he was Regional Coordinator for South East Asia. Before joining the Centre, Meine was a senior research officer in the Root Ecology Section at the DLO Institute for Soil Fertility Research in Haren, the Netherlands, concentrating on models of the relationships between soil fertility, nutrient use efficiency and root development of crops and trees. He also worked for two years as a lecturer in botany and ecology at the University of Juba in Sudan. Meine has a PhD in Agricultural Science from the University of Wageningen, the Netherlands and an MSc in Biology from the Rijksuniversiteit Utrecht, the Netherlands.

Patricia Santa Maria is a forestry engineer graduate from Universidad Nacional Agraria La Molina in Lima, Peru. Her professional experience has touched upon different fields related to development and conservation. She has carried out activities related to environmental impact assessments, forest certification, natural protected areas, as well as projects associated to ecosystem services and forest carbon (CDM and REDD+). During 2009–2010 she joined the Peruvian National Service of Protected Areas where she worked as specialist on ecosystem services and sustainable economic activities for one of the major projects curbing deforestation called *Reducing emissions from deforestation and forest degradation through Protected Areas in the Amazon*. In 2010 she moved to the Netherlands where she joined the Environment and Resource Management MSc Programme at the Institute for Environmental Studies at VU University Amsterdam, graduating in 2011. For her thesis she researched economic alternatives to deforestation in Ucayali, Peru, and how these contribute to a REDD+ mechanism.

Dr Vu Tan Phuong is a Director of Research Centre for Forest Ecology and Environment (RCFEE) of the Forest Science Institute of Vietnam. Much of his work has been devoted to sustainable forest and environment management. His assignments in Vietnam have also given him very good knowledge of forest valuation, carbon accounting and climate change in the forestry sector. He is also a member of the National Working Group on Climate Change in Vietnam and contributed to the development of Vietnam's second national communication report regarding the climate change impacts assessment and greenhouse gases inventory in the forestry sector. He is now a key contributor to the development of a scientific base for REDD implementation in Vietnam and climate change impacts assessment in the forestry sector. He is also active in developing policies on payment for forest environmental services and forest valuation. He is author or co-author of 13 books and 17 articles published in Vietnamese and internationally as well as a number of consultancy reports related to AR CDM, REDD, forest management and carbon trading.

Hye Young Shin is presently assistant Deputy Director at the Ministry of Environment in the Republic of Korea. Between 2009 and 2011 she had educational leave to study and work at the Institute for Environmental Studies in the Netherlands. During this period she graduated from the Environment and Resource Management MSc Programme at the Institute for Environmental Studies at VU University Amsterdam in 2010 and worked on this project.

Acknowledgements

This book is part of a larger project on Reducing Emissions from Deforestation and Degradation through Alternative Land uses in Rainforests of the Tropics (REDD Alert). This project was financed under the European Union's Seventh Framework Programme for Research (FP7). We would like to thank the European Union for its generous funding of this research project. We would also like to thank our project coordinator, Dr Robin Matthews, for his continuing support for our research work.

The team of editors (Professor Joyeeta Gupta, Dr Nicolien van der Grijp, Dr Onno Kuik) and authors (Fahmuddin Agus, Mairon Bastos Lima, Channah Betgen, Léa Bigot, Felix von Blücher, Constanze Haug, Jonathan Y.B. Kuiper, Dr Robin Matthews, Dr Patrick Meyfroidt, Dr Peter Minang, Dr Meine van Noordwijk, Patricia Santa Maria, Dr Vu Tan Phuong and Hye Young Shin) would like to thank the following people in relation to their insights and time as partners, interviewees and reviewers in the extensive research process that has finally led to this book.

From the rest of the REDD-ALERT consortium, we would like to thank Harro van Asselt, Emmy Bergsma, Madhu Subedi, Edzo Veldkamp, Eric Lambin, Jim Gockowski, Glen Hyman, Jose Alloy Cuéllar Bautista, Kristell Hergoualc'h, Martin Tchienkoua and Valentina Robiglio, for their inspiration, support and advice on this project.

We have tested out the materials in this book in teaching sessions and we would like to thank the following students for their work in MSc theses: Azio Lijcklama à Nijeholt, Ngan Bui, Dr Karen Meijer, Marie Louise Filippini, Hanny Newball Hoy, Patricia Santa Maria, Onno Prillwitz and Lisa Kishawi. Valentina Gianni contributed with research on international organizations and REDD during her research stay in Amsterdam. We would also like to thank the following students for their research papers on various aspects for REDD and for contributing in many ways to the intellectual development of this book: Angela Nichols, Arianne van der Wal, Merijn Bas, Gerard Cowan, Ferdinand Goetz, Line van Kesteren, Pim van Berlo, Marleen Schoemaker, Nitsuhe Wolanios, Sandra Megens Santos, Kaja Volker, Omnia Abbas, Rick Assendelft, Gila Merschel, Vincent Merme, Wouter Koene, Lidia Muresan, Arno van Akkeren, Mark Groenhuijzen, Timo Brinkman, Jonathan Cauchi,

Thea Renner, Tara Geerdink, Fabi Fliervoet, Laura Jungmann, Ninya den Haan, Samantha Scholte, Adam Ertur, Annamaria Hajdu, Channah Betgen, Londile Dlamini, Casper B. de Lange, Frank Hemmes, Raluca Alexandru, Lena Modzelewska, Edyta Struzik, Karin de Vries and Riana van der Werf.

We would like to thank the workshop participants who participated in the IVM organized policy exercise on REDD financing, namely: Juan Pablo Castro, Ernestine Meijer, Maria Nijnik, Benoit Morel, Alexandra Morel, Innocent Bakam, Herry Purnomo, Cordula Epple, Jan Fehse, Harro van Asselt, George Dyer, Stefanie von Scheliha, Herman Savenije, Deepak Rughani, Angelica Mendoza and Harko Koster.

We are very grateful to the following interviewees for sharing their time and views with us:

From Cameroon, we thank Christophe Bing, George Akwah, Guillaume Lescuyer, James Acworth, John Oben, Joseph Amougou, Kirsten Hegener, Mendomo Daniel, Michel Ndjatsana, Nankam Appolinaire, Paolo Cerutti, Peter Minang, Samuel Assembe, Valentina Robligio, Vincent Beligne, Yemefack Martin, and the Mesondo Council and Technical Center for Council Forests.

From Indonesia, we thank Agus Suratno, Ahmad Dermawan, Ai Dariah, Andi Basri, Anwar, Asro, Dr Aswanti, Professor Bambang Saharjo, Bandung Sahani, Bang Damsir, Basah Hernowo, Doni Iskandar, Dri Handoyo, Fahmuddin Agus, H. Ahmad Palloge, Haji Napi, Herry Purnomo, Iman Budisetiwan, Iwan Tricahyo Wibisono, Jerry Darli, Jasnari, Jubaedah, Khairiyah, Maswar Bahri, Mawan, Meine van Noordwijk, Muhammad Ali Rohmawan, Muhammad Amin, Muhammad Razi, Muhammad Yakub, Muhammad Yani, Muklis, Ngaloken Gintings, Niken Sakuntaladewi, Norman Jiwan, Nur Milis, Pardi, Pablo Pacheco, Petrus Gunarso, Prihasto Setyanto, Ratna Akiefnawati, Retno Maryani, Rudi Syaf, Saipul Rahman, Sonya Dewi, Suharto, Suharto, Tarman, Wahyu Widodo, Widodo, Winarto, Yasli, and Yusuf Supardi.

From Peru, we thank Diego V. Kau, Eloy Cuellar, Elvira Gómez, Hugo Che Piú, Ildefonzo Riquelme, Jean Pierre Araujo, Johana Garay, Jorge Torres, José Luis Capella, Jose Pilco, Julio Ugarte, Lily Rodriguez, Luis Lanfranco, Manuel Soudre, Marcos Tito, Mauro Scavino, Max Silva, Michael Pollmann, Miguel Sánchez, Milagros Sandoval, Napoleón Jeri, Octavio Galván, Oscar Gil, Sandra Olivera, Shery Tolentino, Violeta Colán and Ymber Flores.

From Vietnam, we thank Akiko Inoguchi, Professor Bao Huy, Mr Chinh Ha Dang, Mr Do Anh Tuan, Ms Do Thi Ngoc Bich, Mr Do Trong Hoan, Mr Duong The, Mr Ha, Mr Ha Cong Dung, Mr Hao, Dr Ho Van Cu, Ms Hoang, Mr Hoang Thanh, Mr Hung, Mr Le Cong Uan, Ms Le Thi Nguyen, Dr Lung, Ms Ly Thi Minh Hai, Mr Nguyen Manh Cung, Dr Nguyen Ton Quyen, Mr Patrick Van Laake, Mr Pham Manh Cuong, Mrs Pham Minh Thoa, Mr.Pham Ngoc Dung, Mr Pham Ngoc Thanh, Mr Pham Quy Anh, Ms Pham Thi Thuy Hanh, Mr Pham Van Dong, Mr Pham Xuan Hoan, Mr Quan, Ms Quyen, Mr Richard McNally, Mr Tien Hung, Ms To Thi Thu Huong, Mr To Xuan Phuc, Mr Tran Quoc Vinh, Mr Tran Van Hung and Mr Vu Tan Phuong.

We thank Andrea Brock, Michael Huettner, Simone Lovera and Till Neeff for sharing insights on REDD with us. We thank the International Institute for Applied Systems Analysis (IIASA) in Vienna, especially Michael Obersteiner and Florian Kraxner, for hosting Constanze Haug as a fellow of the Young Scientist Summer Programme 2009, during which an important part of the research on REDD was carried out.

Earlier drafts of various chapters or cross-cutting issues have been presented at the Colorado Conference on Earth System Governance, 'Crossing Boundaries and Building Bridges', Colorado State University, 17–20 May 2011; Water Governance Meeting the Challenges of Global Change, 7 June 2011, European Science Foundation, Obergurgl, Austria, the Planet Under Pressure Conference, London 2012, the Earth System Governance Conference, Lund. We thank those who participated at and gave comments on our contributions at these conferences.

Finally, we thank the reviewers of the various chapters in this book, namely: Harro van Asselt, Frank Biermann, Andrea Brock, H. Carolyn Peach Brown, George Dyer, Aarti Gupta, Marjan Hofkes, Michael Huettner, Eric Lambin, Simone Lovera, Patrick Meyfroidt, Meine van Noordwijk, Philipp Pattberg, Thu Thuy Pham, Tom Rudel, Heike Schroeder, Sandra Velarde and Fariborz Zelli.

An endeavour such as this book is the result of the intellectual growth that takes place through continuous interactions with people who are all experts on forests in their own way. Without their generous support in terms of time and intellectual input, this book would not have been possible.

Abbreviations

A/R	Afforestation and Reforestation
ASEAN	Association of Southeast Asian Nations
BASIC	Group made up of Brazil, South Africa, India and China
CBD	Convention on Biological Diversity
CCD	Convention to Combat Desertification
CDM	Clean Development Mechanism
CER	Certified Emissions Reduction
CITES	Convention on the International Trade in Endangered Species
COMIFAC	Central African Forests Commission
COP	Conference of the Parties
CPF	Collaborative Partnership on Forests
CSR	Corporate Social Responsibility
DPSIR	Driving forces-Pressures-States-Impacts-Responses
DRC	Democratic Republic of Congo
EU	European Union
FAO	Food and Agriculture Organization of the United Nations
FCPF	Forest Carbon Partnership Facility
FLEG	Forest Law Enforcement and Governance
FLEGT	Forest Law Enforcement, Governance and Trade
FSC	Forest Stewardship Council
GDP	Gross Domestic Product
GEF	Global Environment Facility
GHG	Greenhouse Gas
IFF	Intergovernmental Forum on Forests
ILO	International Labour Organization
IPCC	Intergovernmental Panel on Climate Change
IPF	Intergovernmental Panel on Forests
ITTA	International Tropical Timber Agreement
ITTO	International Tropical Timber Organization
IUCN	International Union for Conservation of Nature
KP	Kyoto Protocol
LULUCF	Land-Use, Land-Use Change and Forestry

MARD	Ministry of Agriculture and Rural Development
MEA	Millennium Ecosystem Assessment
MONRE	Ministry of Natural Resources and Environment
MRV	Measurement, Reporting and Verification
NAMA	Nationally Appropriate Mitigation Actions
NGO	Non-Governmental Organization
NLBI	Non-Legally Binding Instrument (on All Types of Forests)
ODA	Official Development Assistance
(N)PA	(National) Protected Area
P(F)ES	Payment for (Forest) Ecosystem Services
REDD	Reducing Emissions from Deforestation and Forest Degradation (refers to RED, REDD, REDD+ and REDD++ in this book as an umbrella term)
RRI	Rights and Resources Initiative
SAP	Structural Adjustment Program
SFM	Sustainable Forest Management
TFAP	Tropical Forest Action Plan
UNCED	United Nations Conference on Environment and Development
UNCCD	United Nations Convention to Combat Desertification
UNCTAD	United Nations Conference on Trade and Development
UNDP	United Nations Development Programme
UNEP	United Nations Environment Programme
UNESCO	United Nations Educational, Scientific and Cultural Organization
UNFCCC	United Nations Framework Convention on Climate Change
UNFF	United Nations Forum on Forests
UNPFII	United Nations Permanent Forum on Indigenous Issues
VPA	Voluntary Partnership Agreement
WHC	World Heritage Convention
WTO	World Trade Organization
WWF	World Wide Fund for Nature

1 Climate change and forests

From the Noordwijk Declaration to REDD

Joyeeta Gupta, Robin Matthews, Peter Minang, Meine van Noordwijk, Onno Kuik and Nicolien van der Grijp[1]

1.1 Introduction

Forests and climate change interlock in many ways. Many scientists and policymakers feel that a climate perspective provides added energy and resources to resolve the challenges of deforestation and forest degradation. Addressing these challenges can also buy time in dealing with the climate change challenge, as the climate negotiations have reached a deadlock on other issues.

During the past twenty-five years political perceptions about how issues are linked have evolved. In the period 1987–92, the climate change–forests link was actively discussed.[2] However, in the Kyoto Protocol (KP) to the United Nations Framework Convention on Climate Change (UNFCCC)[3] and the subsequent Marrakesh Agreement clarifying the scope of its Clean Development Mechanism (CDM),[4] the forest issue was put on the backburner. This was due to the complexity of getting agreement on forests between developed and developing countries, the wide uncertainty margins around forest-related emissions and the political importance and perceived simplicity of focusing on greenhouse gas (GHG) emissions reduction from industrial sources. Twenty years later, by 2007–12, the relative political complexity of focusing on emissions reduction from industrial sources and the relevance of the presumed 12 to 17 per cent of global GHG emissions related to forests and land use change[5] has led to a return to an emphasis on the role of forests in climate change; and a hope that forest-related emissions will be simpler to reduce since it may be more cost-effective than diminishing industrial emissions (see 1.4). This has led to the new discussions on REDD – Reducing Emissions from Deforestation and Forest Degradation. It seems that the global community has come full circle in the discussion. It appears as if the political situation has changed so much over the last twenty years that the seeming complexity of twenty years ago has evaporated over time; that a focus on forests can help buy time in the climate change regime; and that the knowledge gained in the last twenty years on forests can help to make a better forest regime with a higher compliance pull.

This book addresses these complex political, policy and legal issues, and their relation to the uncertainty in data and attribution of emissions and emissions reduction. It goes beyond analysing the role of forest policy in the context of climate change policy to explore the dimensions of multi-level forest governance. The ultimate objective is to identify the lessons of governance from historical efforts to develop appropriate norms, rules and instruments to deal with forests and to see how best these lessons can be incorporated into the process of designing a new system to deal with forests and related land use issues. This book assesses the different policy instruments used in developing countries to deal with deforestation and current activities to reduce GHG emissions from deforestation and forest degradation under the international climate regime, as well as the incentives flowing from them at the national and subnational levels, to analyse how these policy levers change human behaviour, and how they interface with the global, national and local drivers and pressures of land use change in tropical forests.

The idea of REDD is to develop a mechanism that aims at halting deforestation through financial incentives. In simple words: if a country agrees to reduce deforestation below a baseline that is agreed to with others, it could receive financial compensation for this effort. Other countries would financially support this effort because of its contribution to addressing global deforestation trends and climate change in general and, in particular, in return for emissions reduction credits. There are some basic principles that underlie this idea. These include that countries participate voluntarily in these agreements; that payments are made only when performance can be demonstrated through Measuring, Reporting and Verification (MRV) processes and are in relation to agreed baselines; that the measured reductions are in relation to halting deforestation or forest degradation or conserving/ enhancing carbon stocks or sustainably managing forests; that the processes and procedures are equitable; that they would build on demonstration projects and be consistent with the provisions of relevant United Nations (UN) bodies; and that emphasis is paid on safeguarding other ecosystem services and equity, and achieving co-benefits. The acronym REDD keeps changing from RED through REDD to REDD+ and REDD++ (see 4.3). For the sake of simplicity, this book refers to 'REDD' throughout, using it as an umbrella term for all approaches aimed at financially compensating developing countries for the maintenance of their forest sinks.

This book follows a methodological framework (see 1.5) to explore the generic drivers of deforestation and potential policy instruments at multiple levels of governance (see Chapter 2), examines the global policy evolution on forests (see Chapters 3 and 4), examines forest policy implementation in four case study countries (see Chapters 5–9), presents some economic quantitative analysis in Chapter 10 before drawing conclusions in Chapter 11. This chapter first explores key elements of the climate change problem (see 1.2), the forest problem and the kinds of services forests provide (see 1.3), and the climate change–forests nexus (see 1.4). It elaborates on the problem definition and main research questions, the methodology and structure of this book (see 1.5) and draws some inferences in Section 1.6.

1.2 Climate change

1.2.1 The physical problem

The physical problem of climate change refers to the anthropogenic emissions of carbon dioxide and other gases which are leading to a rapid increase in the trapping of heat ('greenhouse effect') in the atmosphere relative to pre-industrial levels. This increased greenhouse effect leads to rising average global temperatures and altering precipitation patterns, in turn leading to sea-level rise, melting glaciers and ice-caps, and to increased uncertainty of local climates. These changes can drastically affect the hydrological cycle and ocean circulation and can have serious impacts on all biota.

The Fourth Assessment Report of the Intergovernmental Panel on Climate Change (IPCC AR-4) concluded that current atmospheric concentrations of greenhouse gases (GHGs), which include carbon dioxide (CO_2), methane (CH_4), and nitrous oxide (N_2O), are higher than at any time over the last 650,000 years. During the last hundred years, the global average temperature increase has been 0.74° C, the sea level is rising annually by about 3.1 mm since 1993, the global snow and ice cover is receding, there are regional changes to precipitation (more in the Northern hemisphere, less in Africa), and extreme weather events are on the rise. With increased atmospheric concentrations of CO_2, the oceans absorb more CO_2 and their acidity increases. These changes are in line with the predictions through modelling efforts that anthropogenic emissions have increased by 70 per cent over three decades as of 2004. Such emissions are likely to grow by 25–90 per cent (CO_2-eq)[6] between 2000 and 2030 leading to additional warming.[7]

1.2.2 The governance process

Nearly a hundred years after the first calculations by Svante Arrhenius of the impact of increasing CO_2 emissions on the greenhouse gas effect, and after the effective control of industrial sulphur dioxide emissions that masked the effects of CO_2, the issue of climate change was put on the global scientific agenda in 1979 at the first World Climate Conference. Following an *ad hoc* effort to consolidate the science under the umbrella of the Advisory Group on Greenhouse Gases, and a decision of the United Nations General Assembly, the IPCC was established in 1988. It produced four assessment reports, in 1990, 1995, 2001 and 2007 respectively and is currently working on its fifth report. In 2007, the IPCC (along with climate change lobbyist Al Gore) received the Nobel Prize for Peace, thereby achieving recognition for its collaborative efforts at contributing to the science of climate change and recognizing the relationship between climate change and the potential for conflict. However, IPCC has not been entirely without controversy itself. It operates as a 'boundary organization' on the interface of science and public policy[8] and is open to criticism from the scientific community and policymakers. Despite these controversies, the bottom line remains that the climate change problem is a serious one and the impacts have begun to be felt and will continue to materialize.

Ten years after the first scientific conference on climate change in 1979, the issue had been discussed in several political fora at global level, including at the level of heads of state at the Hague Conference on Climate Change in March 1989. Following a decision of the United Nations General Assembly, an Intergovernmental Negotiating Committee was set up to negotiate a treaty on climate change. Two years of negotiations between countries led to the adoption of the UNFCCC;[9] five years later the Kyoto Protocol[10] was adopted. Although negotiations on a post-Kyoto regime began early and were expected to lead to legally binding targets, in 2009 a Copenhagen Accord was 'noted' and not 'adopted'. An Annex to this document notes country offers to take on commitments if others also do the same and recommends Nationally Appropriate Mitigation Actions (NAMAs). In 2010 and 2011 the disappointment over this failure led to the 'Cancun Agreement' with consensus in principle on reducing deforestation (REDD) but not on implementation modalities. The world is now anxiously waiting for a legally binding follow-up to the Kyoto Protocol, but expectations are mixed. Although there has been progress made on forests, the lack of progress on GHG emissions reduction targets may have an impact on the mechanism. The governance problems are further elaborated in Chapter 4.

1.2.3 The key political challenges

The politics of climate change can be explained in terms of four challenges: the definition of climate change; the dominant paradigm for addressing the climate change problem; the mechanism for sharing responsibilities between countries; and methodological challenges.

Climate change was initially defined in the early 1990s as a technical and economic issue, but the true political nature of the problem only became explicit by the end of the decade.[11] Initially, the problem was seen as an abstract, technocratic, sectoral, mitigation issue, but over time it was defined as an urgent, developmental, political, systemic, adaptation and mitigation issue. The evolution from sectoral to systemic implies a critique of the current development paradigm. A search for simple solutions, for example, with a reference to the Environmental Kuznets Curve, proved unsuccessful. This curve graphically illustrates a hypothesis that claims that as societies become richer they pollute more; but after a 'turning point' of per capita income they start to pollute less. Politically, the curve implies that a transfer of modern technologies from advanced countries to the developing countries could help them 'leapfrog' to a low carbon economy. However, research suggests that the Environmental Kuznets Curve does not hold for global problems; hence it is not very likely that countries can simply 'grow' out of their environmental problems.[12] Although discussions in the run-up to the 2012 UN Summit for Sustainable Development reveal that countries hope that the 'green economy' concept[13] will yield a systemic solution, the UN Secretary General sees it as a multi-interpretable concept[14] – which could aim for complete sustainability, but perhaps only achieves a minimal integration of external costs in existing processes.

The dominant paradigm adopted for addressing the climate change problem was that of 'leadership'. This implied that the developed countries should lead by reducing their own emissions and by helping developing countries. It was adopted in preference to a liability paradigm.[15] However, the leadership paradigm morphed in 1996–2001 into a conditional leadership paradigm – where the US was waiting for key developing countries to take action, the EU for the US and Japan to take action, and the developing countries for the EU and US to take action. Subsequently, there was leadership competition with the US promoting several international agreements with relevance for climate change. Following the global recession of 2008, the world is waiting for new leaders to shape the future.

The mechanism for sharing responsibilities built upon the North–South divide. When the third report to the Club of Rome was published under the title *Reshaping the international order*,[16] the world was interpreted as a triangle between 'West', 'East' and 'South', with consequences for the poverty–environment nexus. In the early 1990s, following the fall of the Berlin Wall, this had simplified into a 'North' versus 'South' dichotomy, with most of the global environmental issues caused by the 'North' (specified as 'Annex-I' countries in the UNFCCC and as 'Annex-B' in the Kyoto Protocol) and most of the consequences suffered by the 'South' ('Non-Annex-I' countries). In fact, climate change is a classic North–South issue for four reasons: (1) average per capita Northern emissions is much higher than average per capita emissions from the South; (2) current and immediate future impacts are caused primarily by Northern emissions of the past; (3) efforts to stabilize concentrations at safe levels might mean exhaustion of the total emissions budget by around 2030, leaving no room to grow for developing countries;[17] and (4) climate impacts are likely to be worse in the South because of the geographical location and the greater vulnerability of Southern countries. Hence, the Climate Convention includes the principle of 'common but differentiated responsibilities and respective capabilities'[18] as a way to differentiate between the responsibilities of the poor and rich countries. Initially there was considerable optimism and countries felt that this principle would serve to allocate responsibilities to countries because it flows from justice principles.[19] However, some scholars argue that territorial variation[20] and per capita approaches[21] are unjustified because they tend to emphasize relative emissions (e.g. emission per capita), while what matters for the environment is absolute levels of emissions. In addition, gross national emissions are more in line with State sovereignty principles.[22] The application of the principle is sometimes problematic since some of the so-called developing countries are actually richer (e.g. South Korea) than some of the so-called developed countries. Furthermore, in the future the emerging economies of China and India will have to take on some responsibilities for their emissions because the sheer size of these emissions may negate the reduction efforts of other countries.[23] In any case, the US is less supportive of the rule of law at global level and sees national interests as overriding any global legal principle and this has led to their unwillingness to ratify and implement the Kyoto Protocol; a position that Canada and Russia support with their decision in 2011 not to participate in a post-Kyoto agreement. While differentiation between countries is essential from a justice perspective, such differentiation needs to take into account the changing development status of countries.[24] The 2011 Durban Conference of the Parties (COP) to the Climate Convention may have seen a breakthrough in acknowledging the limits of the dichotomy framework of countries, and the start of a more open and comprehensive round of negotiations.

Methodological challenges relate to the divergence between area- and consumption-based estimates of GHG emissions, with the 'emissions embodied in trade' as a challenge to fair and efficient emissions reduction rules. For example, deforestation is calculated directly into emissions from the deforesting country, even though the wood may have been traded to another destination. Similarly, in production for exports, energy-related emissions are only attributed to the country using the energy and not to the importer of the goods that are produced with the help of the energy that caused the emissions.[25]

The negotiation processes at the COP in Copenhagen and the Copenhagen Accord marked a watershed in climate politics. During the Conference, a new Group of BASIC (Brazil, South Africa, India and China) countries and powerful Annex-I countries negotiated and reached agreement on the Copenhagen Accord, and presented it for adoption by the COP. It marked a new mode of parallel (two-tier) negotiation processes in the Climate Convention (as opposed to the consensus-based approach) with the rest of the world on the one hand and the BASIC group of countries and Annex-I countries on the other. This may be the beginning of a trend in such negotiations. The concept of Nationally Appropriate Mitigation Actions (NAMA) that was first mentioned in the Bali Road Map alongside REDD was endorsed in the Copenhagen agreement also allowing countries to undertake nation or sector-wide emission reductions purely supported by national funds (voluntary NAMAs), with potential for internationally supported NAMAs. One outstanding question in the current negotiations is how REDD (hence forests) should relate to the potentially more encompassing NAMAs and what this implies for funding streams. From the Bali COP in 2007 onwards, it seemed that 'forests' were the only area on which there was consensus (see 1.4 and Chapter 4).

1.3 Forests

1.3.1 Current status

At four billion hectares, forests cover about 31 per cent of the Earth's land area.[26] Of this, primary forests account for 36 per cent, although these have been decreasing at the rate of 0.4 per cent (40 million hectares) annually since 2000. On the other hand, the planted forest area is increasing and now accounts for around seven per cent of the total forest area. Such data are contingent on the operational definition of forests and the types of planted and natural vegetation that are lumped under the 'forest' category in the statistics.

Natural forests are important to the planet's ecological health and to humanity. The Millennium Ecosystem Assessment[27] identified four categories of ecosystem services that are derived from natural, managed and agroecosystems: (1) supporting services such as contributing to nutrient cycles and crop pollination; (2) provisioning services that refer to the production of 'goods' that can be used as food, fuel and fibre; (3) regulating services that include processes controlling the climate and the flows of water; and (4) cultural services such as meeting spiritual and recreational needs. Innovation services include the knowledge generated by the forests that can be used to create future knowledge (see Table 1.1). These services contribute to human well-being by enhancing security, the quality of life, health and good social relations.

The world's forests provide a habitat for nearly 90 per cent of the world's terrestrial biodiversity.[28] Deforestation poses serious threats to the conservation of this biodiversity, particularly in biodiversity 'hotspots'.[29] Forests are also an important store of carbon (C), with an estimated 289 gigatonnes (Gt) of carbon stored in their biomass alone,[30] about 86 per cent of the planet's terrestrial above-ground carbon. This compares with about 1,500 Gt C in the world's soils, and about 730 Gt C in the atmosphere.[31] In terms of ecosystem protection, around eight per cent of the world's forests have protection of soil and water resources as their primary objective.[32] Similarly, managing forests for social and cultural functions, such as recreation, is also increasing, although the areas involved are difficult to quantify.

In addition, forests, although it is unclear what forest definition is used for this assessment, support the livelihoods of about one billion people[33] and provide goods for trade. Around 30 per cent of the world's forests are primarily used for production of wood and non-wood forest products which constitute about two per cent of world trade;[34] other land uses, including oil palm and rubber plantations, contribute one or two orders of magnitude more to trade, per unit area. These figures, however, conceal important trends reflecting the changing perceived value of forests. The number of people working in public forest institutions, for example, declined by about 10 per cent between 1990 and 2005 (probably due to increases in labour productivity),

Table 1.1 The ecosystem services of forests

Ecosystem Services	Examples		Human well-being	Examples
Supporting	Soil formation and erosion prevention; nutrient cycling through atmosphere, plants and soils; primary production, habitat for predator–prey relationships and ecosystem resilience		Security	Access to resources and security from disasters
Provisioning	Food (fruits, nuts, forest animals); fresh water; fuel, wood and fibre; non-timber forest products: silk, rubber, bamboo; genetic resources; biochemical or natural medicines; habitat for humans	Thus contributing to	Good life	Better local climate; drinking water; food; fish; livelihoods; travel; energy
Regulating	Water purification; climate regulation; flood and drought regulation; disease regulation; pollination; carbon sequestration		Health	Reduced vulnerability to diseases; access to water and food (so better health)
Cultural	Aesthetic; spiritual; educational; recreational; tourism; inspirational		Good social relations	Cohesion; mutual respect; (transboundary) cooperation

Source: based on Millennium *Ecosystem Assessment (2005) Ecosystems and human well-being: General synthesis*.

particularly in Europe, North America and East Asia. However, with the increase of protected areas in many countries, 10 million people are employed in forest management.[35] Significant forest-related activity outside the formal sector also contributes to local livelihoods and national economies.

With forests providing so many services, it is no surprise that around 24 per cent of all forests (949 million hectares) are designated for multiple use, with the production of goods, protection of soil and water, conservation of biodiversity and provision of social services all coming from the same forested areas with none of these alone being considered as the predominant function.[36]

1.3.2 Deforestation and degradation

Although forests are valuable for the global ecosystem and humanity, much of this value is 'non-market' and tends to be ignored in economic considerations. Only those goods and services that have a monetary market value are usually considered (e.g. timber), and often these do not compete favourably with other land uses, such as agriculture. Thus, forests are converted to agricultural land. About 13 million hectares of forest is lost annually, with net losses (allowing for afforestation and reforestation) of about 5.2 million hectares per year.[37] Deforestation and degradation together release an estimated 4.4 Gt CO_2 per year into the atmosphere,[38] from biomass burning and the oxidation of carbon stored in the soil under the trees during cultivation and in peatlands under drainage. Degradation, defined as decrease of density or increase of disturbance in forest classes, may represent up to 20 per cent of this loss,[39] or close to 100 per cent if the internationally accepted and very weak forest definition is used (compare Figure 1.1), where 'deforestation' rates are reduced without reducing emissions; the deforestation and degradation terms are best used in combination, where they effectively refer to carbon stock accounting. Other GHGs, such as CH_4 and N_2O, may also be emitted during slash-and-burn and subsequent land use. This represents an estimated 15 per cent of total anthropogenic GHG emissions,[40] greater than that from the whole global transport sector.[41] (See 11.4.2)

Regionally, South America had the highest deforestation rates between 2000 and 2010 at about 4.0 million hectares per year, followed by Africa, at 3.4 million hectares per year.[42] Nationally Brazil and Indonesia have the highest net deforestation rates, losing respectively 2.1 and 0.7 million hectares of forest every year.[43] Total per capita annual CO_2 emissions in Indonesia, for example, may thus be quite high. However, in recent years, Brazil and Indonesia have reduced their rates of deforestation significantly.[44] The official numbers depend on the operational forest definition used and this tends to change with time. Figure 1.1 shows deforestation definitions on the x axis and deforestation rates on the y axis. Depending on the definitions used, the rates of deforestation differ. Comparisons over time of forest statistics are noted for inconsistencies. North America, Europe and China are actually increasing their forest cover through large-scale plantations and natural regeneration of abandoned farmland. Consequently, recent assessments indicate that global deforestation rates are decreasing, although they are still high.

1.3.3 The governance process

During colonial times, colonizers actively traded in timber products. Post-colonial governments often promoted colonial policies in the timber area. At global level, the Food and Agriculture Organization of the United States (FAO) began to register the land under forests in the 1950s. In the 1960s, the international community turned its attention to the protection of rainforests

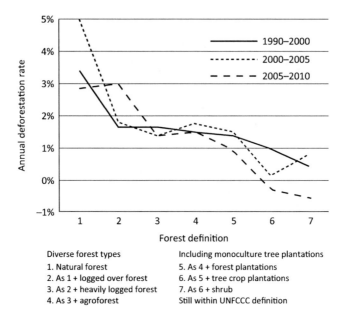

Figure 1.1 Dependence of the 'deforestation rate' for Indonesia in three time periods on the operational forest definition used; negative deforestation rates indicate increase in forest

and species. The 1973 Convention on the International Trade in Endangered Species ranked vulnerable tree, plant and animal species that were to be either banned from international trade or accompanied by strict export rules.[45] The 1975 World Heritage Convention established the protection of sites with key natural or cultural values, including over 76 million hectares of forest land.[46] The 1973/75 Ramsar Convention (on Wetlands of International Importance) prevented the conversion of forested wetlands into other forms of land use.[47] In 1983, the FAO launched the International Tropical Timber Agreement (ITTA) to combat tropical deforestation among International Tropical Timber Organization (ITTO) members who were responsible for 90 per cent of the international tropical timber trade. The ITTA aimed to 'promote the expansion and diversification of international trade in tropical timber from sustainably managed legally harvested forests'.[48] While the ITTO promotes sustainable forest management and deals with illegal commercial logging, it is still criticized as being dominated by forest exploitation interests rather than by conservation interests.[49] And while the World Trade Organization's General Agreement on Tariffs and Trade and other such agreements aim to liberalize trade of, amongst others, forest products, exclusions are allowed based on environmental grounds.[50]

In 1987, the Tropical Forest Action Plan was established by the World Resources Institute together with the FAO, the United Nations Development

Programme and the World Bank to halt tropical forest destruction.[51] This Plan aimed to help countries, especially in Africa, in developing their national plans for forest conservation and management. However, it overemphasized support for the international timber industry and ignored other more important drivers of deforestation (e.g. agriculture), and did not consult indigenous peoples and local communities living in and around the forests. Deforestation rates actually increased over the five years that the Plan was in existence, largely from the building of roads for logging access, which stimulated migration by smallholders.[52]

Subsequently, international forest governance began to include all kinds of forests as well as equity issues. Several arrangements, directly or indirectly related to forests, highlighted the rights of indigenous peoples.[53] The 1992 United Nations Conference on Environment and Development (UNCED) aimed to develop an international forest convention, but the parties were only able to agree on some general principles. The principles include the need for sustainable forest management, but also the sovereign right to exploit natural resources and the right to socio-economic development on a sustainable basis, as well as the notion that the benefits that result out of an international effort to conserve and sustainably manage forests should be distributed equitably among members of the international community.[54]

UNCED also led to the 1992 Convention on Biological Diversity, the most important international hard law on forests.[55] This Convention calls on parties to develop national policies to conserve and sustainably use biodiversity. The Convention established a financial mechanism to support developing countries in meeting their obligations. The 1992 Convention on Climate Change includes forests as carbon sinks and parties are committed to promote the conservation and enhancement of these and other carbon stocks.[56] The 1997 Kyoto Protocol treats forests as sinks that may be used to meet part of parties' emission reduction commitments. Certain afforestation and reforestation activities may be included under the CDM, distinguishing between temporary and long-term credits. Notably, however, few afforestation/reforestation CDM projects have been initiated up to now because of different types of complexity surrounding these projects, such as public resistance, environmental integrity concerns and required technological knowledge.[57]

Following UNCED, the Intergovernmental Panel on Forests and the Intergovernmental Forum on Forests were established, which merged into the United Nations Forum on Forests in 2000. The forum advises and supports the implementation of international and national forestry programmes and in 2007 adopted a 'non-legally binding instrument on all types of forests' and a 'multi-year programme of work' which sets out global objectives for 2015 and national measures to achieve these objectives. The establishment of a Forest Fund discussed at the 2009 forum meeting in the Eighth Session of the United Nations Forum on Forests in New York showed again that the diversity of interests in relation to forests prevents agreement (see Chapter 3).

1.3.4 The key political challenges

Forest politics are linked to colonial and neo-colonial approaches to managing forest resources and the issue of whether deforestation is per se a problem or not, and disagreement about the level at which forests should be governed.

The North's historical power relied heavily on acquisition of natural resources.[58] Timber to make warships (the first treatise on forests in England was commissioned by the Navy) and land to feed people and allow a settler economy to develop were the greatest concerns. Fears of timber scarcity drove European countries to either begin reforestation or secure supplies from their colonies, if they had any. After independence, many of the formerly colonized countries continued this policy and established forest ministries (e.g. Indonesia and India). The exploitation continued, with developing countries cashing in on the demand for timber by industrialized countries for reconstruction after the war.

The colonial idea of using forests as resources was reflected in the initial policies of the FAO that estimated in 1948 that there were nearly four billion hectares of forests globally. For the first time, the concept of a 'world forest' began to develop. However, forests were seen only in terms of their timber resources and, indeed, the four billion hectares were described as being 66 per cent 'productive'. Deforestation was seen largely as desirable, contributing to progress, and scant attention was paid to the other forest ecosystem services (see Table 1.1). Only in specialist circles was there some concern about the adverse environmental effects that deforestation was having, and in the 1960s more than 19 global 'regimes' were developed with specific roles to play in dealing with forests. Some of these roles are synergetic, many are contradictory or confrontational – attempts by developed countries, international agencies and non-governmental organizations (NGOs) are often viewed with suspicion and as hypocrisy by developing countries (see Chapter 3). However, it was not until the mid 1980s that deforestation was seen as a problem in the public consciousness.[59] By that time, the impact of GHGs on the climate was well established. The role of land use change, particularly the clearing of forests for agriculture, was recognized as a major contributor to GHG emissions. The other driver of concern about deforestation was the realization in the late 1980s that deforestation led to loss of biodiversity.

Thus, the record of trying to reduce deforestation over the second half of the twentieth century is mixed, and the impact of all these initiatives at the global level is questionable.

A key challenge has been that forests fall under the national territories of states and state sovereignty has always been a dominating discourse in the forest discussions. There are many arguments to scale up the issue of forests to global level or to scale it down to local level – this is referred to as the politics of scale and discussed further in Chapter 11. Clearly forests need to be protected for their own intrinsic value as an ecosystemic medium and for the ecosystem services they provide. But equally evident is that deforestation serves the immediate needs of people from local to global level – otherwise it would not be happening! Thus there is clearly something wrong with the incentive framework in these previous policies. This calls for a clear diagnosis of the forest governance challenge.

In the 2000s, there has been a shift in thinking from exploitation to the other ecosystem services that forests supply besides timber only, with an emphasis on sustainable forest management. Significant progress has been made in developing forest policies, laws and national forest programmes, with more than 1.6 billion hectares of forest having a management plan.[60] Eighty per cent of the world's forests are still publicly owned, with governments generally spending more on forest than they collect in revenue. Protected areas cover an estimated 13 per cent of the world's forests, but ownership and management of forests by communities, individuals and private companies is increasing, although often contested (see Chapters 5–8).[61]

1.4 Climate change and forests

The discussions on forests in the climate change regime have passed through four phases. In the first phase, forests were seen as important; in the second phase, forests were seen as less critical than emissions reduction from industrial sources; in the third phase, forests were included to a limited extent; and in the fourth phase forests have emerged as their own specific field under the name of REDD.

During the preparations for one of the earliest political conferences on climate change, the Noordwijk Conference on Climate Change, five background documents were prepared. One document focused on the role of forests in climate change.[62] It discussed the many services that forests provide and then continued that 'it inevitably leads to recognition that the world's forests represent a legitimate interest of all nations, for all manner of interdependency reasons apart from the predominant factor of climate regulation; and that this recognition need not be construed to conflict with sovereignty considerations'.[63] The document argues that there is need for the developed countries to support developing countries in reducing their deforestation 'in a spirit of true partnership'.[64] The Noordwijk Declaration on Climate Change[65] stated that the participating ministers agree 'to pursue a global balance between deforestation on the one hand and sound forest management on the other. A world net forest growth of 12 million hectares a year in the beginning of the next century should be considered a provisional aim.' The Declaration requested the IPCC to consider this goal and welcomed the ongoing work of the Tropical Forest Action Plan (TFAP) and the ITTO in this field.

The second phase began in 1990 with the subsequent negotiations on climate change where efforts were made to negotiate separate forest, biodiversity, desertification, and climate change treaties. The negotiations ended with some relevant results in 1992. First, forests were treated in a very limited fashion in the Climate Convention. Second, the forests treaty did not materialize – only a set of forest principles were finally agreed upon. Third, the Biodiversity Convention was adopted and this incorporated forest-related concerns to the extent that they were relevant to biodiversity. Fourth, the Rio Principles on Environment and Development were adopted – however, the relationship between these principles and the forest principles are not very clear. The moment was not ripe to promote global consensus on forest issues; and a more single-minded focus on reducing emissions from industrial sources was seen as more constructive and feasible. The adoption and entry into force of the Kyoto Protocol in 1997 indicated that industrial emissions reduction would be prioritized.

The third phase began with an opening in the Kyoto Protocol that resulted from the framing of the quantitative emission targets in terms of 'net' emissions. This implied that countries with targets could include the change in carbon stocks into their emission accounting process. Furthermore, the Marrakesh Agreement of 2001 on the modalities of the Clean Development Mechanism treated afforestation and reforestation projects in developing countries as eligible, but protection or enhancement of existing forest carbon stocks was not seen as eligible. This Mechanism allowed developed countries to invest in developing countries in return for carbon credits (certified emission reductions). Avoided deforestation was excluded in this Mechanism because of the challenges at project scale to deal with 'additionality' (what would have occurred without intervention), 'leakage' (emission reductions in one location causing emission increases in another) and 'permanence' issues. This Mechanism could not operate at the national scale where such issues would, in principle, be tractable. However, afforestation and reforestation activities have not been very prominent in the CDM portfolio (see Chapter 4).

In the fourth phase, forests are dominating the climate agenda under the acronym REDD. Ten years after Rio, the difficulties in getting the US to ratify the Kyoto Protocol, and the complexities in negotiating post-Kyoto quantitative commitments for countries, led to prioritizing forest issues. A key reason was that about 12–17 per cent of all GHG emissions come from the forest sector. The *Stern Review* emphasized the possible cost-effectiveness of forest policy;[66] while a group of forest scientists concluded that appropriate carbon pricing could help address deforestation by 2020.[67] This led to renewed optimism about the ability of the global community to deal with forests and thereby contribute to GHG mitigation. Furthermore, the new global driving force of biofuel production led perversely to more land conversion and calls for a more integrated perspective. Deforestation and land degradation discussions were launched in 2005 at the 11th COP. The 13th COP in Bali in 2007 stimulated experiments with REDD and decided that all countries should work towards improving data collection, estimation of emissions from deforestation and forest degradation, monitoring and reporting, and addressing the institutional needs of developing countries. The drivers of deforestation should also be addressed. In contrast with the project scale of the Clean Development Mechanism, REDD is perceived as operating at national scales, with only under exceptional situations an option to have subnational entities as accounting units for the assessment of emissions and success rate of emission reduction. At the same time, afforestation and reforestation did not become a key part of the CDM programme[68] and this may have implications for REDD design. Other operational difficulties in REDD relate to social and environmental issues, the scope of an operational forest definition (see Figure 1.1 and Box 1.1), the degree to which such issues can be managed by 'safeguards', and to the financing mechanism. While fund-based mechanisms are non-controversial, involvement of an 'offset' carbon market is controversial, especially in the absence of additionality in commitments to reduce emissions. A growing number of countries and stakeholders are interested in actual reduction of global emissions and not in trading emission 'rights'. Other issues include leakage, additionality and permanence; the difficulties of monitoring; the problem of the multiple levels at which the drivers of deforestation operate; the institutional dimensions of application and verification rules; the loss of sovereignty for individual countries; the potential of flooding the market with credits; and how global policies can be translated through the different administrative levels to finally help shape new behavioural patterns without exploiting or marginalizing the poorest in the system. Further progress on REDD is therefore contingent upon success in negotiating emission cuts at the global scale (see Chapter 4 for details).

Box 1.1 **Defining deforestation**

Forests are defined differently in different contexts:

The UNFCCC defined forests in the context of afforestation and reforestation CDM as 'A minimum area of land of 0.05–1.0 hectares with tree crown cover (or equivalent stocking level) of more than 10–30 per cent with trees with the potential to reach a minimum height of 2–5 metres at maturity *in situ*. A forest may consist either of closed forest formations where trees of various storeys and undergrowth cover a high proportion of the ground or open forest. Young natural stands and all plantations which have yet to reach a crown density of 10–30 per cent or tree height of 2–5 metres are included under forest, *as are areas normally forming part of the forest area which are temporarily unstocked as a result of human intervention such as harvesting or natural causes but which are expected to revert to forest*' [emphasis added].[69]

A number of counter-intuitive implications emerge from the UNFCCC definition including:[70]

- Conversion of forests to oil palm plantations is not considered 'deforestation';
- The current transformation of natural forest, after rounds of logging, into fastwood plantations occurs fully within the 'forest' category, out of reach of REDD policies;

> - Most tree crop production and agroforestry systems do meet the minimum requirements of forest, e.g. unpruned coffee can easily reach a height of 5 m, *but* are not 'forests' as long as they are not under the institutional frame and jurisdiction;
> - There is technically no deforestation in a country like Indonesia, as land remains under the institutional control of forest institutions, and is only 'temporarily unstocked'; and
> - Swiddening and shifting cultivation is not a driver of deforestation, as long as the fallow phase can be expected to reach minimum tree height and crown cover.

1.5 The research questions and the analytical framework

Climate change, thus, is clearly a serious problem; but governance efforts at reducing industrial emissions of GHGs are flagging and global attention has now turned to reducing deforestation and forest degradation as a way to address part of the GHG emissions. However, if past forest governance aimed at addressing deforestation at global level had been successful there would have been no need now for such action. This implies, prima facie, that there are serious social, economic, political and institutional challenges that hinder effective forest governance at global (see Chapters 2, 3 and 4) through to local level (see Chapters 5–9).

This implies that the development and implementation of any new instrument such as REDD will not take place in a vacuum. It must learn from past lessons and build on the existing framework of rules on forests at the differing levels of governance. Hence, the key research question is: Based on an assessment of the existing forest institutional arrangements at global to local level, how can a better policy framework (i.e. a combination of norms, principles and regulatory, market and suasive instruments) be designed to effectively and equitably govern the challenges of reducing deforestation and forest degradation at national to local levels under the global climate change regime?

In order to answer this question, an analytical framework has been developed. This framework draws upon six prominent methods for analysing responses to environmental problems. The research framework developed by the International Human Dimensions Programme's Project on the Institutional Dimensions of Global Environmental Change is the starting point.[71] This framework is very appropriate for the research question as it analyses whether individual policy response instruments can change the behaviour of social actors given drivers or contextual reasons towards reducing the impact on the state of the environment and enhancing social welfare. Based on an assessment of which instruments work, when and where, the model tries to determine how the instruments can be improved. This model has been made more dynamic by including an evolutionary aspect to it.

This model matches well with the Driving forces-Pressures-States-Impacts-Responses (DPSIR) method but provides more emphasis to the social science analysis of the instruments and their redesign.[72] This model is made multi-level by building on ideas emerging from multi-level governance theory.[73] It also draws on some elements from the Earth System Governance Project which focuses on issues of policy architecture (how is the policy field organized (A1)), agency (who are the key actors and how do they influence policy processes (A2)), and access and allocation (how is equity taken into account in the process (A3))[74] and looks at the cross-cutting themes of knowledge, scale, power and norms. A cosmopolitan touch has been added by drawing on insights from Third World Approaches to International Law[75] and their North–South Goals and counter goals (G, CG), Arguments and counter arguments (A, CA), Patterns and counter patterns (Pa, CPa) approach[76] to see if the analysis would change if specific attention is paid to what developing countries are saying. This global-to-local multi-layered analysis is primarily qualitative in nature. It is complemented by quantitative analysis using the economic model GTAPdyn,[77] a dynamic version of the well-known GTAP (Global Trade Analysis Project) model, a so-called Computable General Equilibrium (CGE) model to simulate the production, consumption and trade of multiple products across multiple countries.[78] The land market representation of the GTAPdyn model was improved for this study (see Figure 1.2).

The methodological implication of the integrated model is as follows. The first step is to create a conceptual framework based on the literature to understand the concept of the forest transition (i.e. what patterns emerge from an analysis of how forests have been managed in the past and what does this mean, if anything, for the future); relate it to the generic drivers of deforestation and forest degradation and the 26 response instruments that can be used to tackle these drivers (see Chapter 2).

The second step is to identify four case study countries for analysis at the national level that represent the three major basins of tropical rainforest in the world (see Figure 1.3). Cameroon is situated in the Congo Basin, Peru in the Amazon Basin, and Indonesia and Vietnam in South East Asia. These countries have more than 40 per cent of their land area covered by forests and are at different stages in terms of deforesting these areas. Multi-level layered case studies deal with forest governance issues at the national to local level. The case studies are based on a literature review, a content analysis of policy and legal documents, and stakeholder interviews. Their aim is to draw lessons for institutional design in general and in relation to REDD in particular.

The third step is to describe and analyse evolving forest institutions at multiple levels of governance in terms of the architecture, actors and processes. Chapters 3 and 4 cover the global level, Chapters 5–9 the national to local level.

18 *Joyeeta Gupta et al.*

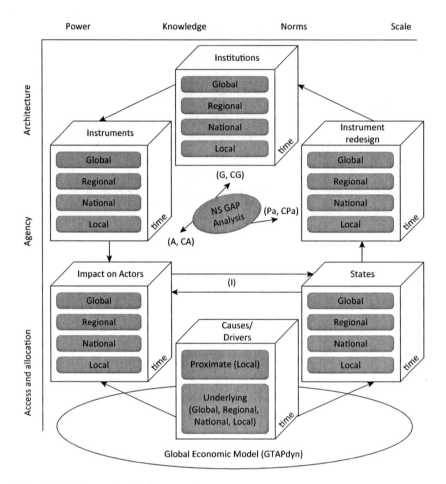

Figure 1.2 The analytical framework

The fourth step is to identify the proximate and underlying drivers of deforestation and forest degradation in the case study countries based on the literature and interviews (see Chapters 5–9). The fifth step is to identify the key policy instruments respectively, used at global to local level (see Chapters 3–9) in relation to the generic list and analysis in Chapter 2.

The sixth step is to assess the impact of forest instruments in changing actor behaviour, given the drivers that affect them (Chapters 5–9). Here, a rough overall estimate is made on the impact on the environment and on social issues. The case studies examine the most relevant of these instruments in their particular context. Based on the literature and interviews with stakeholders, these instruments have been qualitatively assessed regarding their strengths and weaknesses. Where the instruments are seen as achieving their

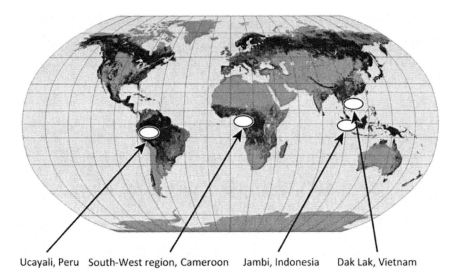

Ucayali, Peru   South-West region, Cameroon   Jambi, Indonesia   Dak Lak, Vietnam

Figure 1.3   The case study countries and regions

goals and forest conservation in an equitable manner, they get a maximum of + +. Here emphasis is given to the instrument's goal as well as forest conservation because some instruments ostensibly aim at forest conservation, but their design allows for forest conversion. Where they are seen as weak in terms of achieving their goals and forest conservation in an equitable manner they receive a maximum of – –. Where the instrument is quite successful but has some weaknesses, this is depicted as + + –. Where the instrument is primarily weak but has some strengths, this is depicted as + – –. Where the strengths and weaknesses appear evenly matched, it is represented as + –. Where the instrument is still being developed a question mark has been added.

The seventh step is to discuss the equity dimensions in an integrated and comparative manner in Section 9.4. The eighth step is to redesign instruments based on the lessons learned from the analysis of the existing instruments and an understanding of what works and what does not work in a specific context and why. The ninth step is the quantitative analysis which addresses the question of whether solving problems in one country leads to policy leakage elsewhere (see Chapter 10). Developing country perspectives and positions are emphasized throughout the research as part of the analysis and included in a story line in Chapter 11 which makes an effort to integrate the information gained in the different chapters.

The method is systematic and very useful for structuring the analysis. However, in analysing drivers and instruments, the research had to rely significantly more on interviews than on existing scientific analysis of these drivers and instruments; as such data was often not available in terms of effectiveness, equity and gendered implications. The qualitative assessments of the instruments is thus more to give an impression of how these instruments are perceived than an analysis of the exact effect of each instrument. Furthermore, this book focuses more on the potential of REDD to build on existing instruments than on the shifting discussions on REDD at national and international levels, including technical discussions on defining avoided emissions and how it should be financed internationally.

1.6 Inferences

This chapter has set the stage for the forthcoming discussions in this book. It has elaborated on the relationship between climate change, forests and REDD, and on the role of forests in providing ecosystem services. It has argued that the science and politics of forests are closely linked and contested. Forest definitions and statistics at global level are contested. It has further elaborated on the role of sovereignty in relation to forests and how this affects the framing of the forest problem, an issue discussed in the last chapter. It has elaborated on the diverging perceptions of the new instrument of REDD. In the climate change regime, REDD is seen as a young, low hanging fruit that helps to enlarge the negotiation pie in the international negotiations and to buy time, especially in the context of flagging negotiations. In the forest regime, which is used to incremental approaches and learning, this is a new way to infuse the forest negotiations with enthusiasm, resources and political commitment. Finally, it has explained the theoretical method that is used in this book.

Notes

1 The authors thank Heike Schroeder and Aarti Gupta for their extensive comments on a former draft of this chapter.
2 Williams, M. (2006) *Deforesting the earth, from prehistory to global crisis: an abridgement*, Chicago: The University of Chicago Press; Noordwijk Conference (1989) *Background paper on forestry and afforestation: The role of forests in relation to climate change*, Noordwijk: Noordwijk Conference.
3 KP (1997) *Kyoto Protocol to the United Nations Framework Convention on Climate Change*, 37 ILM 1998.
4 The Clean Development Mechanism allows developed country investors to invest in projects in developing countries in return for certified emission reduction credits. These projects also need to meet the criteria of sustainable development.
5 Agrawal, A., Nepstad, D. and Chhatre, A. (2011) Reducing emissions from deforestation and forest degradation, *Annual Review of Environmental Resources* 36, 11.1–24.
6 CO_2-equivalent: The amount of global warming a given type and amount of greenhouse gas may cause (over a specified period), using the functionally equivalent amount or concentration of carbon dioxide (CO_2) as the reference.
7 IPCC (2007) *Climate change 2007: The physical science basis, contribution of Working Group 1 to the fourth assessment report of the Intergovernmental Panel on Climate Change*, Cambridge: Cambridge University Press.

8 Miller, C. (2001) Hybrid management: Boundary organizations, science policy, and environmental governance in the climate regime, *Science Technology and Human Values* 26, 478–500.
9 UNFCCC (1992) *United Nations Framework Convention on Climate Change*, 31 ILM 849 (1992).
10 KP (1997) op. cit.
11 Gupta, J. and van der Grijp, N. (2010) Climate change, development and development cooperation, in Gupta, J. and van der Grijp, N., *Mainstreaming climate change in development cooperation: Theory, practice and implications for the European Union*, Cambridge: Cambridge University Press.
12 The Environmental Kuznets Curve describes a relationship between environmental pressure and per capita wealth in a country, suggesting that environmental pressure will decline once a certain level (threshold) of wealth has been reached. See for explanation and critique: Opschoor, J.B. (1995) Ecospace and the fall and rise of throughput intensity, *Ecological Economics* 15 (2), 137–41; De Bruyn, S.M. and Opschoor, J.B. (1997) Developments in the throughput–income relationship: Theoretical and empirical observations, *Ecological Economics* 20 (3), 255–69; Dinda, S. (2004) Environmental Kuznet curve hypothesis: A survey, *Ecological Economics* 49 (4), 431–55.
13 UNEP (2010) *Towards a green economy: Pathways to sustainable development and poverty eradication*, online. Available at http://www.unep.org/greeneconomy (accessed 13 February 2012).
14 UN Secretary General (2010) *Report of the Secretary General*, A.Conf./216/PC/2, 1 April 2010, para 44.
15 See Gupta, J. (1998) Leadership in the climate regime: Inspiring the commitment of developing countries in the post-Kyoto phase, *Review of European Community and International Environmental Law* 7 (2), 178–88; Gupta, J. (2009) Climate Law: Gap between normative rhetoric and politics, in Capaldo, G.Z. et al., *Yearbook of International Law and Jurisprudence*, Oxford: Oxford University Press.
16 Tinbergen, J. (1976) *Rio: Reshaping the international order – A report to the Club of Rome*, Boston: Dutton.
17 UNDP (2007) *Fighting climate change: Human solidarity in a divided world, Human Development Report 2007–2008*, New York: Palgrave Macmillan.
18 UNFCCC (1992) op. cit., Art. 3.
19 Rajamani, L. (2000) The principle of common but differentiated responsibility and the balance of commitments under the climate regime, *Review of European Community and International Environmental Law* 9 (2), 120–31; Anand, R. (2004) *International environmental justice: A North–South dimension*, London: Ashgate Publishing.
20 Weisslitz, M. (2002) Rethinking the equitable principle of common but differentiated responsibility: Differential versus absolute norms of compliance and contribution in the global climate change context, *Colorado Journal of International Environmental Law & Policy* 13 (2), 473–509.
21 Adams, T.B. (2003) Is there a legal future for sustainable development in global warming? Justice, economics and protecting the environment, *Georgetown International Environmental Law Review* 16, 77–132.
22 Bafundo, N.E. (2006) Compliance with the ozone treaty: Weak states and the principle of common but differentiated responsibilities, *American University International Law Review* 21 (3), 461–95.

23 Honkonen, T. (2009) The principle of common but differentiated responsibilities in post-2012 climate negotiations, *Review of European Community and International Environmental Law* 18 (3), 257–9.
24 Gupta, J. (2003) Engaging developing countries in climate change: (KISS and Make-Up!), in Michel, D., *Climate policy for the 21st century: Meeting the long-term challenge of global warming*, Washington DC: John's Hopkins University Press.
25 Gupta, J. (1997) *The climate change convention and developing countries – From conflict to consensus?* Dordrecht: Kluwer Academic Publishers.
26 FAO (2011) *State of the world's forests 2011*, Rome: FAO.
27 Millennium Ecosystem Assessment (2005) *Ecosystems and human well-being: synthesis*, Washington DC: Island Press.
28 World Bank (2004) *Sustaining forests: A development strategy*, Washington DC: The World Bank.
29 Schmitt, C.B., Belokurov, A., Besançon, C., Boisrobert, L., Burgess, N.D., Campbell, A., Coad, L., Fish, L., Gliddon, D., Humphries, K., Kapos, V., Loucks, C., Lysenko, I., Miles, L., Mills, C., Minnemeyer, S., Pistorius, T., Ravilious, C., Steininger, M. and Winkel, G. (2009) *Global ecological forest classification and forest protected area gap analysis: Analyses and recommendations in view of the 10% target for forest protection under the Convention on Biological Diversity (CBD)*, 2nd revised edition, Freiburg, Germany: Freiburg University Press.
30 FAO (2011) op. cit.
31 IPCC (2007) op. cit. Aboveground carbon stocks in the natural forest lands around the world amounted to about 1,146 Gt (giga ton), while in the 1st, 2nd and 3rd metres of the soil profile, soils globally contain 1,502, 491 and 351 Petagram Carbon (Pg C), respectively. See Jobbágy, E.G. and Jackson, R.B. (2000) The vertical distribution of soil organic carbon and its relation to climate and vegetation, *Ecological Applications* 10, 423–36.
32 FAO (2011) op. cit.
33 FAO (2011) op. cit.; World Bank (2004) op. cit., although it is unclear what forest definitions are used for these assessments.
34 Kanninen, M., Murdiyarso, D., Seymour, F., Angelsen, A., Wunder, S. and German, L. (2007) *Do trees grow on money?* Bogor: CIFOR.
35 FAO (2011) op. cit.
36 FAO (2011) op. cit.
37 FAO (2011) op. cit.
38 Van der Werf, G.R., Randerson, J.T., Giglio, L., Collatz, G.J., Kasibhatla, P.S. and Arellano, A.F. (2006) Interannual variability in global biomass burning emissions from 1997 to 2004, *Atmospheric Chemistry and Physics* 6, 3423–41.
39 Putz, F.E., Zuidema, P.A., Pinard, M.A., Boot, R.G.A., Sayer, J.A. and Sheil, D. (2008) Improved tropical forest management for carbon retention, *PLoS Biology* 6, e166.
40 Van der Werf, G.R., Morton, D.C., DeFries, R.S., Olivier, J.G.J., Kasibhatla, P.S., Jackson, R.B., Collatz, G.J. and Randerson, J.T. (2009) CO_2 emissions from forest loss, *Nature Geoscience* 2 (11), 737–8.
41 Stern, N. (2006) *Stern review: the economics of climate change*, London: HM Treasury.
42 FAO (2011) op. cit.
43 FAO (2011) op. cit.
44 FAO (2011) op. cit.
45 Convention on International Trade in Endangered Species of Wild Fauna and Flora (Washington), 3 March 1973, in force 1 July 1975, 12 ILM (1973), 1085.

46 Convention for the Protection of the World Cultural and Natural Heritage (Paris), 16 November 1972, in force 17 December 1975, 11 ILM (1972), 1358.
47 1971 Convention on Wetlands of International Importance especially as Waterfowl Habitat (Ramsar), 11 ILM (1972), 963.
48 UNCTAD (2006) *International Tropical Timber Agreement*, United Nations Conference on Trade and Development, TD/TIMBER.3/12., Art.1.
49 E.g. Nagtzaam, G.J. (2008) *The international Tropical Timber Organization and conservationist forestry norms: A bridge too far?* Victoria: Monash University, online. Available: http://works.bepress.com/gerry_nagtzaam/1 (accessed 13 February 2012).
50 See, for example, Tallontire, A. and Blowfield, M.E. (2000) Will the WTO prevent the growth of ethical trade? Implications of potential changes to WTO rules for environmental and social standards in the forest sector, *Journal of International Development* 12 (4), 571–84.
51 Lyke, J. and Fletcher, S.R. (1992) *Deforestation: An overview of global programs and agreements*, Washington DC: CRS Report 92-764 ENR.
52 Colchester, M. and Lohman, L. (1990) *The Tropical Forestry Action Plan: What progress?* The World Rainforest Movement.
53 The International Labour Organization's (ILO) Conventions Nos. 107 and 169 establish legally binding rights for indigenous people to the lands that they have traditionally occupied, to use the natural resources on these lands, and to be involved in the decision-making procedures in their country. The United Nations Permanent Forum on Indigenous Issues (UNPFII 1982/2000) adopted the non-binding but highly influential Declaration on the Rights of Indigenous Peoples, upholding the rights of indigenous people to 'own, use, develop and control the lands, territories, and resources that they possess' (Article 26).
54 Non-Legally Binding Authoritative Statement of Principles for a Global Consensus on the Management, Conservation and Sustainable Development of all Types of Forests, adopted at UNCED 1992, 31 ILM (1992), 1333.
55 Humphreys, D. (2006) *Logjam – Deforestation and the crisis of global governance*, Earthscan forestry library, London: Earthscan; Humphreys, D. (1996) *Forest politics: The evolution of international cooperation*, London: Earthscan.
56 United Nations Framework Convention on Climate Change (New York), 9 May 1992, in force 24 March 1994; 31 ILM 1992, 822.
57 E.g. Hunt, C.A.G. (2009) *Carbon sinks and climate change: Forests in the fight against global warming*, Cheltenham: Edward Elgar Publishing; Streck, C. and Scholz, S.M. (2006) The role of forests in global climate change: whence we come and where we go, *International Affairs* 82 (5), 861–79; Haupt, F. and von Lupke, H. (2012) *Obstacles and opportunities for afforestation and reforestation projects under the Clean Development Mechanism of the Kyoto Protocol*, Rome: FAO.
58 E.g. Beinart, W. (2000) African history and environmental history, *African Affairs* 99, 269–302; Greenblatt, S. (1991) *Marvelous possessions: The wonder of the New World*, Chicago: University of Chicago Press.
59 Williams (2006) op. cit.
60 FAO (2011) op. cit.
61 RRI and ITTO (2009) *Tropical forest tenure assessment*, Washington DC and Yokohama: Rights and Resources Initiative and International Tropical Timber Organization.
62 Noordwijk Conference (1989) op. cit.
63 Noordwijk Conference (1989) op. cit., pp.1–2.

64 Noordwijk Conference (1989) op. cit., p. 3.
65 Anon. (1989) *The Noordwijk declaration on climate change*, Leidschendam, the Netherlands: VROM, UNEP, WMO, Art. 23.
66 Stern (2006) op. cit.
67 Rokityanskiy, D., Benítez, P., Kraxner, F., McCallum, F., Obersteiner, M., Rametsteiner, E. and Yamagata, Y. (2007) Geographically explicit global modeling of land-use change, carbon sequestration, and biomass supply, *Technological Forecasting and Social Change* 74 (4), 1057–82.
68 Van Noordwijk, M. and Minang, P.A. (2009) If we cannot define it, we cannot save it, in van Bodegom, A.J., Savenije, H. and Wit, M., *Forests and climate change: adaptation and mitigation* (pp. 5–10), Wageningen: Tropenbos International.
69 FCCC/CP/2001/13/Add.1, Decision 11/CP.7, Land use, land-use change and forestry, Annex A. Definitions.
70 Van Noordwijk and Minang (2009) op. cit.
71 Young, O.R., Agrawal, A., King, L.A. and Sand, P.H.W.M. (1999) *Institutional dimensions of global environmental change (IDGEC) Science Plan, IHDP Report No.9*, Bonn.
72 The method postulates that drivers lead to pollutants that put pressure on the system and this changes the state of the environment, which has impacts on health and ecosystems. This calls for response strategies that can focus on drivers, pressures, state, or impact. Atkins, J.P., Gregory, A.J., Burdon, D. and Elliott, M. (2011) Managing the marine environment: Is the DPSIR framework holistic enough? *Systems Research and Behavioral Science* 28 (5), 497–508; Svarstad, H., Petersen, L.K., Rothman, D., Siepel, H. and Wätzold, F. (2008) Discursive biases of the environmental research framework DPSIR, *Land Use Policy* 25 (1), 116–25.
73 Hooghe, L. and Marks, G. (2003) Unraveling the Central State, but how? Types of multi-level governance, *The American Political Science Review* 97 (2), 233–43; Termeer, C.J.A.M., Dewulf, A. and van Lieshout, M. (2010) Disentangling scale approaches in governance research: comparing monocentric, multilevel, and adaptive governance, *Ecology and Society* 15 (4), 29,online. Available at: http://www.ecologyandsociety.org/vol15/iss4/art29/.
74 Biermann, F., Betsill, M.M., Gupta, J., Kanie, N., Lebel, L., Liverman, D., Schroeder, H., Siebenhüner, B. and Zondervan, R. (2010) Earth system governance: a research framework, *International Environmental Agreements: Politics, Law and Economics* 10 (4), 277–98.
75 Chimni, B.S. (2006) Third World approaches to international law: A manifesto, *International Community Law Review* 8 (1), 3–27; Khosla, M. (2007) The TWAIL discourse: The emergence of a new phase, *International Community Law Review* 9 (3), 291–304.
76 Gupta, J. (2011) Climate Change: A GAP analysis based on Third World approaches to international law, *German Yearbook of International Law* 53, 341–70.
77 Ianchovichina, E. and McDougall, R.A. (2001) *Theoretical structure of Dynamic GTAP*, GTAP Technical Paper 17, West Lafayette, Indiana, USA: Purdue University.
78 Hertel, Th.W. (1997) *Global trade analysis: modeling and applications*, Cambridge: Cambridge University Press.

2 The forest transition, the drivers of deforestation and governance approaches

Joyeeta Gupta, Hye Young Shin, Robin Matthews, Patrick Meyfroidt and Onno Kuik[1]

2.1 Introduction

Any effort at understanding the global forest challenge has to begin with an evaluation of how societies evolve over time. Transition analysis looks at the underlying processes influencing past empirical patterns as a way of understanding changes that may occur in the future. For example, the demographic transition approach[2] shows that as societies develop, the death rate falls first, then the birth rate, leading to a rapid rise in population and eventually stabilizing at a newer and higher equilibrium. Although global population growth rates have peaked, population is expected to rise to 9 billion by 2050.[3] The economic take-off concept,[4] with all its inherent limitations, argues that societies evolve from subsistence economies, through specialization, trade and industrialization to maturity with a dominant service sector. Similarly, the Environmental Kuznets Curve shows that as societies pass beyond a critical point in income per capita, they will pollute less and less in relation to their income.[5]

2.2 The forest transition

The forest transition concept[6] describes empirical patterns of change in forest cover resulting from the exploitation by human societies of their forest resources. Essentially a reversed Environmental Kuznets Curve, the concept has had some success in describing the patterns in North America, Europe and some tropical countries,[7] but due to the multitude of factors, both existing and emerging, affecting deforestation, the likelihood of a pattern becoming reality depends on the interplay of the contextual circumstances, the driving factors and the policy regime in each specific locality, as well as other transitions occurring in society (see Figure 2.1). As such, the concept has only limited predictive power. Nevertheless, it does provide a framework of how and why transition processes come into play. These transition processes can be explained by two factors: first, the contextual issues and drivers that lead social actors to deforest or degrade their forest resources (see 2.3). Such drivers can include policy processes – such as policies that stimulate economic

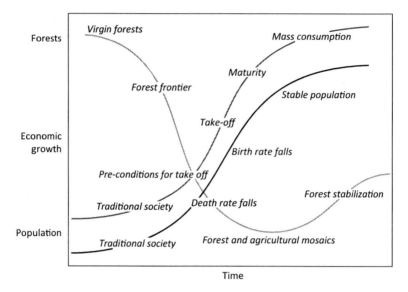

Figure 2.1 Stylized representations of the demographic, economic and forest transitions

development. Second, deliberate forest policies adopted by state and non-state actors can influence actors to change their behaviour (see 2.4). The effectiveness of these policies depends on their contextual relevance and their ability to neutralize the effect of the drivers of deforestation. This may even require a social and economic transformation that modifies the structure of land use at a national scale. This chapter presents a typology of forest instruments; the evolutionary adoption of these instruments in specific contexts and in relation to the forest transition is explored in subsequent chapters. The focus on drivers and instruments is in line with the analytical framework (see 1.5). It shows that REDD (see Chapters 1 and 4) will not fall on barren ground; there is already much where it can, and perhaps *should*, build on.

Understanding the trajectory that a country is on in terms of deforestation is essential for understanding the role of deforestation drivers and forest policies and equitable baselines against which to measure progress. The forest transition describes forest cover changes that occur as societies industrialize and urbanize, with forest cover declining to a minimum, then slowly rising and eventually stabilizing and increasing in some cases. Initially, the forest cover is relatively undisturbed, with poor infrastructure and market access making it inaccessible for commercial exploitation.[8] Deforestation is triggered by several drivers and accelerated by positive feedback loops leading to a possible *forest frontier* period. Eventually, high deforestation rates bring about forest scarcity which slows deforestation rates, leading to a third *forest and agricultural mosaics stage*. Finally, forest cover may begin to increase again.

Several broad typical pathways of forest transitions can be described.[9] The *economic development* path is associated with increasing urbanization and higher earning capacity, drawing labour from the surrounding rural areas. This relative rural depopulation means that fewer people are making their livelihoods from the land, and landowners gradually switch from labour-intensive agriculture to less labour-demanding activities such as forestry, so that forest cover increases, either through plantations or through regeneration of indigenous forest due to farm abandonment. The *forest scarcity* route is taken when declining forests leads to a shortage of forest products, hence rising prices, which in turn leads to increased tree planting and protection of remaining forests. A third related route is the combination of multiple pathways and pressures from environmental movements and local communities and state motivation and sensitivity to such pressure leading to reductions in deforestation as in Costa Rica and Brazil.[10]

Forest transitions have a contingent character.[11] They have occurred in European and North American countries over the past two centuries, and recent forest transitions have occurred in developing countries like Vietnam, India, China, Costa Rica, Cuba, El Salvador, Chile and Uruguay (the latter two only when including plantations).[12] The slopes (deforestation and reforestation speed) and the turning points (minimum forest cover) of the Forest Transition Curve across countries may vary,[13] and policies may influence the slopes and turning points.[14] Yet, the multiple trends that may result in a forest transition – natural regeneration of forests, forest plantation, adoption of agroforestry, continuing deforestation – combine in various ways through time and space. The causes of *reforestation* also vary geographically.[15] Reforestation of abandoned land is more common in Central America and the Caribbean, mostly associated with economic changes and globalization. Large, industrial tree plantations dominate in subtropical and temperate South America. South and East Asia contain the largest share of the world's tree plantations; driving forces include decentralization and state-sponsored programmes. Land-use policies and agricultural changes also contribute to forest regrowth in Asia. In sub-Saharan Africa, forest plantations and agroforestry expand locally in countries with high population densities and supportive forest policies.[16]

Box 2.1 **An Environmental Kuznets Curve for forest cover**

The Forest Transition Curve (see 2.2) plots forest cover against time. Transforming the Forest Transition Curve into an Environmental Kuznets Curve requires plotting on the horizontal axis some measure of income (or wealth) instead of time over a sufficiently long period of time. One could then compare the individual country curves to statistically check whether one dynamic pattern emerges. A recent study employing this method found evidence of inverted

U curves for Latin America and Africa, but not for Asia.[17] A simpler approach is to take a cross-section of many countries at different stages of development at only one small period in time. The idea is that if the correlation between income and forest cover would be strong enough, the cross-section would show some statistical correlation, given a large enough sample. The annual deforestation rates in the period 1990–2010 of a panel of 215 countries and regions show a decreasing trend with per capita income (Top panel).[18] But the variation in deforestation rates is very large and the trend only explains about nine per cent of the variation ($R^2 = 0.088$). By plotting the share of land under forest cover as a function of per capita income (Bottom panel), the variation is even larger. The trend vaguely resembles the forest transition curve, but it explains only three per cent of the variation ($R^2 = 0.032$). These plots suggest that there must be more factors explaining rates of deforestation and shares of forest cover, and that per capita income only explains a small part. This confirms the findings of earlier literature.[19]

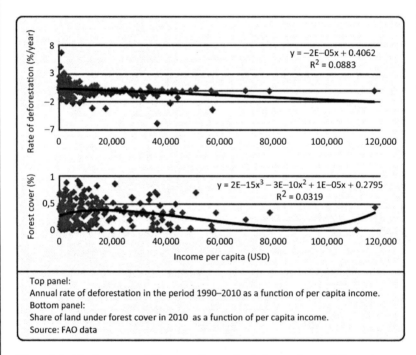

Top panel:
Annual rate of deforestation in the period 1990–2010 as a function of per capita income.
Bottom panel:
Share of land under forest cover in 2010 as a function of per capita income.
Source: FAO data

Figure 2.2 Environmental Kuznets Curves for deforestation and forest cover

Countries do not necessarily experience a forest transition with economic development (see Box 2.1). Multiple causes and effects can be associated with forest transitions depending on the geographic, social and environmental contexts. In several countries, rising urban wage rates did not lead to rural–urban migration, nor was there a sufficient demand for forest products, so that the resulting poverty trap aggravated deforestation.[20] In some countries, civil wars increased deforestation rates, while in others (e.g. Indonesia, Cameroon), weak forest governance and strong vested interests prevented a transition from occurring. Per capita income and the extent of national forest cover (as indicators of the economic development and forest scarcity pathways, respectively) explain some country variation, but other factors such as corruption are also important. Thus, there remains uncertainty about the causes of forest loss and forest poverty and about effective policy responses.[21]

2.3 The drivers of deforestation

2.3.1 *Generic drivers and forest transitions*

'Drivers' refer to the causes of a problem. The generic drivers of deforestation include proximate (PD) and underlying drivers (UD).[22] Proximate drivers directly affect the behaviour of a local actor to engage in deforestation and/or forest degradation. Underlying drivers indirectly influence the proximate drivers and via that have an effect on deforestation. Proximate drivers are mostly local. Underlying drivers can be local to global in scope (see Table 2.1). The literature also refers to other drivers including civil war and biophysical factors. This book subsumes biophysical factors under direct drivers. Each driver is also associated with specific actors – for example, shifting agriculture is associated with subsistence farmers, while cash crops are associated with commercial farmers.

These drivers change depending on the geographic and historic contexts. In an ideal-typical sense, these drivers can be linked to the forest transition, *ceteris paribus*. When the virgin forests are intact, local communities have no reason to deforest or degrade the forests, enjoying instead their customary usufructuary rights. However, such forests can still be affected by climate variability and change. As societies develop, practise shifting cultivation, wood extraction, mining and other such activities, they begin to deforest the lands. In the third stage, urbanization, industrialization, technology development, and economic progress, stronger development policies, weak forest policies and unclear property rights are key drivers of deforestation. In the fourth stage, economic growth leads to more advanced technology and stability in the use of land resources; the land rights are clearer and the institutions are stronger, leading to stable forest cover (see Figure 2.3).

Ideal-typical representations simplify complex realities. However, there are context-specific differences across countries. For example, Papua New Guinea has shifted from 'virgin forests' to industrial logging virtually without any

Table 2.1 Proximate and underlying causes of deforestation and forest degradation

Drivers of deforestation and forest degradation

Proximate drivers (PD)	Underlying drivers (UD)
Local to national	*Local*
Agriculture: permanent (cash crops), shifting agriculture, animal husbandry	Economic: poverty
	Local and national
Extraction: commercial, charcoal, fuelwood, polewood, non-timber forest products	Cultural: attitudes, behaviour
	National
Infrastructure: transport (roads), markets, settlements, large infrastructure (e.g. dams)	Demographic: growth, migration, density, distribution, urbanization
	Economic: markets, economic structures
Industry: mining	Technological: agro-technical change, agricultural productivity, productivity in the processing sector
Biophysical: drought, floods, fire	Non-forest policies: agriculture, infrastructure, development policy
	Other institutional factors: policy climate including lack of resources/ capacity, corruption, property rights
	Global
	Demand/trade/consumption at global level for food, fibre, fuel, timber and other forest products; minerals; international debt; climate change

Source: Adapted from Geist, H.J. and Lambin, E.F. (2002) Proximate causes and underlying driving forces of tropical deforestation, *Bioscience* 52 (2), 143–50.

form of 'society development' or shifting cultivation. The following sections discuss context-specific forest developments in different regions, further explored in Chapters 5–9.

2.3.2 Drivers of deforestation in different regions

Forests in South and South East Asia are mainly tropical, and spread across India, Nepal, Sri Lanka, Bangladesh, Myanmar, Thailand, Cambodia, Lao PDR, Vietnam, Indonesia, Malaysia, Philippines and New Guinea. These forests account for 18 per cent of global tropical forests,[23] shelter millions of people and many rare plant and animal species. They include four of the 25 world biodiversity hotspots.[24] Colonization led to increased deforestation in many of these countries, but after independence, deforestation rates dropped

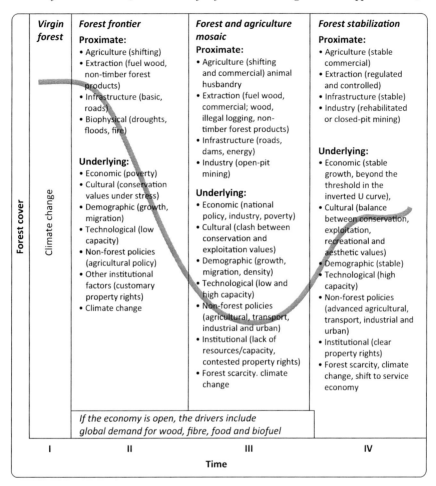

Figure 2.3 Drivers and the forest transition: an ideal typical representation

(Sources: adapted from Geist, H.J. and Lambin, E.F. (2002) Proximate causes and underlying driving of tropical deforestation, *Bioscience* 52 (2), 143–50; Mather, A.S. (1992) The forest transition, *Area* 24 (4) 367–79; Angelsen, A. (2007) Forest cover change in space and time: combining the von Thünen and forest transition theories, *World Bank Policy Research Working Paper*, Washington DC: The World Bank.

slightly or increased due to, amongst other factors, national development policies, implementation challenges and global resource demand. In recent decades, tropical deforestation rates in Asia were twice those of Latin America or Africa.[25] Forest-rich countries (Indonesia, Malaysia, Myanmar, Lao PDR, Cambodia and Papua New Guinea) face greater threats of deforestation than forest-poor countries (except Nepal and Sri Lanka). Annual deforestation rates are highest in Philippines, Indonesia and Cambodia. In *absolute terms*, annual deforestation is highest in Indonesia (18,710 km^2 per year). Some countries have similar drivers (e.g. expanding oil palm in Malaysia and Indonesia; institutional factors in Papua New Guinea and Indonesia). The proximate drivers are agricultural expansion, logging, oil palm and other cash crops (e.g. rubber, coffee), and timber, pulp and paper driven by the economic interests of private companies (driven by rising global demand). Underlying drivers include government development policies, global demand for forest commodities, poor governance, population growth and density, poverty and forest-dependence (particularly for indigenous groups). Forest fires (natural, especially in El Niño years, and human-induced) open up forest areas for economic activity.

The South and Central American region has the largest area of contiguous tropical forest on Earth, the Amazon Basin, and, as such, has global significance in terms of biodiversity and climate regulation. The continent also includes seasonally dry tropical, mountain and Atlantic forests. Between 2000 and 2005 the highest annual rates of deforestation were in Honduras, Nicaragua and Guatemala. Brazil lost the largest area of forest annually (31,030 km^2 per year). Up until 1990, deforestation resulted from shifting cattle ranching and cultivation, often driven by government programmes and subsidies.[26] Today, mechanized cash crop agriculture (e.g. soybean) is another – smaller – driver. Timber extraction and mining play a minor role. In Argentina, cattle production in the seasonally dry Chaco forest[27] and agriculture (especially for soybean production) are key drivers.[28] Bolivia and Paraguay are also experiencing conversion of forest to crop land with their seasonally dry forests with flat terrains being most threatened.[29] In less politically stable states, illegal crop (e.g. coca and opium poppies) production is also a significant driver, especially in the humid slopes of the Andes.[30] Twenty-four per cent of deforestation in the Peruvian Amazon can be attributed to coca production.[31] In Mexico the key driver is cash crop expansion.[32] In many cases small-scale farmers, out-competed by large-scale commercial growers, sell cleared land and then move on to clear new areas. Roads (e.g. the BR163 in Brazil from Cuiaba to the port of Santarem in Pará) are a key driver of deforestation. The Initiative for Integration of Regional Infrastructure in South America to link transport, energy and telecommunications in 12 countries could accelerate deforestation considerably.[33] Timber extraction, including illegal extraction, in Brazil,[34] and selective logging cause forest degradation[35] and loss of species.[36] Gold mining and oil exploration,[37] natural, climate-change-related[38] and human-induced fires[39] are also significant drivers of deforestation in South and Central America. Since 2004, the Brazilian government and smallholder organizations have reduced deforestation by 50 per cent by, inter alia, establishing 20 million hectares of parks, indigenous reserves[40] and protected areas, particularly in Pará.

African forests, unevenly distributed across the continent, accounted for 16.7 per cent of global forests in 2010. The Congo Basin in Central Africa is the world's second largest dense tropical rainforest. Since 1990, African forest loss accounts for almost 55 per cent of global deforestation. However, the 'low forest cover' countries of Northern Africa are quite stable.[41] Agricultural activities including shifting cultivation (Democratic Republic of Congo, Zambia, Cameroon and Madagascar)[42] and cash crops (e.g. sugar cane in Uganda, coffee and cocoa in Cameroon, maize in Madagascar, *khat* in Ethiopia and oil palm in Congo) are important drivers of deforestation. Logging (legal, illegal and selective) has a significant impact as well. Although selective logging removes few trees, it may trigger large-scale deforestation by providing access to previously remote areas, resulting in migration into the area, and leading ultimately to further clearing for agriculture (e.g. Congo and Gabon). The dependence of Africans on fuel wood (e.g. Congo) also increases pressure on forests. Such practices are associated with population growth, migration due to war, and structural adjustment policies imposed from outside the country.

2.4 Instruments of forest governance

2.4.1 Introduction

This section links the forest transition story and the drivers of deforestation to forest policy (see 2.4.2). It presents a typology of forest-policy instruments (see 2.4.3), which is subsequently elaborated in subsections on regulatory (see 2.4.4), economic and market (see 2.4.5), suasive, information and research (see 2.4.6), and voluntary management instruments (see 2.4.7). The last subsection links these instruments to the drivers they address (see 2.4.8). This typology is subsequently applied in the following chapters both in terms of how individual instruments have evolved in specific contexts and how they may have influenced the forest transition in these countries.

2.4.2 Forest transitions, drivers and forest policy

Designing policy instruments that address deforestation requires understanding of the complex contextual feedbacks between drivers and policies. Policies will only be effective if they address the relevant drivers of deforestation and land degradation.[43] Policy measures may not always have the desired effect due to a partial understanding of the forces at work.[44] High timber prices can stimulate sustainable management of plantations and secondary forests but can also provoke mining of old-growth forests. Improving agricultural technology and crop yields can relieve pressure on forests, or encourage more deforestation if the surpluses generated are used for forest clearing.[45] Higher farm incomes may also attract more migrants to the forest region, resulting in more land cleared.

Furthermore, policies can be designed in a static or dynamic manner. Policies can focus on simply halting deforestation (see Figure 2.4 (a)) or on accelerating and shifting the forest transition curve upwards (see Figure 2.4(b)), which is possibly more practical and sustainable.[46]

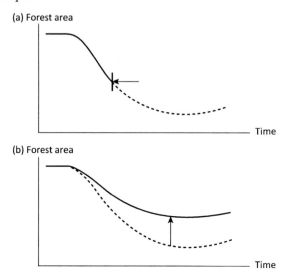

Figure 2.4 Halting deforestation (a) or moving the forest transition curve upwards (b)

Source: Lambin, E.F. and Meyfroidt, P. (2010) Land use transitions: Socio-ecological feedback versus socio-economic change, *Land Use Policy* 27 (2), 108–18.

2.4.3 Classifying governance instruments

Policy refers to a 'course of action or principle adopted or proposed by a government, party, business or individual'.[47] Policy instruments are 'tools' used by state or non-state actors to implement the policy that 'are designed to cause people to do things, refrain from doing things, or continue doing things they would otherwise not do'.[48] They can be classified as regulatory (command-and-control and coercive, which include permission, prohibition, and standard setting and enforcement), economic and market-based (which promote the incorporation of environmental costs and benefits into the budgets of households and enterprises) and suasive/research/information instruments. Furthermore, in the case of forests, there are also 'management' options that form a separate category (see Table 2.2). A choice can be made between the above different classes of instruments, different types within a class (e.g. a tax or a subsidy within the economic class), and combinations of the above (instrument mixes). A measure is a specific instrument that is adopted in a specific situation (e.g. fire control measures). This section classifies and lists policy instruments. It is not ideal as there are often overlaps. However, it is functional and allows for further systematic analysis.

2.4.4 Regulatory instruments

Regulatory instruments operate at different levels of governance. At the international governance level are (i) trade restrictions, which prohibit or limit the

Table 2.2 Classifying policy instruments on forest management

Category	Definition (including actors)	Links to literature
Regulatory	Includes policies that are binding, do not have a market component; mostly adopted by state authority	Weiss 2000; Bemelmans-Videc et al. 1998; 'compulsory' – Rivera 2002; World Bank 1997
Economic, financial and market incentives	Includes public and private options to encourage actors to adopt a specific kind of behaviour; is often backed by regulatory measures;	'economic' – Weiss 2000; 'carrots' – Bemelmans-Videc et al. 1998; mixed – Rivera 2002; 'using markets' and 'creating markets'– World Bank 1997
Suasive, informational and research	Includes public and private information tools, research, public awareness	Weiss 2000; 'sermons' – Bemelmans – Videc et al. 1998; 'engaging the public'– World Bank 1997
Management	Includes mostly private actors, self-management; voluntary management and hybrid management processes	'voluntary' – Rivera 2002

Sources: Weiss, G. (2000) Evaluation of policy instruments for protective forest management in Austria, *Forest Policy and Economics* 1, 243-255; Bemelmans-Videc, M., Rist, L. and Vedung, R.C. (eds.) (1998) Carrots, sticks, and sermons: Policy instruments and their evaluation, Transaction Publisher, Inc., New Brunswick, NJ, p.30; Rivera, J. (2002) Assessing the voluntary environmental initiative in the developing world: the Costa Rican certification for sustainable tourism, *Policy Science* 35, 333–360; World Bank (1997) Five years after Rio: Innovations in environmental policy, *Environmentally sustainable development studies and monograph series*, no. 18, Washington DC, USA: World Bank, p.6.

trade in specific types of products. These restrictions can be unilateral – when a government decides it will not import products that do not meet specific standards (e.g. the US Lacey Act);[49] or bilateral – when two countries decide jointly on standards for export and import (e.g. Peru–US Free Trade Agreement).[50] These instruments also focus on the demand for forest products in the developed countries. Important at the national level is (ii) decentralization, the process by which authority to manage resources is shifted to lower government authorities in the hope that they have better and more relevant local interests and capabilities to deal with local problems. In the forest arena, many regard this as a critical instrument.[51] (iii) Policymakers often adopt a principle that influences policies. This is a tool used to shape other tools; e.g. sustainable forest management (SFM), once adopted, returns in different policy instruments. (iv) States often use binding forest rules and regulations which can include policies on reforestation, methods for fire prevention and management especially in relation to climate change; current and anticipated pest and disease control; creation of forest habitat corridors; improved tree species selection; the introduction of irrigation and/or fertilizers to compensate for changing climatic conditions; and the use of biotechnology to introduce adaptive species. Enhancing the carbon content of forests could lead to rules regarding the number of plantations and their intensity and agroforestry. Such rules may be incidental or integrated within comprehensive rules regarding SFM.[52]

Box 2.2 **Protected Areas**

A Protected Area (PA) is a spatial planning tool to reserve certain areas for limited purposes such as the conservation of forests and their biodiversity.[53] There is an international classification of protected areas[54] with different regimes, ranging from very strict (only conservation-related use allowed) to multi-use (sustainable use by local inhabitants is allowed).[55] PAs have expanded rapidly in tropical forest countries in recent years,[56] and it is arguably the most prominent and best-funded conservation instrument.[57] In the 1990s, PAs were seen as ineffective, 'paper parks'.[58] In the 2000s, improved monitoring techniques (especially remote sensing) and studies suggested that PAs are effective in protecting the forests.[59] Recent studies have suggested, however, that problems with establishing counter-factuals (i.e. would these lands have otherwise been deforested?) might have biased this evidence.[60] This bias would be even stronger if protection would lead to leakage, i.e. if deforestation outside the PAs would increase *because of* the protection.[61] New research confirms that in many countries PAs are biased to higher elevations, steeper slopes and greater distances to roads and cities;[62] that the overall effectiveness of PAs is less than suggested by earlier studies, and that 'location' matters.[63] Protection has the highest impact for flat and less remote areas that face the highest deforestation pressure, even if it is not strictly focused on conservation, but allows for multiple-use.[64] Other factors with a positive impact on the effectiveness of PAs include adequate funding, enforcement[65] and community participation in forest governance.[66]

(v) Spatial planning can be used for different land use options. Spatial planning refers to a state public policy instrument to assign certain spaces/areas to certain uses. Thus, for example, the government can allocate land to agriculture, cities, forests and so on. Social actors need planning permission before they can undertake activities in specific areas and this permission is given based on the allocation of land. A key spatial planning tool is Protected

Areas[67] (see Box 2.2). (vi) In order to regulate access to forest areas, concessions and permits may be allocated, as well as rules regarding 'allowable cut' which states what amount of forests can be cut down[68] and the provision and retiring of concessions/rights to industry.[69]

(vii) Land rights are a critical aspect. Forest land rights are generally bundles of formal and informal property rights in relation to forests.[70] Forest 'owners' include the state, the community, the user groups and individuals. The bundle of ownership rights include: (a) Usufructuary rights including the right of access (e.g. for birdwatching; allowing animals to cross the land), use (e.g. using fruits and medicines of the forests for family and household needs), and exploitation (e.g. harvest specific products even for commercial goals); (b) Control rights including the right to manage or co-manage a forest and the right to exclude the participation of others); and (c) Alienation rights (i.e. the ability to rent or sell a forest).[71] In forests, as in fresh water and fish ownership,[72] there are generally no exclusive ownership rights. The different rights are spread among the different 'owners'; and the recognition of these rights is also difficult to identify. Sometimes, the state may own the forest in terms of the right of alienation, but the community may hold the right of use. Sometimes, the community and the state may have joint rights of control. Sometimes, the community may not be aware of the state's claim on the land and vice versa.

Regulatory approaches include (viii) reporting by domestic actors on their various activities as a way to monitor how policies are being implemented, (ix) monitoring and surveillance of the above approaches through, for example, inspection teams and remote sensing surveillance;[73] and (x) improving law enforcement through setting up an administrative and judicial system to ensure that civil and criminal law procedures are in place to discourage the violation of the law.[74]

Box 2.3 **Payment for ecosystem services (PES)**

Payment for ecosystem services (PES) is an instrument that is based on voluntary transactions between buyers and sellers in which payments are conditional on maintaining an ecosystem use that provides well-defined environmental services (see Table 1.1).[75] Payments can, for example, be made by government agencies, companies, international non-governmental organizations (NGOs), international banks and development organizations.[76] Interest in PES schemes in developing countries has rapidly increased in recent years,[77] not least because it is believed that they can achieve a win–win situation by simultaneously protecting ecosystems *and* alleviating poverty *and* attracting donor funds.[78] But how effective is PES in protecting ecosystem services or in avoiding deforestation? The challenge of assessing PES effectiveness is similar to that of Protected Areas (see Box 2.2). One needs to determine what would have happened in the absence of a PES scheme. Only few PES evaluations meet certain minimum methodological requirements.[79] The (limited) available evidence suggests small environmental effects of PES so far, partly because suppliers appear to prefer to offer land without alternative uses (without a deforestation threat) to the PES scheme.[80] Economists call this 'adverse selection', which leads to inefficiencies. There are ways to improve the design of PES schemes to avoid adverse selection (and moral hazard).[81] How effective is PES in reducing poverty? Here, a distinction must be made between poor service providers and poor service users. Preliminary evidence suggests that the effect on the welfare of service providers is positive but small, provided that the resources are not subject to elite capture. The welfare effects on poor ecosystem users (that do not participate in the PES scheme) are mixed: some may benefit from the better protection of ecosystem services; others may lose because of use restrictions that are implemented to protect the ecosystems. In the aggregate, in most cases PES may have positive but small poverty-reduction effects.[82] On a somewhat different note, the fear has been expressed that the introduction of PES schemes might change the mindset of ecosystem providers: from doing what is considered appropriate for the community to calculating what the most lucrative action is from a private perspective.[83]

2.4.5 Economic and market instruments

The next category of policy instruments includes economic and market incentives, generally adopted by public and private actors. These include (xi) import and export tariffs for forest products. Those States that are party to the World Trade Organization should respect its rules and disciplines in this policy area and are therefore somewhat restricted in their use. (xii) International funds, grants and loans such as those provided by the Global Environment Facility are important international tools to help bridge the financial, technological and capacity gap in many developing countries. (xiii) Taxes on forest products, severance stumpage and (xiv) subsidies[84] can serve as important incentives to promote change in human behaviour through financial carrots and sticks. (xv) A relatively new instrument is that of payment for ecosystem services (PES)[85] (see Box 2.3). (xvi) Micro-credit is one way to provide financial resources to local farmers to reduce their need to exploit forests.[86] However, in recent years, there has been considerable discussion of the very high rate of interest charged to these farmers because of the transaction costs of making such loans that reduce the actual attractiveness of this instrument in specific circumstances. (xvii) A relatively popular instrument in the 1990s was debt-for-nature swaps (see Box 2.4). Although this instrument is less popular today, there are important lessons to be learnt from it. (xviii) Carbon offset funds such as the Clean Development Fund[87] (see Chapter 4) are also ways to raise financial resources for additional and verifiable emissions reduction from projects that also meet the criteria of sustainable development. (xix) Corporate social responsibility (CSR) (where companies adopt more socially conscious behaviour) and ecotourism (where companies charge a premium price to tourists and promote more sustainable sites of tourism) are voluntary programmes of individual companies who try to meet certain social and environmental standards, while certification schemes attempt at promoting sustainable production processes in producer companies in order to sell these products at a premium price to conscious consumers (see Box 2.5).

Box 2.4 **Debt-for-nature swaps**

Debt-for-nature swaps allow a country's foreign debt to be cancelled or forgiven, if the debtor government invests an (almost) equivalent amount in domestic currency in domestic environmental projects.[88] A debt-for-nature swap involves a debtor and national bank, an investor, a lending bank and a nature organization. A foreign actor (environmental organization/ aid agency) pays off the loan from the lending bank at a discounted price directly or indirectly via the secondary market. The debt purchaser converts this debt into local currency through the debtor country's national bank often at lower than face value which is then invested directly or indirectly in environmental projects.[89] The first swap in 1987 between Bolivia and Conservation International (US-INGO) involved cancelling Bolivia's foreign debt of USD 650,000 in exchange for USD 100,000 worth of local currency to protect the Beni Biosphere Reserve.[90]

Foreign debt is a driver of deforestation when it is paid back through exports of tropical wood products[91] and of agricultural and livestock commodities often encouraged through structural adjustment programmes (see Chapter 7). However, other studies suggest no clear link between deforestation and debt.[92]

Debt-for-nature swaps can increase domestic environmental awareness[93] and reduce debt as a driver of deforestation.[94] However, in practice, they exacerbate inflation,[95] address only a small part of the debt (e.g. Costa Rica)[96] thus not always addressing foreign debt as a driver; cannot address other deforestation drivers, have small direct impacts,[97] and there is potential for leakage.[98] Debtors benefit through debt reduction, environmental protection and a better credit rating. They face the risk of increased inflation, 'overpaying' for redeeming the debt, public discontent and ignoring the opportunity costs of creating PAs. International investors can purchase/subsidise greater conservation efforts through fewer resources, improve their reputation and their links with financial institutions; however there is a risk of high transaction costs involved and project failure. The creditor banks benefit by dealing with bad debts, improving their relations with debtor countries, and acquiring goodwill. However, they have to accept a loss on the balance sheet and the risk of moral hazard (that others don't pay back loans) free-riding.[99]

> *Box 2.5* **Forest certification**
>
> Forest certification implies that forest management is inspected and certified by an independent body based on assessing the management processes and verifying the chain of custody of forest products from the producer to the end user.[100] A certification system involves a producer who meets forest management standards, an independent certifier, and a consumer who pays a premium price for wood labelled as certified.
>
> Certification creates management standards and promotes best practices for logging companies, raises the quality of the forest products, generates public awareness in producer and consumer countries, leads to dialogue on forest land tenure, generates resources for SFM, assesses the willingness to pay for SFM,[101] and addresses the driver of industrial logging.[102] However, international consumers may be unwilling to pay the premium price, certifying companies may not be effective, the process of developing standards non-inclusive, forest certification in tropical areas can be quite expensive, requiring capacity building and training[103] (most certification takes place in the temperate zones)[104] and certification has no effect on local markets. International certification schemes are discussed in Chapter 3.

2.4.6 *Suasive, information and research instruments*

Suasive instruments are those that aim to convince actors to change their behaviour. This includes, for example, (xx) principles that have a normative force. The adoption of international and/or national principles often has a learning effect on individual actors leading to the converging adoption of common concepts and ideas. (xxi) Soft law targets can also provide a common aspirational direction for policies by all social actors. For example, the Millennium Development Goals were soft law targets that have provided a benchmark against which country achievements can be measured. (xxii) A key instrument is research[105] and monitoring and remote sensing activities which help to develop new knowledge about emerging problems (such as on drought-resistant seedlings that can help deal with climatic impacts on forests), why they are important and how these can be dealt with.[106] (xxiii) Such knowledge can also be used to feed into educational and information sharing tools.[107] International policy guidance can also be used to promote information and education.

2.4.7 *Management measures*

Management measures refer to voluntary management by forest dwellers alone or in collaboration with other actors (e.g. non-governmental organizations, local municipalities). It excludes management that falls under the Protected Areas scheme. Increasingly, states are trying to incorporate local

The forest transition, the drivers of deforestation and governance approaches 41

management efforts into national governance schemes and thus the distinction between a regulatory and non-regulatory measure becomes increasingly blurred.

These measures include community-based schemes, NGO-based schemes and joint schemes. (xxiv) Community forest management, community-based forest management, common property resource management and community natural resource management all more or less refer to the approaches developed by local communities to manage their own resources (see Box 2.6). There are also forests that are managed by (xxv) non-governmental organizations.[108] Finally, there is (xxvi) joint forest management where different local actors combine their expertise and interests to manage forest resources. Sometimes, such management is referred to as stewardship.

Box 2.6 **Community-based forest management**

Community-based forest management (CBM) implies self-government of forests by the communities and other stakeholders (e.g. local to international NGOs and local government) directly or indirectly dependent on the forest.[109] CBM is generally financed by the communities themselves in cash or in kind, but sometimes (aided) by local government or NGOs. Most forests fall under state jurisdiction. However, forests have traditionally been managed by local communities and community-based organizations, and this continues to be the case in many places.[110] But how effective is community-based management in avoiding deforestation? The seminal work of Elinor Ostrom focuses on the potential efficiency of CBM of common pool resources, such as forests.[111] There is no guarantee, however, that local communities will not deforest.[112] There is also no guarantee that CBM will benefit the poorest forest dwellers.[113] The returns to the local communities may be low and alternative land use options may be more profitable,[114] local groups are not always harmonious communities,[115] local social justice issues may remain unaddressed, and national and international drivers of deforestation cannot be addressed. Research suggests that, on balance, CBM may be effective. Research across 80 forests in ten countries in Asia, Africa and Latin America suggests that greater rule-making autonomy at the local level is associated with less deforestation (more carbon storage) and livelihood benefits.[116] With this in mind, some scholars fear that REDD (with its potentially high financial stakes) may pose a threat to the trend of decentralization and community-based forest management.[117]

2.4.8 *Forest instruments and drivers*

We now turn to establish the relationship between key forest instruments and drivers. Based on an examination of the literature, Table 2.3 sums up the key instruments, and how they deal with specific measures and drivers.

Some of these instruments are possibly more likely to emerge in different phases of the forest transition. For example, one can hypothesize that parks

Table 2.3 Forest policy instruments, measures and drivers

	Instrument	Counters driver(s)	And promotes sustainable practices by (for example)
Regulatory	i) Trade restrictions	Global demand for timber and non-timber forest products	Banning trade in illegally*/ unsustainably harvested timber
	ii) Decentralization/ subsidiarity	Policy and institutional factors	Tuning policies to local needs and circumstances
	iii) Principles (including SFM)	Wood extraction	Prescribing sustainable management
	iv) Binding forest rules, reforestation, fire rules	Wood extraction, other factors	Regulatory requirements, prohibitions
	v) Spatial planning – Protected Areas	Infrastructure extension, agricultural expansion, demographic factors (migration, distribution)	Zoning regulations, land use planning
	vi) Sustainable forest logging/ concessions	Wood extraction	Sustainable forest logging/ concessions
	vii) Land rights	Policy and institutional factors	Establishing land rights
	viii) Reporting	Infrastructure extension, agricultural expansion, wood extraction	Reporting illegal* practices
	ix) Monitoring/ surveillance	Infrastructure extension, agricultural expansion, wood extraction	Monitoring illegal* practices
	x) Law enforcement	Infrastructure extension, agricultural expansion, wood extraction	Punishing illegal* practices

(*Continued*)

Table 2.3 (Contd)

		Instrument	Counters driver(s)	And promotes sustainable practices by (for example)
Economic and market	xi)	Export and import tariffs	Global demand	Export tariff on logs or biofuels
	xii)	International funds, grants, loans	Policy and institutional factors, poverty	World Bank loans
	xiii)	Taxes (on forest products)	Economic factors, global demand	Timber tax
	xiv)	Subsidies (for alternative work, etc.)	Economic factors, poverty	Subsidy on alternative employment
	xv)	PES	Economic factors, poverty	Payments for maintaining forest ecosystem services
	xvi)	Micro-credit	Economic factors, poverty	Giving access to credit
	xvii)	Debt-for-nature swap	International debt	Supporting conservation
	xviii)	Carbon offsetting funds (CDM, voluntary)	Economic factors, poverty	Payments for storing carbon in forests
	xix)	CSR/certification/ ecotourism	Economic factors, poverty, tourism demand	Promoting sustainable forest management
Suasive research	xx)	(Soft international law) principles	Policy and institutional factors	Inviting government to adopt principles
	xxi)	(Soft international law) targets	Policy and institutional factors	Inviting government to adopt these targets
	xxii)	Research/ monitoring	Technological factors, cultural factors, (policy and institutional factors, demographic factors)	By encouraging research institutes and local communities to engage in research and/or participatory monitoring
	xxiii)	Information and education	Cultural factors	Information campaigns
L Mgt	xxiv)	Community based management	Wood extraction, agricultural expansion	Generally voluntary action by local communities, can be stimulated by government
	xxv)	NGO-based management		Generally voluntary action by NGOs
	xxvi)	Joint management		State promotes joint management

NB* Illegality is a contested term when it refers to customary access activities that have been labelled as illegal (see 9.4) are created when deforestation rates accelerate in a country, that subsequently decentralization of forest policies emerges, and that subsidies for reforestation may emerge at later stages of the forest transition.[118] This will be further explored inductively in Section 11.3.2.

2.5 Inferences

This chapter leads to inferences that are of relevance to the following chapters. First, the literature indicates that like most transition theories, forest transition theory cannot be interpreted as a predictive mechanism but can help to identify ideal typical drivers that may apply at different stages of the forest transition. The confluence of context, drivers and policies will determine the transition path a specific country may follow in dealing with forests. Besides, the forest transition will always occur in the context of other transitions. Second, the literature helps to identify generic proximate and underlying drivers that lead to deforestation and forest degradation that may apply within different countries. The drivers change in accordance with local social and biophysical factors (e.g. population density, income, yields, availability of forestlands and many other factors). In specific contexts the drivers may be different. Where countries are closed to international trade, there will be limited influence of global drivers. The degree to which the economy is open and affected by globalization (e.g. increasing demand, integration in global financial system, international debt) may expose a country to greater global drivers.

Third, four classes of instruments have been identified (regulatory, economic, suasive, management) that promote the adoption of measures to deal with local through to global proximate drivers. Some of these instruments are linked to international mechanisms and governance processes. Instruments to mitigate and adapt to climate change have been integrated into the above framework. The adoption of instruments may be linked to specific stages in a country's forest transition, but may also be exogenously influenced by international policies as will be discussed in subsequent chapters. The relatively new instrument of REDD (see Chapter 4) will need to build on the experiences with these instruments, and the institutions established to implement them.

Fourth, the bulk of the available policy instruments tend to focus on local and proximate drivers. Some address institutional drivers. There are very few instruments that address global underlying (e.g. world demand) and national underlying drivers (e.g. population growth, the perceived need for economic growth) (see also Chapters 5–9 and 11).

Fifth, each instrument has a specific task and deals possibly only with few specific drivers (e.g. debt-for-nature swap with foreign currency debt; certification with international demand). These instruments need to be used in combination to deal with the multiple drivers that may operate within a specific context.

Sixth, the box items (Box 2.3, Box 2.4 and Box 2.5) suggest that economic instruments might be potentially promising policy instruments for avoiding deforestation; one main challenge is raising funds for implementing them (see Chapters 5–9 and 11).

Seventh, although instruments aim to benefit local people and communities, there are risks that the schemes are only cost-effective and environmentally effective when they protect large tracts of land and may thus exacerbate local inequities.

What does this chapter imply for the new instrument of REDD? Although REDD is explored in further detail in Chapter 4, this chapter shows that REDD has not emerged in a historical vacuum and designing REDD may need to take into account the lessons learnt from the transition analysis, the drivers of deforestation and degradation, and REDD implementation can build on the existing types of policy instruments to deal with forests.

Notes

1 The authors would like to thank Tom Rudel and Eric Lambin for their extensive comments on a previous draft of this chapter.
2 Thompson, W.S. (1929) Population, *American Journal of Sociology* 34, 959–75.
3 United Nations (2009) World population prospects: The 2008 revision.
4 Rostow, W.W. (1960) The stages of economic growth: A non-communist manifesto, Cambridge: Cambridge University Press.
5 Grossman, G.M. and Krueger, A.B. (1995) Economic growth and the environment, *Quarterly Journal of Economics* 110 (2), 353–77.
6 Mather, A.S. (1992) The forest transition, *Area* 24 (4), 367–79.
7 Rudel, T.K., Coomes, O.T., Moran, E., Achard, F., Angelsen, A., Xu, J. and Lambin, E.F. (2005) Forest transitions: towards a global understanding of land use change, *Global Environmental Change* 15, 23–31.
8 Angelsen, A. (2007) Forest cover change in space and time: combining the von Thünen and forest transition theories, *World Bank Policy Research Working Paper*, Washington DC: The World Bank.
9 Rudel *et al.* (2005) op. cit.; Meyfroidt, P. and Lambin, E.F. (2011) Global forest transition: Prospects for an end to deforestation, *Annual Review of Environment and Resources* 36, 343–71.
10 Boucher, D. (2008) What REDD can do: the economics and development of reducing emissions from deforestation and forest degradation, Paper presented at the World Bank workshop: The costs of reducing carbon emissions from deforestation and forest degradation, Washington DC: The World Bank.
11 Meyfroidt and Lambin (2011) op. cit.
12 Meyfroidt and Lambin (2011) op. cit.
13 Rudel, T.K. (1998) Is there a forest transition? Deforestation, reforestation and development, *Rural Sociology* 65, 533–52.
14 Angelsen (2007) op. cit.
15 Meyfroidt and Lambin (2011) op. cit.
16 Rudel, T.K. (2009) Tree farms: Driving forces and regional patterns in global expansion of forest plantations, *Land Use Policy* 26, 545–50.

17 Culas, R.J. (2012) REDD and forest transition: Tunneling through the environmental Kuznets curve, *Ecological Economics* 79, 44–51. Doi:10.1016/j.ecolecon.2012.04.015. The author used a number of conditioning variables (such as absolute forest area, share of forest land, population density, indebtedness, export price index, etc.) to explain for additional heterogeneity across countries.
18 FAO (2011) State of the world's forests 2011, Rome: FAO.
19 Meyfroidt, P., Rudel, T.K. and Lambin, E.F. (2010) Forest transitions, trade, and the global displacement of land use, *Proceedings of the National Academy of Sciences* 107 (49), 20917–22; Meyfroidt and Lambin (2011) op. cit.; Culas (2012) op. cit.
20 Rudel *et al.* (2005) op. cit.
21 Chomitz, K.M., Buys, P., De Luca, G., Thomas, T.S. and Wertz-Kanounnikoff, S. (2006) *At Loggerheads? Agricultural expansion, poverty reduction and environment in the tropical forests*, Washington DC: World Bank.
22 Geist, H.J. and Lambin, E.F. (2002) Proximate causes and underlying driving forces of tropical deforestation, *Bioscience* 52 (2), 143–50.
23 Laurance, W.F. (2007) Forest destruction in tropical Asia, *Current Science* 93 (11), 1544–50.
24 Myers, N., Mittermeier, R., Mittermeier, C.G., Fonseca, G.A. and Kent, J. (2000) Biodiversity hotspots for conservation priorities, *Nature* 403, 853–8.
25 Laurance (2007) op. cit.
26 Grau, H.R. and Aide, M. (2008) Globalization and land use transitions in Latin America, *Ecology and Society* 13 (2), 1–16.
27 Grau, H.R., Gasparri, N.I. and Aide, T.M. (2008) Balancing food production and nature conservation in the Neotropical dry forests of northern Argentina, *Global Change Biology* 14 (5), 985–97.
28 Zak, M.R., Cabido, M., Caceres, D. and Diaz, S. (2008) What drives accelerated land cover change in central Argentina? Synergistic consequences of climatic, socioeconomic, and technological factors, *Environmental Management* 42 (2), 181–9.
29 Grau and Aide (2008) op. cit.
30 Grau and Aide (2008) op. cit.
31 Moran, E.F. (1993) Deforestation and land use in the Brazilian Amazon, *Human Ecology* 21 (1), 1–21.
32 Bray, D.B., Ellis, E.A., Armijo-Canto, N. and Beck, C.T. (2004) The institutional drivers of sustainable landscapes: a case study of the 'Mayan Zone' in Quintana Roo, Mexico, *Land Use Policy* 21, 333–46.
33 McCormick, S. (2007) Troublesome construction: IIRSA and public–private partnerships in road infrastructure, *Journal of Latin American Studies* 39, 917–18.
34 Barreto, P., Souza Jr, C. and Noguerón, R. (2006) Human pressure on the Brazilian Amazon forests, World Resources Institute, Belem, Brazil (accessed on 24 September 2008 at www.globalforestwatch.org/common/pdf/Human_Pressure_Final_English.pdf)
35 Asner, G.P., Broadbent, E.N., Oliveira, P.J.C., Keller, M., Knapp, D.E. and Silva, J.N.M. (2006) Condition and fate of logged forests in the Brazilian Amazon, *Proceedings of the National Academy of Sciences of the United States of America* 103 (34),12947–50.
36 Escobal, J. and Aldana, U. (2003) Are non-timber forest products the antidote to rainforest degradation? Brazil nut extraction in Madre De Dios, Peru, *World Development* 31 (11), 1873–87.

37 Kirby, K.R., Laurance, W.F., Albernaz, A.K., Schroth, G., Fearnside, P.M., Bergen, S., Venticinque, E.M.and da Costa, C. (2006) The future of deforestation in the Brazilian Amazon, *Futures* 38 (4), 432–53; Finer, M., Jenkins, C.N., Pimm, S., Keane, B. and Ross, C. (2008) Oil and gas projects in the Western Amazon: threats to wilderness, biodiversity and indigenous peoples, *PLoS ONE* 3 (8), e2932.
38 Cox, P.M., Betts, R.A., Collins, M., Harris, P.P., Huntingford, C. and Jones, C.D. (2004) Amazonian forest dieback under climate–carbon cycle projections for the 21st century, *Theoretical and Applied Climatology* 78, 137–56.
39 Cochrane, M.A. and Barber, C.P. (2008) Climate change, human land use and future fires in the Amazon, *Global Change Biology* 15 (3), 601–12.
40 Nepstad, D., Schwartzman, S., Bamberger, B., Santilli, M. and Ray, D. (2006) Inhibition of Amazon deforestation and fire by parks and indigenous lands, *Conservation Biology* 20 (1), 65– 73.
41 FAO (2010) Global forest resources assessment 2010, *FAO Forestry Paper 163*, Rome: Food and Agriculture Organisation of the United Nations.
42 Jarosz, L. (1993) Defining and explaining tropical deforestation: shifting cultivation and population growth in colonial Madagascar, *Economic Geography* 64 (9), 366–80; Ickowitz, A. (2006) Shifting cultivation and deforestation in tropical Africa: Critical reflections, *Development and Change* 37 (3),599–626.
43 Forner, C. *et al.* (2006) Keeping the forest for the climate's sake: avoiding deforestation in developing countries under the UNFCCC, *Climate Policy* 6 (3), 275–94; Young, O.R., Agrawal, A., King, L.A. and Sand, P.H.W.M. (1999) *Institutional dimensions of global environmental change (IDGEC) Science Plan, IHDP Report No. 9*, Bonn; Miles, L. and Kapos, V. (2008) Reducing greenhouse gas emissions from deforestation and forest degradation: Global land-use implications, *Science* 320, 1454–5.
44 Chomitz *et al*. (2006) op. cit.
45 Angelsen, A. and Kaimowitz, D. (1999) Rethinking the causes of deforestation: Lessons from economic models, *The World Bank Research Observer* 14 (1), 73–98.
46 Lambin, E.F. and Meyfroidt, P. (2010) Land use transitions: Socio-ecological feedback versus socio-economic change, *Land Use Policy* 27 (2), 108–18.
47 OECD Glossary of Statistical Terms, http://stats.oecd.org/glossary/ (accessed 18 May 2012).
48 Anderson, J.E. (2010) *Public policymaking: An introduction*, Hampshire UK: Cengage Learning, p. 242.
49 See Chapter 3 of this book.
50 See Chapter 8 of this book.
51 Edmunds, E. and Wollenberg, E. (2003) Whose devolution is it anyway? Divergent constructs, interests and capabilities between poorest forest users, in: Edmunds, E. and Wollenberg, E., *Local forest management: The impacts of devolution policies*, 150–66, London: Earthscan.
52 Fischlin, A. and Midgley, G.F. (2007) Ecosystems, their properties, goods and services, in: M. L. Parry, O.F. Canziani, J. P. Palutikof, P. J. van der Linden and C. E. Hanson (eds), *Climate change 2007:Impacts, adaptation and vulnerability*, Contribution of Working Group II to the Fourth Assessment Report of the Intergovernmental Panel of Climate Change (IPCC), Cambridge, UK, Cambridge University Press, 211–72; Cubbage, F. and Sills, E. (2007) Policy instruments to enhance multi-functional forest management, *Forest Policy and Economics*

9, 833–51; MacDonald, M.A. (2003) The role of corridors in biodiversity conservation in production forest landscapes: a literature review, *TASFORESTS* 14, 41–52; Spittlehouse, D.L. and Stewart, R.B. (2003) Adaptation to climate change in forest management, *BC Journal of Ecosystems and Management*, 4 (1); Nabuurs, G.J., Masera, O., Andrasko, K., Benitez-Ponce, P., Boer, R., Dutschke, M., Elsiddig, E., Ford-Robertson, J., Frumhoff, P., Karjalainen, T., Kurz, W.A., Matsumoto, M., Oyhantcabal, W., Ravindranath, N.H., Sanz Sanchez, M.J. and Zhang, X. (2007) Forestry, in Metz, B., Davidson, O.R., Bosch, P.R., Dave, R. and Meyer, L.A., Climate Change 2007: *Mitigation. Contribution of Working Group III to the Fourth Assessment Report of the Intergovernmental Panel on Climate Change* (pp. 542–84), Cambridge, UK and New York, NY, USA: Cambridge University Press; Sterner (2003) Policy instruments for environmental and natural resource management, Washington DC: Resources for the Future; Nepstad, D., Lefebvre, P., da Silva, U.L., Tomasella, J., Schlesinger, L., Solórzano, L., Moutinho, P., Ray, D. and Benito, J.G. (2004) Amazon drought and its implications for forest flammability and tree growth: A basin-wide analysis, *Global Change Biology* 10 (5), 704–17.
53 IUCN (1994) Guidelines for protected area management categories, IUCN: Cambridge, UK and Gland, Switzerland.
54 IUCN (1994) op. cit.
55 Nelson, A. and Chomitz, K.M. (2011) Effectiveness of strict vs. multiple use protected areas in reducing tropical forest fires: a global analysis using matching methods, *PLoS ONE* 6 (8), e22722.
56 Coad, L., Campbell, A., Granziera, A., Burgess, N. and Fish, L. (2008) State of the world's protected areas in 2007 – An annual review of global conservation progress, Cambridge, UK: UNEP-WCMC.
57 Nelson and Chomitz (2011) op. cit.
58 Nelson and Chomitz (2011) op. cit.
59 Clark, S., Bolt, K. and Campbell, A. (2008) *Protected areas: an effective tool to reduce emissions from deforestation and forest degradation in developing countries?* Working Paper, Cambridge, UK: UNEP World Conservation Monitoring Centre.
60 Joppa, L.N. and Pfaff, A. (2009) High and far: Biases in the location of protected areas, *PLoS ONE* 4 (12), e8273; Nelson and Chomitz (2011) op. cit.
61 Discussion on leakage is not settled yet. Oliveira, P.J.C., Asner, G.P., Knapp, D.E., Almeyda, A., Galván-Gildemeister, R., Keene, S., Raybin, R.F. and Smith, R.C. (2007) Land-Use allocation protects the Peruvian Amazon, *Science* 317, 1233–6 find, for a sustainably managed timber concession in Peru, leakage rates of 100 per cent, while Andam, K.S., Ferraro, P.J., Pfaff, A., Sanchez-Azofeifa, G.A. and Robalino, J.A. (2008) Measuring the effectiveness of protected area networks in reducing deforestation, *PNAS* 105 (42), 16089–94 find 'negligible' leakage due to protected areas in Costa Rica. For more discussion, see Ewers, R.M. and Rodrigues, A.S.L. (2008) Estimates of reserve effectiveness are confounded by leakage, *Trends in Ecology and Evolution* 23 (3), 113–16 and Chapter 10 of this book.
62 Joppa and Pfaff (2009) op. cit.
63 Pfaff, A., Robalino, J.A., Sanchez-Azofeifa, G.A., Andam, K.S. and Ferraro, P.J. (2009) Park location affects forest protection: Land characteristics cause differences in park impacts across Costa Rica, *The BE Journal of Economic Analysis and Policy* 9 (2), Art. 5.

64 Nelson and Chomitz (2011) op. cit.
65 Clark *et al.* (2008) op. cit.
66 Persha, L., Agrawal, A. and Chhatre, A. (2011) Social and ecological synergy: Local rulemaking, forest livelihoods, and biodiversity conservation, *Science* 331, 1606–8.
67 Gullison, R.E. (2003) Does forest certification conserve biodiversity? *Oryx* 37 (2), 153–65.
68 Sterner (2003) op. cit.
69 Gullison (2003) op. cit.
70 Sterner (2003) op. cit.; Padgee, A., Kim, Y.-S. and Daugherty, P.-J. (2006) What makes community forest management successful: a meta-study from community forests throughout the world, *Society and Natural Resources* 19, 33–52.; Nabuurs *et al.* (2007) op. cit.; Stern, N. (2006) *Stern review: the economics of climate change*, London: HM Treasury.
71 Barry, D. and Meinzen-Dick, R. (2008) The invisible map: community tenure rights, *Proceedings of the 12th Biennial Conference of the International Association for the Study of Commons*, Cheltenham, England.
72 Bavinck, M. (2005) Understanding fisheries conflicts in the south – a legal pluralist perspective, *Society and Natural Resources* 18, 805–20.
73 Oestreicher, J.S., Benessaiah, K., Ruiz-Jaen, M.C., Sloan, S., Turner, K., Pelletier, J., Guay, B. and Clark, K.E. (2009) Avoiding deforestation in Panamanian protected areas: an analysis of protection effectiveness and implications for reducing emissions from deforestation and forest degradation, *Global Environmental Change* 19, 279–91.
74 Struhsaker, T.T., Struhsaker, P.J. and Siex, K.S. (2005) Conserving Africa's rain forests: problems in protected areas and possible solutions, *Biological Conservation* 123, 45–54.
75 Wunder, S. (2007) The efficiency of payments for environmental services in tropical conservation, *Conservation Biology* 21 (1), 48–58.
76 Pagiola, S. and Platais, G. (2005) *Introduction to payments for environmental services*, Presentation, Washington DC: The World Bank.
77 Pattanayak, S.K., Wunder, S. and Ferraro, P.J. (2010) Show me the money: do payments supply environmental services in developing countries? *Review of Environmental Economics and Policy* 4 (2), 254–74.
78 Ferraro, P.J., Lawlor, K., Mullan, K.L. and Pattanayak, S.K. (2012) Forest figures: Ecosystem services valuation and policy evaluation in developing countries, *Review of Environmental Economics and Policy* 6 (1), 20–44.
79 Pattanayak *et al.* (2010) op. cit.
80 Arriagada, R., Sills, E., Pattanayak, S.K. and Ferraro, P. (2009) Combining qualitative and quantitative methods to evaluate participation in Costa Rica's program of Payments for Environmental Services, *Journal of Sustainable Forestry* 28 (3), 343–67.
81 Ferraro, P. (2008) Asymmetric information and contract design for payments for environmental services, *Ecological Economics* 65 (4), 810–21.
82 Wunder, S. (2008) Payments for environmental services and the poor: concepts and preliminary evidence, *Environment and Development Economics* 13 (3), 279–97.
83 Vatn, A. (2010) An institutional analysis of payments for environmental services, *Ecological Economics* 69 (6), 1245–52.
84 For alternative work, reforestation activities, etc.; Chomitz *et al.* (2006) op. cit.

50 *Joyeeta Gupta et al.*

85 Forner *et al.* (2006) op. cit.
86 Sterner (2003) op. cit.; Cubbage and Sills (2007) op. cit.
87 Nabuurs et al. (2007) op. cit.; Stern (2006) op. cit.; Brown, K. and Corbera, E. (2003) Exploring equity and sustainable development in the new carbon economy, *Climate Policy* 3 (S1), S41–S56; Chomitz *et al.* (2006) op. cit.; Karousakis, K. and Corfee-Morlot, J. (2007) Financing mechanisms to reduce emissions from deforestation: issues in design and implementation, COM/ENV/EPOC/IEA/SLT(2007)7, Paris: OECD.
88 Thapa, B. (2000) The relationship between debt-for-nature swaps and protected area tourism: a plausible strategy for developing countries, *USDA Forest Service Proceedings* RMRS-P-0, 268–72.
89 Asiedu-Akrofi, D. (1991) Debt-for-nature swap: Extending the frontiers of innovative financing in support of the global environment, *International Lawyer* 25 (3), 557–86; Thapa, B. (1998) Debt-for-nature swaps: an overview, *International Journal of Sustainable Development and World Ecology* 5 (4), 249–62; Hansen, S. (1989) Debt for nature swaps – Overview and discussion of key issues, *Ecological Economics* 1 (1), 77–93, p.78.
90 Thapa (1998) op. cit., p.249.
91 Kahn, J.R. and McDonald, J.A. (1995) Third-world debt and tropical deforestation, *Ecological Economics* 12 (2), 107–23, pp.107, 122.
92 Pfaff, A., Sills, E.O., Amacher, G.S., Coren, M. J., Lawlor, K. and Streck, C. (2010) *Policy impacts on deforestation: Lessons learned from past experiences to inform new initiatives,* Nicholas Institute for Environmental Policy Solutions, Duke University.
93 Asiedu-Akrofi (1991) op. cit., p.577.
94 Kahn and McDonald (1995) op. cit., p.122.
95 Asiedu-Akrofi (1991) op. cit., p.578.
96 Thapa (1998) op. cit., p.260.
97 Chambers, P. E., Jensen, R. and Whitehead, J. C. (1996) Debt-for-nature swaps as noncooperative outcomes, *Ecological Economics* 19 (2), 135–46., p.144; Pfaff *et al.* (2010) op. cit., p.6.
98 Hansen, S. (1989) op. cit., p.86; cited in Asiedu-Akrofi (1991) op. cit., p.563.
99 Dogse, P. and von Droste, B. (1990) Debt-for-nature exchanges and biosphere reserves: Experiences and potential, MAB Digest 6, Paris: UNESCO; Thapa (1998) op. cit., p.259.
100 Burger, D. and von Kruedener, B. (1998) *Forest certification: Status report and overview,* Forest Certification- Working Paper No. 1, Programme Office for Social and Ecological Standards, Deutsche Gesellschaft für Technische Zusammenarbeit (GTZ) GmbH., p.4.
101 Molnar, A. (2003) Forest certification and communities: looking forward to the next decade, in *Forest Trends 2003,* Washington DC: Forest Trends, p.4.
102 Dudley, N., Jeanrenaud, J. and Sullivan, F. (1995) *Bad harvest? The timber trade and degradation of the world's forests,* Earthscan, London, UK; cited in Gullison (2003), p.153.
103 Ibid: 32.
104 Leslie, A.D. (2004) The impacts and mechanics of certification, *International Forestry Review* 6 (1), 30–39, p.31.
105 Cubbage, F., Moore, S., Henderson, T. and Araujo, M. (2009) Costs and benefits of forest certification in the Americas, in J.B. Pauling (ed.), Natural resources:

Management, economic development and protection, New York: Nova Science Publishers.
106 Barret, T.M., Andersen, H.E. and Winterberger, K.C. (2009) Integrating field and lidar data to monitor Alaska's boreal forest, available online at blue.for.msu.edu/meeting/proceed.php; Chung, S.Y., Yim, J.S., Cho, H.K. and Shin, M.Y. (2009) Comparison of forest biomass estimation methods by combining satellite data and field data, available online at http://blue.for.msu.edu/meeting/proceed.php; Ducey, M.J., Zheng, D. and Heath, L.S. (2009) Creating compatible maps from remotely sensed and forest inventory data using radial basis functions, available online at blue.for.msu.edu/meeting/proceed.php
107 Cubbage *et al.* (2009) op. cit.; Scherr, S.J, White, A. and Kaimowitz, D. (2003) Making markets work for forest communities, *International Forestry Review* 5 (1), 67–73.
108 Edmunds and Wollenberg (2003) op. cit.
109 Carter, J. and Gronow, J. (2005) *Recent experience in collaborative forest management: A review paper*, Bogor: Center for International Forestry Research (CIFOR).
110 Local communities and organizations govern almost 200 million hectares of forest more than in the 1980s: Agrawal, A., Chhatre, A. and Hardin, R. (2008) Changing governance of the world's forests, *Science* 320 (5882), 1460–62.
111 Ostrom, E. (1990) *Governing the Commons: The evolution of institutions for collective action*, Cambridge: Cambridge University Press; Ostrom, E. (1999) *Self-governance and forest resources*, Occasional Paper No. 20, Bogor: Center for International Forestry Research (CIFOR).
112 Tacconi, L. (2007) Decentralization, forests and livelihoods: Theory and narrative, *Global Environmental Change* 17, 338–48.
113 Sikor, T. and Nquyen, T.Q. (2007) Why may forest devolution not benefit the rural poor? Forest entitlements in Vietnam's Central Highlands, *World Development* 35 (11), 2010–25.
114 Carter and Gronow (2005) op. cit.
115 Radachowsky, J., Ramos, V.H., McNab, R., Baur, E.H. and Kazakov, N. (2012) Forest concessions in the Maya Biosphere Reserve, Guatemala: A decade later, *Forest Ecology and Management* 268, 18–28.
116 Chhatre, A. and Agrawal, A. (2009) Trade-offs and synergies between carbon storage and livelihood benefits from forest commons, *PNAS* 106 (42), 17667–70.
117 Phelps, J., Webb, E.L. and Agrawal, A. (2010) Does REDD+ threaten to recentralize forest governance? *Science* 328 (5976), 312–13.
118 We are indebted to Tom Rudel for this insight.

3 Global forest governance

Constanze Haug and Joyeeta Gupta[1]

3.1 Introduction

This chapter focuses on the architecture of global forest governance by examining the institutions that make up the international 'forest regime complex'[2] and the principles and policy instruments that emanate from it. It presents a brief history of global forest governance, examines existing institutions related to forests at the global and regional level (see 3.2) and discusses the related principles and policy instruments (see 3.3 and 3.4).

3.2 Institutions

3.2.1 *A brief history*

Tropical forests and deforestation have been on the international agenda for the past 50 years, although it took some decades until forest-specific global institutions were established. A global forest convention, which was to provide a comprehensive legal framework for forest issues, had been expected from the 1992 United Nations Conference on Environment and Development (UNCED), yet parties failed to agree on its content and scope (see Box 3.1).[3] Subsequent efforts to arrive at a comprehensive global framework were also unsuccessful, leaving little prospect for a legally binding international forest agreement.[4] A fragmented mosaic of issue-specific efforts has evolved in its place. Among those, the function of forests as carbon sinks has received increasing attention since the emergence of REDD in the global climate arena (see 1.4 and Chapter 4).

Institutions in global forest governance include international organizations, UN agencies, treaty secretariats, policy forums and processes working on forest-related issues, as well as private arrangements dealing for instance with forest certification. They all conform to Keohane's definition of an institution in that they 'involve persistent and connected sets of rules (formal or informal) that prescribe behavioural roles, constrain activity and shape expectations' and are embedded into practices.[5] Figure 3.1 provides a – non-exhaustive – overview of the landscape of international forest governance.

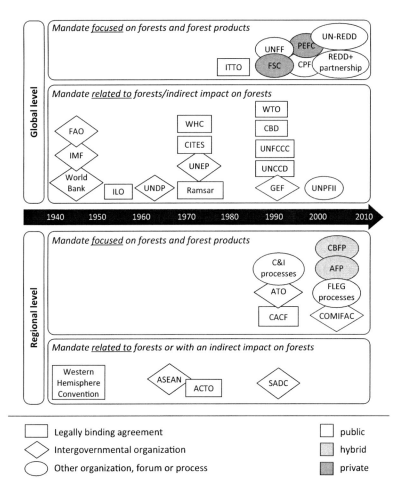

Figure 3.1 Institutions in international forest governance

ACTO = Amazon Cooperation Treaty Organization; AFP = Asia Forest Partnership; ASEAN = Association of South East Asian Nations; ATO = African Timber Organization; CACF = Central American Forest Convention; CBD = Convention on Biological Diversity; CBFP = Congo Basin Forest Partnership; C&I = Criteria and Indicators; CITES = Convention on International Trade in Endangered Species; COMIFAC = Central African Forest Commission; CPF = Collaborative Partnership on Forests; FAO = Food and Agriculture Organization of the United Nations; FLEG = Forest Law Enforcement and Governance; FSC = Forest Stewardship Council; GEF = Global Environment Facility; ILO = International Labour Organization; IMF = International Monetary Fund; ITTO = International Tropical Timber Organization; PEFC = Programme for the Endorsement of Forest Certification; SADC = Southern African Development Community; UNCCD = United Nations Convention to Combat Desertification; UNDP = United Nations Development Programme; UNEP = United Nations Environment Programme; UNFCCC = United Nations Framework Convention on Climate Change; UNFF = United Nations Forum on Forests; UNPFII = United Nations Permanent Forum on Indigenous Issues; UN-REDD = United Nations Collaborative Programme on Reducing Emissions from Deforestation and Forest Degradation in Developing Countries; WHC = World Heritage Organization; WTO = World Trade Organization.

The discussion below follows the logic of the figure. It first treats global-level and then regional institutions, distinguishing between arrangements whose mandate focuses exclusively and explicitly on forest-related issues and other arrangements that produce indirect impacts on forest and the various services they provide.

As Figure 3.1 shows, institution-building related to forests has proceeded in fits and starts. Some institutions founded in the immediate aftermath of the Second World War were of major significance for global forests, although the latter were not at the heart of their mandate. The 1970s brought on several forest-related international environmental agreements, as did the 1992 UNCED summit. Forest-specific institutions did not emerge until the 1980s. The mid to late 1990s and the 2000s saw much international activity focused on forests, but no more global or regional legally binding agreements were adopted. This period also brought the onset of private and hybrid governance initiatives, globally and regionally.

3.2.2 Global institutions with a forest-focused mandate

Legally binding agreements

Relatively few institutions have forests and forest products as their sole focus. The earliest and also the only legally binding forest-specific agreement is the International Tropical Timber Agreement (ITTA) negotiated in 1983 under the UN Conference for Trade and Development. The resulting International Tropical Timber Organization (ITTO) aims to regulate the international trade in tropical timber between producer and consumer countries. Over time, the ITTO has adopted a broad vision of forest management, aiming to 'promote the expansion and diversification of international trade in tropical timber from sustainably managed legally harvested forests'.[6] While the ITTO increasingly addresses controversial issues such as illegal logging, certification and the concept of sustainable forest management (SFM), it continues to be criticized for being dominated by forest exploitation rather than conservation interests.[7]

UN forest negotiations

A second forest-specific global arrangement is the United Nations Forum on Forests (UNFF), an intergovernmental policy forum that continues the global dialogue on forests which began at the UNCED summit in 1992. Instead of a forest convention, UNCED had only adopted a chapter on 'Combating Deforestation' within its 'Agenda 21', and a list of principles on SFM (the Forest Principles[8]), applicable to 'all kinds of forests'. To promote international forest dialogue, the UN Commission for Sustainable Development established the Intergovernmental Panel on Forests in 1995, which later merged into the Intergovernmental Forum on Forests. Together, the two forums adopted more than 270 non-binding 'Proposals for Action' on global forest issues. Since 2000, forest negotiations continue under the auspices of the UNFF. The UNFF advises and supports the implementation of international forest-related agreements and national forest programmes. The adoption of a 'Non-Legally Binding Instrument on All Types of Forests' (NLBI)[9] constituted a breakthrough of sorts in 2007, but its effect remains doubtful.[10] Recent UNFF meetings were dominated by discussions on forest financing, but have yet to lead to the Forest Fund called for by developing countries. The reasons behind the difficulties of reaching a comprehensive agreement on forests at the global level are explored in Box 3.1.

To streamline international efforts in forest governance, the Collaborative Partnership on Forests (CPF) was established in 2001 as an informal

---

Box 3.1 **Why is there no global forest convention?**[11]

The failure to negotiate a legally binding global treaty on forests at the UNCED summit and its aftermath has been widely discussed in the academic literature.[12] Key reasons for the absence of consensus are developing country concerns about national sovereignty and national development priorities, including their sovereign right to exploit forest resources.[13] Many countries in the global south rejected the notion of stewardship, i.e. international responsibility for forests as a global commons, and consequently opposed the view that deforestation was a global problem, and not a national or local one.[14] This attitude was especially pronounced among forest-rich countries, whereas forest-poor nations tended to be more supportive of global forest conservation efforts.[15] Countries with strong forest industries also tended to oppose a forest convention, though this pattern was less consistent.[16] A key issue since the forest negotiations in Rio is the question of financing.[17] Developing nations requested the provision of 'new and additional funds' over official development assistance for sustainable forest management through a 'green fund' as compensation for committing themselves to a binding regime.[18] When it became apparent that the requested finance would not be forthcoming, developing countries blocked the convention.[19]

After Rio, prospects for a global forest treaty became increasingly dim. Convention fatigue and a growing dominance of a neo-liberal discourse in global politics favoured softer approaches over legally binding agreements.[20] Strong leadership was also lacking: the United States withdrew its support when it became clear that limiting the scope of the agreement to tropical deforestation would be unacceptable to developing countries.[21] And while the EU officially supported a convention, differing member state positions weakened its stance on the issue.[22]

Structural reasons underlying the difficulties of reaching a comprehensive forest agreement relate to the complexity and the cross-sectoral, cross-scale character of forests (see 11.2) along with the multiple services they provide (see Table 1.1). Forest-related issues involve and affect many actors (forest dwellers, indigenous peoples, logging companies, transnational corporations), multiple levels of policymaking, and intersect with other land-uses, particularly agriculture. Federal systems with devolved forest management responsibilities have further complicated the adoption and implementation of international forest commitments.[23]

56 *Constanze Haug and Joyeeta Gupta*

coordination mechanism between 14 international forest-related organizations. It aims to support the UNFF and enhance collaboration between its member organizations. So far its work has been considered quite effective by member organizations and countries alike.[24] Finally, the emergence of REDD in the international climate negotiations since 2005 has prompted the establishment of two new institutional arrangements focused on lesson-learning and capacity-building for REDD: the UN-REDD programme, a collaborative effort of the United Nations Environment Programme (UNEP), the United Nations Development Programme (UNDP) and the UN Food and Agriculture Organization (FAO), and the REDD+ partnership (see Chapter 4).

Transnational governance arrangements

International rule-making on forest issues has increasingly shifted to private actors, with the rise of certification schemes being a prime example.[25] The Forest Stewardship Council (FSC), initiated in 1993 as a consequence of the disappointment of environmental NGOs with the ITTO's failure to promote SFM practices,[26] was the first significant global private arrangement. Over time, competitors emerged such as the PEFC (Programme for the Endorsement of Forest Certification), which started out as a European scheme, but became a global organization in 2003 (see Box 3.2; and Box 2.5).

Box 3.2. **Certification bodies**

Forest certification is a popular forest instrument introduced in the 1990s (see Box 2.5). The two biggest certification standard setting organizations globally are the Forest Stewardship Council (FSC) and the Programme for the Endorsement of Forest Certification (PEFC). The FSC is credited with representing the more ambitious standards in environmental and social terms and enjoys most support from NGOs.[27] But its ambitious standards impose a high burden on small landholders.[28] By contrast, the producer-backed PEFC is a framework organization based on the mutual recognition of national schemes.

There is heavy competition between FSC, PEFC and others for dominance in the certification market. In 2000, the FSC opposed an initiative by the International Forest Industry Roundtable aiming at mutual recognition of all 'credible' certification schemes as it was reluctant to see its own normative pull weakened in a 'race to the bottom' of rule harmonization.[29] Global forest labelling may, however, have entered a phase of consolidation now, with the FSC and the PEFC remaining the two central players in the global 'certification game'.[30]

In recent years, the area of certified forest has grown rapidly, covering several hundred million hectares of forests in over 80 countries, including the case study countries.[31] Currently, the area certified by FSC and PEFC corresponds to almost 10 per cent of global forests.[32] However, only a small percentage of certified forests are tropical or subtropical; for the FSC, they make up only 11 per cent of the total certified area.[33]

3.2.3 *Global institutions with a forest-related mandate and/or an indirect impact on forest services*

Apart from the arrangements which focus exclusively on forests, there are many intergovernmental organizations and agreements whose mandate relates to forests in one way or another. This section outlines how those institutions impact on the services that forests provide.

Intergovernmental organizations

The Food and Agriculture Organization (FAO), established in 1945 as an intergovernmental organization with currently 194 member states, promotes food security and agricultural development. Its Forest Department supports SFM by providing states with information (e.g. the five-yearly Forest Resource Assessment), policy advice and technical assistance. The 2010 FAO Strategy on Forests and Forestry lays out the organization's vision on the sector and emphasizes a 'broad view of forestry' as a 'multi-disciplinary concept that encompasses social, economic and environmental aspects'.[34]

The World Bank has been engaged in forest activities within the context of its larger mandate to help reduce poverty.[35] It has funded a significant share of forest activities in developing countries over the past decades. Whereas 'early projects financed industrial operations to the detriment of tropical native forests',[36] the Bank's strategy later shifted towards forest conservation and since the 2000s towards a sustainable management paradigm.[37] The World Bank has been slow to adapt to changing discourses in international forest discussions in the past.[38] However, recently it has adopted a more active role, for instance in the Forest Law Enforcement and Governance (FLEG, see 3.2.5) processes and in the development of forest carbon projects.

The involvement of the International Monetary Fund (IMF) in forest issues is more indirect. Its lending policy is concerned with issues that impact the macroeconomic situation of a country. However, the prescriptions it makes in its structural adjustment programmes can have an indirect impact on forests and/or a country's forest policy, for instance by promoting lower export taxes for timber or by prompting cuts in government budgets, with adverse effects on forest law enforcement.[39]

The United Nations Development Programme (UNDP)'s forest work consists of on-the-ground technical assistance and project implementation in developing countries. UNDP is one of the three implementing agencies of the Global Environment Facility (GEF) and manages many forest-related projects within the GEF's climate change and biodiversity focal areas. It collaborates on the UN-REDD programme with FAO and the United Nations Environment Programme (UNEP). The traditional home of the forest-related activities of UNEP is its biodiversity programme. Moreover, UNEP supports the administration of several forest-related treaties, including the Convention on the International Trade in Endangered Species (CITES), the Convention on Biological Diversity (CBD) and the United Nations Convention to Combat Desertification (UNCCD). UNEP is also involved in forest ecosystem-related projects.

The Global Environmental Facility (GEF) was established by the World Bank, UNDP and UNEP in 1991. GEF provides grants and concessional funding to cover the 'incremental' cost for projects that yield global environmental benefits.[40] Its annual budget increased from USD one billion during its pilot phase in 1991 to USD 4.34 billion in 2011.[41] The GEF administers specific funds for environmental agreements including the CBD, the UNCCD, and the UNFCCC and its Kyoto Protocol. GEF funding for forest initiatives has grown over time; in total, the GEF has committed USD 1.6 billion forest-related funding to more than 300 projects, leveraging another USD 5 billion in co-financing from other sources.[42]

Legally binding agreements

The first legally binding environmental agreements relevant to forests were adopted in the 1970s (see Figure 3.1). The 1973 CITES treaty was the first multilateral treaty focusing on species conservation. Its goal is to control international trade in endangered species of wild flora and fauna. It operates via the inclusion of certain species and subspecies in one of its three appendices that impose different levels of control in international trade. However, very few tree species are actually listed under the treaty, mainly due to economic concerns of exporting countries.[43] The 1971 Ramsar Convention on Wetlands of International Importance strives for 'the conservation and wise use of all wetlands through local and national actions and international cooperation'.[44] This Convention is particularly relevant to forests since the increasingly threatened mangrove forests around the world fall under its scope. The 1975 Convention Concerning the Protection of the World Cultural and Natural Heritage (WHC) is of significance to forests due to its mechanism for listing and protecting sites of key cultural and/or natural value.

Even though no forest convention came forth from the 1992 UNCED summit, the three conventions adopted there each have a forest dimension. The CBD's objective of promoting 'the conservation of biological diversity, the sustainable use of its components and the fair and equitable sharing of the benefits arising out of the utilization of genetic resources'[45] is the key global instrument on biodiversity conservation, and arguably also the most important international hard law on forests.[46] It requires Parties to develop national strategies, plans or programmes for the conservation and sustainable use of biodiversity. Forests in the broader context of land-use change also featured in the second Convention agreed in Rio, the UNFCCC. This Convention recognizes the role of forests as both sources and sinks of carbon in the context of climate change and commits Parties to promote the conservation and enhancement of these and other carbon stocks (see Chapters 1 and 4). The 1997 Kyoto Protocol endorsed forests as sinks that may be used to meet part of Parties' emission limitation and reduction commitments, against the initial opposition of a number of Parties including the European Union and Brazil.[47] The modalities of the Protocol later also admitted afforestation and reforestation projects under the scope of the Clean Development Mechanism (CDM), although only few such projects have been realized to date[48] (see Chapter 4). Finally, the UN Convention to Combat Desertification (UNCCD) is relevant for forests that are close to areas prone to desertification. Its implementation has however been hampered by a severe lack of resources and funding.[49] This has also prevented it from developing a stronger focus on forests and deforestation.[50]

In the non-environmental realm, agreements under the World Trade Organization (WTO) are significant for trade in forest products. WTO's governing norm of trade liberalization might clash with attempts to ensure preferential treatment for sustainably sourced timber or timber certification. Moreover, the Agreement on Trade-Related Aspects of International Property Rights (TRIPS) potentially conflicts with rules on access and benefit-sharing of genetic resources under the CBD. However, TRIPS as well as the General Agreement on Tariffs and Trade (GATT) and the Agreement on Technical Barriers to Trade (TBT Agreement) allow for exceptions on environmental grounds, and no forest-related cases have so far been brought before WTO dispute settlement bodies.

Finally, Conventions No. 107 and 169 of the International Labour Organization (ILO) and the UN Permanent Forum on Indigenous issues (UNPFII), which in 2007 adopted the non-binding but influential UN Declaration on the Rights of Indigenous Peoples (UNDRIP), focus exclusively on indigenous and tribal people's rights. Given that many of these groups depend on forests for their livelihoods, both can be considered highly relevant for this dimension of forest governance (see Box 3.3).

Box 3.3 **Policies on indigenous peoples**

Globally, 60 of the 370 million indigenous people depend on forests. They can include local communities, hunter-gatherers, pastoralists, ethnic groups, minorities, tribal groups or even the whole local population (particularly in Africa).[51] They include peoples 'whose social, cultural and economic conditions distinguish them from other sections of the national community, and whose status is regulated wholly or partially by their own customs or traditions or by special laws or regulations' and those who are descendants from the initial local populations at the time of conquest or colonization or the establishment of present state boundaries and who retain some or all of their own social, economic, cultural and political institutions.[52]

Indigenous peoples depend on and guard forests, especially in areas where primary forests have been conserved.[53] Whereas their contribution to climate change is minimal, they are vulnerable to climatic impacts[54] exacerbated by their low incomes, poor information access, no effective access to and participation in the policy processes, and the poor recognition of rights and customs. Indigenous people are increasingly asserting their rights. During colonial and post-colonial times, forests were often listed as owned by states which then proceeded to use them to maximize their incomes. The ILO Convention[55] calls on states to recognize and effectively protect the traditional rights of ownership, possession or access of these peoples to their lands.

The ECOSOC Working Group on Indigenous Populations' draft Declaration on the Rights of Indigenous Peoples, adopted 20 years later in 2007 by the General Assembly, states that indigenous people are free and equal to others, have the right to self-determination and may freely pursue their own development, may strengthen their own institutions, and that states shall take action to ensure that they are not deprived of their lands or resources. Their rights to own, occupy or access resources on their lands should be recognized and protected. Over the years, indigenous peoples have acquired the right to participate in UN conferences, have become the subject of research and policy work by the WHO on health issues, UNESCO on education issues, ILO on employment issues, UNDP on development issues and UNEP on environmental issues. The International Union for the Conservation of Nature has also increasingly studied the relationship between indigenous peoples and forests. Increasingly, their interests in REDD are also being represented (see Box 4.2).

3.2.4 Global governance institutions and ecosystem services

Global institutions can also be characterized in terms of their focus on ecosystem services – provisioning, regulating, supporting and cultural services (see 1.3.1; Millennium Ecosystem Assessment). Figure 3.2 shows that international activity on forests is high for some ecosystem services and lower for others. The least covered seem to be the cultural services (combining the spiritual and recreational functions of forests), though attention for these issues may be increasing as of late. Supporting services include the protection of soils and of forest biodiversity. International activities on biodiversity are for the most part synergistic, but there is a regulatory gap globally with regard to soil conservation.[56] Since the advent of REDD, the carbon sequestration function, which belongs to the regulating services provided by forests, has dominated global-level activity above all others (see Chapter 4), leading existing institutions to refocus their efforts and prompting the establishment of new arrangements. Finally, many institutions focus on the provisioning services of forests, perhaps more on forest products than on forests as habitat for humans.

Global forest governance 61

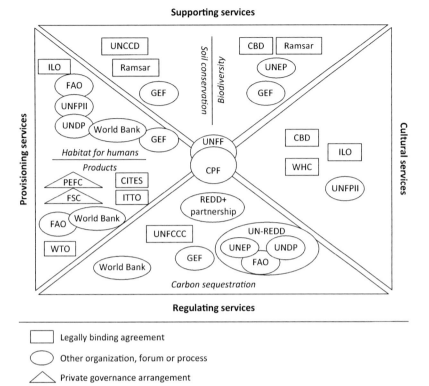

Figure 3.2 Global forest governance arrangements and ecosystem services

3.2.5 *Regional institutions with a forest-focused mandate*

Much forest-related activity has also taken place at regional level. Regional initiatives are arguably more flexible and responsive to region- and ecosystem-specific concerns, and incur lower transaction costs. Yet they may also be more easily dominated by powerful regional actors.[57] In line with this book's focus, the following paragraphs discuss regional institutions relevant to tropical forests.

The first regional forest convention is the 1993 Regional Convention for the Management and Conservation of the Natural Forest Ecosystems and the Development of Forest Plantations (Central American Forest Convention, CAFC). This Convention emphasizes the relationship between forest management, development and poverty through the implementation of sustainable forestry.[58] It requires Parties to consolidate national and regional systems of protected areas, establish dynamic inventories of forest cover, create mechanisms to control illegal trade in flora and fauna, promote public participation and recognize the rights of indigenous peoples and other inhabitants of forested areas (see Box 3.3).

In Africa, several governance arrangements focus on forests. The oldest is the African Timber Organization (ATO) which deals with the sustainable production and commercialization of timber. The Central African Forests Commission (COMIFAC), an intergovernmental commission with ten member states from the Congo Basin, emerged as the operational structure for regional cooperation on forest issues from the first Summit of Central African Heads of State on conservation and sustainable management of forests in Yaoundé in 1999. It is currently engaged in regional REDD efforts, in implementing the Action Plan for Forest Law Enforcement, Governance and Trade (FLEGT, see below) and in the CBD's access and benefit-sharing initiative. However, lack of institutional capacity and of linkages to other regional organizations hampers its effectiveness in working at the interface of these regimes.[59] A hybrid public–private initiative active in the same region as COMIFAC is the Congo Basin Forest Partnership (CBFP), which was launched as a 'type II partnership' at the World Summit on Sustainable Development in Johannesburg in 2002. The CBFP is an informal network of about 40 governmental, non-governmental and international organizations. Its primary role is to work as a 'transmission belt between donors and implementing agencies',[60] providing a forum for consultation of all stakeholders that is not immediately involved in programme implementation. There are several more regional public–private partnerships focusing on forest management in the Tropics, including the Asia Forest Partnership (AFP), which emanated from the 2002 Johannesburg summit, the Heart of Borneo Initiative, the Green Heart of Africa initiative and the Latin American Puemblo initiative.

Arguably one of the most important regional governance arrangements is the Forest Law Enforcement and Governance (FLEG) initiative, a series of regional interministerial processes to combat illegal logging and promote better forest governance. FLEG processes are co-hosted by producer and consumer governments and the World Bank and supported by non-governmental organizations, international agencies and the private sector. Initially conceived because UN institutions were thought to be too rigid, slow and inflexible to address this problem,[61] FLEG initiatives have been most dynamic in Asia and Africa.[62] However, efforts to ensure the compatibility of measures such as import controls and licences with WTO rules have continuously limited the extent of actions taken under FLEG.[63]

3.2.6 Regional institutions with a forest-related mandate and/or an indirect impact on forest services

Finally, there are regional agreements with a more indirect impact on forests. Two of them, the ASEAN Agreement on the Conservation of Nature and Natural Resources and the SADC Forestry Protocol, are embedded into wider regional integration efforts in South East Asia and Southern Africa, respectively. Two agreements on the American continent have an exclusively environmental outlook: the Amazon Cooperation Treaty Organization (ACTO) strives for achieving sustainable development in the Amazon Basin, and the Convention on Nature Protection and Wild Life Preservation in the Western Hemisphere (Western Hemisphere Convention) focuses on species and habitat protection.

3.2.7 Extraterritorial impacts of national governance

In addition to global and regional institutions, in some cases, national legislation in one part of the world may produce extraterritorial effects on forests elsewhere. For instance, both the United States and the EU have implemented measures to counter illegal logging. In the United States, the Lacey Act of 1900 (16 USC §§ 3371–8) is one of the oldest US conservation laws, originally focusing on trade in endangered species. Its scope was extended in 2008 to prohibit the import of illegally harvested forest products including timber from other countries into the US.

The EU's approach to combating illegal logging and associated trade was first defined in its 2003 Forest Law Enforcement Governance and Trade (FLEGT) Action Plan, which covered both demand and supply side measures to address the problem and which prompted two key pieces of legislation. FLEGT Regulation No. 2173/2005 endorsed a licensing scheme for timber imports from countries that had entered bilateral Voluntary Partnership Agreements (VPAs) with the EU. The EU Timber Regulation No. 995/2010 imposes obligations on EU timber importers to exercise 'due diligence' and to keep records of their suppliers and customers. Both measures thus seek to encourage better law enforcement in developing countries as well as greater supply chain transparency.

Of a different nature are the potential effects of the provisions on biofuels in EU Renewable Energy Directive No. 28/2009. This Directive stipulates that by 2020, ten per cent of transport fuel used in the EU has to come from renewable sources. This has stimulated export-oriented biofuel production in many developing countries, including Indonesia. In an attempt to mediate adverse environmental impacts from the policy, the EU has imposed a number of sustainability criteria for biofuel production, which are applicable to both EU member states and imports into the EU. Similarly, there are a number of policies underway in the US to promote biofuels.[64]

3.3 Key principles and concepts in international forest governance

Global forest governance is thus spread across many different institutions, each with its own objectives, focus and normative outlook. This section highlights some overarching principles and concepts that have emerged over time in the international forest regime complex. In Chapter 2, we discussed this ideational dimension of policy-making in the context of suasive policy instruments (see Table 2.3).

3.3.1 Key principles

This section highlights some of the overarching principles, concepts and instruments of forest governance that have emerged over time.

The 1992 UNCED summit was significant for norm development relevant to forest governance. The 1992 Rio Declaration on Environment and Development includes 27 principles, which either promote cooperation between countries, apply to countries individually or draw attention to specific actor groups, like indigenous peoples, women, and youth. Many – for instance the principle of national sovereignty, the need for countries to integrate environmental issues into their national policies, the precautionary, and the polluter pays principles, as well as the principle of common but differentiated responsibilities, and the principles on indigenous peoples and women – are pertinent to forests and have been reiterated later in forest-specific declarations and documents.

UNCED also adopted the 'Forest Principles', the first global consensus document on forests. Their content has been judged rather weak.[65] The assertion of countries' 'sovereign right to exploit their own resources' – as long as their activities 'do not cause damage' to other states – precedes all other principles in the declaration and constitutes a cornerstone of global forest governance until today. The Forest Principles emphasize that forest resources and forest lands 'should be sustainably managed to meet the social, economic, ecological, cultural and spiritual needs of present and future generations' and that national policies should provide a framework to that end. The need for financial support for developing countries to increase their capacity for SFM is also recognized in the Declaration. The NLBI, the non-binding declaration adopted by the United Nations Forum on Forests in 2007, strengthened the sovereignty principle in the forest context even further, by making three out of its four global objectives a state, not international community, responsibility.[66]

General procedural principles of good governance, such as inclusiveness, transparency, and accountability, are equally part of the global forest governance canon.[67] In the forest sector, this also translates into combating illegal practices, including illegal logging.[68] Finally, equity has emerged as a key relevant normative principle. At the international level, it is closely related to the principle of common but differentiated responsibilities, which was first formulated in the Rio Declaration and later also endorsed in the NLBI. In the forest context, international equity 'implies different types and levels of national responsibility for addressing and reversing deforestation'.[69] At the domestic level, equity refers to the participation of marginalized groups such as indigenous communities and forest dwellers in decision-making and resource use, land rights and benefit sharing. A key challenge here is that many customary activities have been subsequently labelled as 'illegal' as they have been delegitimized by top-down neo-liberal policy processes.

3.3.2 Key concepts

Two concepts provide an overarching perspective on the governance of forests – SFM and the ecosystem approach. In line with the evolving sustainable development discourse in global politics, the preservationist discourse in forest governance ended with the rise of the SFM concept[70] since the 1992 UNCED summit, with the Forest Principles stipulating that 'Forest resources and forest lands should be sustainably managed to meet the social, economic, ecological, cultural and spiritual needs of present and future generations.' The SFM terminology quickly gained popularity, though its exact meaning remains contested.[71] The Ministerial Conference for the Protection of Forests in Europe (MCPFE) defined SFM as '[t]he stewardship and use of forests and forest lands in a way, and at a rate, that maintains their biodiversity, productivity, regeneration capacity, vitality and their potential to fulfil, now and in the future, relevant ecological, economic and social functions, at local, national, and global levels, and that does not cause damage to other ecosystems'.[72] The NLBI refrains from giving a precise definition of SFM, merely referring to it as a 'dynamic and evolving concept'. The notion of SFM and its implications for practice have been at the heart of negotiations in the UNFF and its predecessor bodies. Nine regional Criteria & Indicator (C&I) processes, supported by international organizations such as FAO, ITTO, UNEP and others, work on breaking SFM down into specific criteria and elements for implementation across the different types of ecosystems.

The ecosystem approach entered forest governance from the biodiversity regime. While the term is not used in the CBD itself, its Conference of the Parties stipulated that the ecosystem approach – for which no definition was adopted – 'should be the primary framework of action to be taken under the Convention'.[73] As of 2012, there is no universally agreed upon definition of the ecosystem approach. However, it is generally understood to encourage management of human activities that arise from an in-depth understanding of ecological processes and interactions and that aims to preserve an ecosystem with its structure and functions for future generations. The concept thus endorses a holistic, systemic approach to the environment that seeks to appreciate the entirety of implications of human interventions on an ecosystem.

3.4 Instruments of international forest governance

The institutions of international forest governance give rise to many policy instruments that interact with policy frameworks at lower levels of governance. This section distinguishes between three types of instruments – regulatory, economic, and suasive (see Tables 2.2 and 2.3). Local management instruments are not included here because given their local level focus, they are not contained in any global agreement.

3.4.1 Regulatory instruments

Table 3.1 shows that global-level institutions include a limited variety of instruments. One example of regulatory instruments is the restrictions on the international trade in endangered species flowing from the 1975 CITES Convention. Also the designation of protected areas, as stipulated by the WHC and the Ramsar Convention, falls into this group. The WHC obliges Parties to identify, protect and conserve sites of outstanding cultural or natural value. Among the 936 sites on the World Heritage List, 104 are forests, covering a total surface area of over 75 million hectares.[74] Similarly, in order to accede to the Ramsar Convention, a Party must designate at least one wetland site on their territory as a nature preserve which then becomes subject to Ramsar's standard of 'wise use'.

Other agreements, most notably ILO Conventions No. 107 and 169, establish legally binding rights for individuals, in this case indigenous peoples and tribal populations, which are particularly dependent on forests for their livelihoods. These Conventions endorse indigenous peoples' rights to participate in decision-making, to decide on their own development priorities and to exercise control over their economic, social and cultural environment.

Many institutions also impose monitoring and reporting obligations on national governments. The most generic in terms of issue coverage may be the National Forest Programmes (NFPs) that states were required to produce by Chapter 11 on deforestation of UNCED's Agenda 21. National Forest Programmes gained renewed prominence with the adoption of the NLBI in 2007. The NLBI, in keeping with its focus on national sovereignty, considered them as the key tool for countries to keep track of and report on their progress towards achieving SFM.[75] But also international agreements like the CBD, UNFCCC and ITTO, and the regional criteria and indicator processes (C&I) require regular reporting from parties. Taken together, these obligations prove cumbersome especially for least developed countries and are 'inadequately addressed in capacity-building and technical and financial assistance at the intergovernmental level'.[76] The CPF is undertaking efforts to streamline forest-related reporting, and has set up a task force to this effect.[77]

3.4.2 Economic instruments

Turning to economic instruments, international institutions are active primarily with regard to financing as well as forest certification (see Chapter 2). Some organizations issue grants for projects or capacity-building initiatives. The largest source of multilateral finance for forests is the World Bank, with forest-related assistance totalling USD 327 million a year on average between 2009–11.[78] However, many of its loans for infrastructure, mining or agriculture also have large negative impacts on forests in the target countries.[79] Its strategy in forest lending has been highly controversial in the past.[80] While there is still concern about the extent to which forest and conservation issues

Table 3.1 Instruments in forest-related agreements[1]

	Type of policy instrument	Institutions directly mandating the instrument or its implementation at the national level	Institutions encouraging the implementation of the instrument at the national level
Regulatory	(i) Trade restrictions	CITES	
	(ii) Decentralization/ subsidiarity		UNCCD, CBD, Rio Declaration
	(iii) Principles (including SFM)	Agenda 21, Rio Declaration, Forest principles, NLBI	
	(iv) Binding forest rules, reforestation, fire rules		ITTO, FAO
	(v) Spatial planning – Protected Areas	WHC, Ramsar, CBD	NLBI
	(vi) Sustainable forest logging/concessions	—	—
	(vii) Land rights	ILO Conventions No. 107 and 169	—
	(viii) Reporting	CITES, Ramsar, CBD, UNCCD, UNFCCC, ITTO, FAO, UN-REDD, etc.	—
	(ix) Monitoring/ surveillance	—	FAO, ITTO, CBD, UNFCCC, UNCCD, NLBI, etc.
	(x) Law enforcement	—	FLEG, NLBI
Economic and market	(xi) Export and import tariffs	WTO (encouraging trade liberalization)	—
	(xii) International funds, grants, loans	ITTO, World Bank, UN-REDD, GEF (CBD/ UNCCD/UNFCCC)	—
	(xiii) Taxes (on forest products)	—	—
	(xiv) Subsidies (for alternative work, etc.)	—	—
	(xv) PES	—	—
	(xvi) Micro-credit	—	—
	(xvii) Debt-for-nature swap	—	—
	(xviii) Carbon offsetting funds (CDM, voluntary)	UNFCCC, voluntary forest carbon market	—
	(xix) CSR/certification/ ecotourism	FSC, PEFC	—
Suasive and research	(xx) (Soft international law) principles	CBD, UNCCD, FAO, ITTO, Rio Declaration, NLBI, C&I processes, etc.	—
	(xxi) (Soft international law) targets	ITTO, CBD, NLBI	—
	(xxii) Research/ monitoring	FAO	CBD, UNCCD, FAO, ITTO, World Bank, Rio principles, NLBI, C&I processes, etc.
	(xxiii) Information and education	FAO, ITTO, World Bank, CBD, UN-REDD, REDD+ partnership	CBD, UNCCD, FAO, ITTO, World Bank, Rio principles, NLBI, C&I processes, etc.

[1] The information in Table 3.1 draws on the foundational documents of the arrangements as well as recent annual reports and strategic plans outlining the vision of an organization.

have been mainstreamed in Bank business and whether forest-related safeguards are adequately implemented, the 2007 Review of Implementation of the World Bank Strategy acknowledges the Bank's efforts to this effect.[81] More generally, after funding for forest conservation had largely stagnated over the past two decades, the emergence of REDD has prompted a significant shift of multilateral and bilateral financing to the forest sector (see Chapter 4). This is in contrast to the limited success of carbon offset projects under the Clean Development Mechanism of the Kyoto Protocol (see Chapters 1 and 4).[82] However, forest carbon offset projects have proven quite successful on the voluntary markets.[83]

The international trade regime gives rise to a different type of forest-related economic incentive – trade liberalization. The Uruguay Round brought about significant reductions of tariffs for traded forest products and addressed other non-tariff barriers as well.[84] So far no forest-related cases have been brought before WTO dispute settlement bodies. Nonetheless, there may be a chilling effect from the WTO's guiding norm of non-discrimination, for instance on the licensing scheme for timber legality implemented through the EU's Voluntary Partnership Agreements with timber exporting countries in the context of EU FLEGT.[85]

Finally, forest certification – which could also be considered under the suasive category – has become an important instrument for encouraging SFM (see 3.2.2). While certified forest companies have certainly changed their operations, the environmental effectiveness of the schemes, including their role in reducing forest loss, is less clear-cut.[86]

3.4.3 Suasive instruments

Not surprisingly, suasive incentives are the instrument type most commonly found in international forest institutions. Virtually all arrangements promote specific principles relating to sound forest management, good governance, or the rights of forest dwellers (see 3.3). Some have also adopted explicit targets. Thus, the CBD Conference of the Parties recently updated its Biodiversity Target agreed in 2002, which had aimed at reaching a significant reduction of the current rate of biodiversity loss by 2010. The new 'Aichi targets' include a long-term 'vision' for 2050 (Living in harmony with nature) and a short-term 'mission' for 2020 (Taking effective and urgent action to halt the loss of biodiversity in order to ensure that by 2020 ecosystems are resilient and continue to provide essential services). The latter is further specified into 20 headline targets for 2015/2020, including the objective to at least halve and where feasible bring close to zero the rate of loss of natural habitats, including forests, by 2020.[87]

Similarly, ITTO members continue to pursue the ITTO 'Objective 2000' – striving for international trade in timber from sustainably managed sources. The initial deadline for reaching the objective – the year 2000 – was not met, but the International Tropical Timber Council reaffirmed its commitment to achieve the goal 'as rapidly as possible'.[88] Last but not least, the NLBI is organized around four 'shared global objectives' that member states committed themselves to strive towards: first, to reverse forest cover loss worldwide; second, to enhance forest-based economic, social and environmental benefits by improving the livelihoods of forest dwellers; third, to increase significantly the area of protected and sustainably managed forests worldwide; and fourth, to reverse the decline in official development assistance for SFM.[89]

Finally, international institutions, especially the FAO, fulfil an important function of knowledge synthesis and assessment. Various organizations and forums issue reports and guidance documents related to forest policy and management. FAO considers the delivery of direct technical support to countries as its primary mandate on forests. Its National Forest Programme Facility aims to support and coordinate the implementation of national and subnational forest policy.[90] ITTO publishes technical reports, manuals and guidelines on topics ranging from environmental standard-setting to timber procurement policies, and the CBD Secretariat has guidelines and training modules available to support Parties in treaty implementation.

Other suasive measures include the criteria and indicator processes underway in nine major global and regional intergovernmental processes. The first was initiated by ITTO in 1992, to define comprehensive criteria and subsequently indicators for SFM, with a view to harmonizing national-level measurement and monitoring of progress in achieving SFM.

This brief analysis shows that 'harder', regulatory instruments are included in early arrangements that have a focus on the natural value of forests (e.g. CITES, Ramsar). Once additional concerns (e.g. on socio-economic development, sovereignty and equitable benefit sharing) are introduced, agreement seems to be difficult to reach, leading to the adoption of more suasive-type of instruments. Recently, in line with the general trend towards 'new' environmental policy instruments,[91] different kinds of market-based or economic incentives have been introduced to stimulate forest conservation and protection, such as forest certification and forest carbon offsetting. The use of such economic instruments is often triggered by difficult negotiation processes where parties cannot agree on regulatory measures, as market-based instruments can generally be implemented bilaterally or individually, and on a voluntary basis. As international forest negotiations under the UNFF show, financing of measures for forest conservation and SFM remains a critical challenge. While carbon storage or wood products have specific buyers, biodiversity protection and payment for ecosystem services are much more vague concepts and reliable long-term sponsors for these products and services are not necessarily available unless a regulatory system at global level is set up like the Kyoto Protocol which created a market for carbon trading. The prospects for a global REDD mechanism aiming to reduce emissions from deforestation and forest degradation are explored in Chapters 4 and 11.

Institutions			
Global institutions	**Regional institutions**	**Private institutions**	**Extra-territorial effects of national policies**
• Forest-focused: ITTO, UNFF, FSC, PEFC, CPF, UN-REDD, REDD+ partnership • Forest-related: FAO, IMF, World Bank, ILO, UNDP, UNEP, Ramsar, UNEP, CITES, WHC, GEF, CBD, UNFCCC, CBD, WTO, UNPFII	• FLEG and C&I processes • Asia: ASEAN, AFP • Africa: COMIFAC, CBFP, ATO • South America: Western Hemisphere Convention, ACTO, CACF	• FSC • PEFC	

Concepts and principles
Sustainable forest management, ecosystem approach
Sovereignty, precautionary principle, common but differentiated reponsibilities, environmental policy integration, good governance, equity...

Policy instruments		
Regulatory instruments	**Economic and market instruments**	**Suasive instruments**
• Trade restrictions • Protected areas • Reporting • Land rights	• Trade liberalization • Funds, grants and loans • Forest carbon offsetting • Forest certification	• Concepts and principles • Targets • Information and education • Research

Figure 3.3 Global forest governance and its instruments

3.5 Inferences

Two decades after the failed attempt at arriving at a global forest convention in Rio, global forest governance is dispersed,[92] its evolution having been characterized by 'creeping ad hoc incrementalism'.[93] There is reluctance to comprehensively deal with forests at the global level for reasons explored above (concerns of national sovereignty, the contested scaling of forest issues, lack of leadership, the question of finance, convention fatigue and a growing neo-liberal discourse). The absence of a unified global instrument had left a void for others to fill, and this task has been taken up both by existing institutions with a mandate far broader than forests, but also by new – partly private – institutions dedicated specifically to forest issues. As a consequence, rules relating to forests are 'peppered throughout' many agreements;[94] with few legally binding instruments (e.g. CITES, ITTA, CBD), which work together in a sometimes more conflictive, sometimes more synergetic manner. The last legally binding text of relevance to forests was adopted in 1997;

since then, the focus has been primarily on soft law. While some scholars claim that a forest 'regime' exists[95] since a regime does not have to have a core treaty, the term 'regime complex'[96] may be a more appropriate label. Figure 3.3 provides a summary of the key institutions, concepts, principles and policy instruments contained in the international forest regime complex.

Over the years, some key principles and concepts have emerged that make up the global forest governance canon, with the emerging international norm against illegal logging the most recent addition. Similarly, two concepts are now central in forest governance: SFM, whose exact meaning and implementation is still contested, and the ecosystem approach. As for the policy instruments on forest governance at the global level, there is a shift over time from more regulatory types of instruments to more economic and suasive ones, with significant overlaps especially in the area of reporting and provision of information.

Recently, the new interest in tropical deforestation as a result of the discussions in the climate arena has brought fresh momentum to the global forest agenda, prompting a growing focus on the carbon sequestration function of forests. The next chapter will examine the potential promise and pitfalls of REDD in greater detail.

Notes

1 The authors would like to thank Philipp Pattberg and Harro van Asselt for their detailed comments on this chapter.
2 Glück, P., Angelsen, A., Appelstrand, M., Assembe-Mvondo, S., Auld, G. and Hogl, K. (2010) Core components of the international forest regime complex, in Rayner, J., Buck, A. and Katila, P., *Embracing complexity: Meeting the challenges of international forest governance: A global assessment report prepared by the Global Forest Expert Panel on the International Forest Regime* (pp.37–55), Vienna: International Union of Forest Research Organizations (IUFRO).
3 Lipschutz, R.D. (2000) Why is there no international forestry law: An examination of international forestry regulation, both public and private, *UCLA Journal of Environmental Law and Policy* 19, 153; Dimitrov, R.S. (2005) Hostage to norms: states, institutions and global forest politics, *Global Environmental Politics* 5 (4), 1–24.
4 Dimitrov (2005) op. cit.
5 Keohane, R.O. (1988) International institutions: two approaches, *International Studies Quarterly*, 379–96, p.383.
6 ITTO (2006) International Tropical Timber Agreement.
7 Nagtzaam, G.J. (2008) *The International Tropical Timber Organization and conservationist forestry norms: A bridge too far?* Victoria: Monash University, http://works.bepress.com/gerry_nagtzaam/1 (accessed 13 February 2012).
8 Officially entitled the 'Non-Legally Binding Authoritative Statement of Principles for a Global Consensus on the Management, Conservation and Sustainable Development of all Types of Forests'.
9 NLBI (2007) *Non-legally binding instrument on all types of forests*, New York: adopted by the seventh session of UNFF and by the United Nations General Assembly in 2007 (A/RES/62/98)

10 Kunzmann, K. (2008) The non-legally binding instrument on sustainable management of all types of forests – Towards a legal regime for sustainable forest management? *German Law Journal* 1 (8), 981–1005.
11 This section has benefitted considerably from the research undertaken by Aizo Lijcklama à Nijeholt (2010) The forest convention debate: From Rio to REDD, MSc thesis, Institute for Environmental Studies, VU University Amsterdam.
12 Humphreys, D. (2001) Forest negotiations at the United Nations: explaining cooperation and discord 2345, *Forest Policy and Economics* 3 (3–4), 125–35; Humphreys, D. (1996) *Forest politics: The evolution of international cooperation*, London: Earthscan; Humphreys, D. (2005) The elusive quest for a global forests convention, *Review of European Community and International Environmental Law* 14 (1), 1–10; Dimitrov (2005) op. cit.; Tarasofsky, R. (1995) *The international forests regime: legal and policy issues*, Gland: WWF/IUCN; Glück, P., Tarasofsky, R., Byron, N. and Tikkanen, I. (1997) *Options for strengthening the legal regime for forests*, Joensuu: European Forest Institute; Davenport, D.S. (2005) An alternative explanation for the failure of the UNCED forest negotiations, *Global Environmental Politics* 5 (1), 105–30; Reischl, G. (2009) *The European Union and the international forest negotiations: An analysis of influence*, doctoral thesis, Uppsala: Swedish University of Agricultural Sciences.
13 Humphreys (1996) op. cit.; Dimitrov (2005) op. cit.
14 Humphreys (2005) op. cit.
15 Lele, U.J. (2002) *Managing a global resource: Challenges of forest conservation and development*, New Brunswick: Transaction Publishers.
16 Lipschutz (2000) op. cit.; Humphreys (2005) op. cit.
17 Humphreys (1996) op. cit.; Humphreys, D. (2006) *Logjam – Deforestation and the crisis of global governance*, Earthscan forestry library, London: Earthscan.
18 Scholz, I. (2004) *A forest convention – Yes or no*, Bonn: BMZ.
19 Kolk, A. (1996) *Forests in international environmental politics: international organisations, NGOs and the Brazilian Amazon*, Utrecht: International Books.
20 Smouts, M.C. (2003) *Tropical forests, international jungle: the underside of global ecopolitics*, New York: Palgrave Macmillan; Reischl (2009) op. cit.
21 Davenport (2005) op. cit.
22 Reischl (2009) op. cit.
23 Interlaken Workshop (2004) *Decentralization, federal systems in forestry and national forest programs, Report of a worskhop co-organized by the Governments of Indonesia and Switzerland, 27–30 April 2004*, Interlaken, Switzerland.
24 Humphreys (2006) op. cit.; Reischl, G. (2012) Designing institutions for governing planetary boundaries. Lessons from global forest governance, *Ecological Economics*.
25 E.g. Auld, G., Gulbrandsen, L.H. and McDermott, C.L. (2008) Certification schemes and the impacts on forests and forestry, *Annual Review of Environment and Resources* 33 (1), 187–211; Overdevest, C. (2010) Comparing forest certification schemes: the case of ratcheting standards in the forest sector, *Socio-Economic Review* 8 (1), 47–76.
26 Pattberg, P.H. (2005) The forest stewardship council: risk and potential of private forest governance, *The Journal of Environment and Development* 14 (3), 356–74; Auld et al. (2008) op. cit.

27 Gulbrandsen, L. (2004) Overlapping public and private governance: Can forest certification fill the gaps in the global forest regime? *Global Environmental Politics* 4, 75–99.
28 Fry, A. (2000) Sustainable forest management: The role of certification, *Sustainable Development International* 2, 129–31.
29 Humphreys (2006) op. cit.
30 Humphreys (2006) op. cit.
31 http://register.pefc.cz/statistics.asp and http://www.fsc.org/facts-figures.html, accessed 29 March 2012.
32 Own calculation, based on http://register.pefc.cz/statistics.asp, http://www.fsc.org/facts-figures.html, accessed 29 March 2012.
33 http://www.fsc.org/facts-figures.html, accessed 29 March 2012.
34 FAO (2010) *FAO strategy for forests and forestry*, Rome, p.3.
35 World Bank (2004) *Sustaining forests: A development strategy*, Washington DC: The World Bank.
36 Hajjar, R. and Innes, J.L. (2009) The evolution of the World Bank's policy towards forestry: push or pull? *International Forestry Review* 11 (1), 27–37, p.27.
37 Hajjar and Innes (2009) op. cit.
38 Humphreys (2006) op. cit.; Hajjar and Innes (2009) op. cit.
39 Cashore, B., Galloway, G., Cubbage, F., Humphreys, D., Katila, P., Levin, K., Maryudi, A., McDermott, C. and McGinley, K. (2010) Ability of institutions to address new challenges, in Mery, G., Katila, P., Galloway, G., Alfaro, R.I., Kanninen, M., Lobovikov, M. and Varjo, J., *Forests and society – Responding to global drivers of change* (pp.441–85), Tampere: International Union of Forest Research Organizations; Ariell, A. (2010) Forest futures: A causal layered analysis, *Journal of Futures Studies* 14 (4), 49–64.
40 GEF (2011) *Instrument for the establishment of the restructured global environment facility*, Washington DC: GEF.
41 http://www.thegef.org/gef/sites/thegef.org/files/Images/GEF-timeline-1920.png (last accessed 5 May 2012).
42 http://www.thegef.org/gef/SFM (last accessed 7 May 2012).
43 Assembe-Mvondo, S. (2008) A review of States' practice of sustainable forest management with regard to some international conventions, *Miskolc Journal of International Law* 5 (2), 109–131.
44 *Convention on Wetlands of International Importance especially as Waterfowl Habitat*, Ramsar (Iran), 2 February 1971, UN Treaty Series No. 14583, Art 3.1.
45 The Convention on Biological Diversity, 1760 UNTS 79; 31 ILM 818 (1992), Art.1.
46 Humphreys (2006) op. cit.; Humphreys (1996) op. cit.
47 Boyd, E., Corbera, E. and Estrada, M. (2008) UNFCCC negotiations (pre-Kyoto to COP-9): what the process says about the politics of CDM-sinks, *International Environmental Agreements: Politics, Law and Economics* 8 (2), 95–112.
48 Thomas, S., Dargusch, P., Harrison, S. and Herbohn, J. (2010) Why are there so few afforestation and reforestation Clean Development Mechanisms projects? *Land Use Policy* 28, 880–87.
49 McDermott, C.L., O'Carroll, A. and Wood, P. (2007) *International forest policy – the instruments, agreements and processes that shape it*, New York: Department of Economic and Social Affairs; Bauer, S. and Stringer, L.C. (2009) The role of

science in the global governance of desertification, *The Journal of Environment and Development* 18 (3), 248–67.
50 Ruis, B. (2001) No forest convention but ten tree treaties, *Unasylva* 52 (3).
51 IUCN (2010) *Indigenous peoples and climate change/REDD. An overview of current discussions and main issues.*
52 ILO (1989) *Convention concerning indigenous and tribal peoples in independent countries (C169)*, Geneva: International Labour Organisation.
53 Johns, T., Merry, F., Stickler, C., Nepstad, D., Laporte, N. and Goetz, S. (2008) A three-fund approach to incorporating government, public and private forest stewards into a REDD funding mechanism, *International Forestry Review* 10 (3), 458–64.
54 IUCN (2010) op. cit.
55 ILO (1989) op. cit., Art. 14.
56 McDermott *et al.* (2007) op. cit.
57 Martin, R.M. (2004) Regional approaches: bridging national and global efforts, *Unasylva* 55 (3).
58 Aguilar, G. and González, M. (1999) Regional legal arrangements for forests: The case of Central America, in Tarasofsky, R., *Assessing the international forest regime* (pp. 113–23), Geneva: IUCN.
59 Howlett, M., Rayner, J., Goehler, D., Heidbreder, E., Perron-Welch, F., Rukundo, O., Verkooijen, P. and Wildburger, C. (2010) Overcoming the challenges to integration: Embracing complexity in forest policy design through multi-level governance, in Rayner, J., Buck, A. and Katila, P., *Embracing complexity: Meeting the challenges of international forest governance: A global assessment report prepared by the Global Forest Expert Panel on the International Forest Regime*, Vienna: International Union of Forest Research Organizations (IUFRO).
60 http://www.cbfp.org/objectifs_en.html, accessed 25 April 2012.
61 Humphreys (2006) op. cit.
62 Cashore *et al.*(2010) op. cit.
63 Humphreys (2006) op. cit.
64 Ziolkowska, J., Meyers, W.H., Meyer, S. and Binfield, J. (2010) Targets and mandates: Lessons learned from EU and US biofuels policy mechanisms, *AgBioForum* 13 (4), 398–412.
65 Pattberg (2005) op. cit.
66 Glück *et al.*(2010) op. cit.
67 Blaser, J., Contreras, A., Oksanen, T., Puustjarvi, E. and Schmithusen, F. (2005) *Forest law enforcement and governance in Europe and North Asia (ENA)*, Reference paper prepared for the Ministerial Conference, St Petersburg, Russia, 22–25 November 2005; World Bank (2006) *Strengthening forest law enforcement and governance. Addressing a systemic constraint to sustainable management*, Washington DC: The World Bank.
68 UNCTAD (2006) *International Tropical Timber Agreement*, United Nations Conference on Trade and Development, TD/TIMBER.3/12.
69 McDermott, C.L., Humphreys, D., Wildburger, C., Wood, P., Marfo, E., Pacheco, P. and Yasmi, Y. (2010) Mapping the core actors and issues defining international forest governance, in Rayner, J., Buck, A. and Katila, P., *Embracing complexity: Meeting the challenges of international forest governance: A global assessment report prepared by the Global Forest Expert Panel on the International Forest Regime* (pp. 19–36), Vienna: International Union of Forest Research Organizations (IUFRO).

70 Cock, A.R. (2008) Tropical forests in the global states system, *International Affairs* 84 (2), 315–33.
71 Tucker, C.M. (2010) Learning on governance in forest ecosystems: Lessons from recent research, *International Journal of the Commons* 4 (2), 687–706.
72 MCPFE (1993) *Resolution H1: General guidelines for the sustainable management of forests in Europe*, Helsinki, Finland: Second ministerial conference on the protection of forests in Europe, Art. D.
73 CBD (1995) *Decision II/8 Preliminary consideration of components of biological diversity particularly under threat and action which could be taken under the Convention*.
74 http://whc.unesco.org/en/list and http://whc.unesco.org/en/forests/, accessed 25 April 2012.
75 Glück *et al*.(2010) op. cit.
76 Davenport, D., Bulkan, J., Hajjar, R., Hardcastle, P., Assembe-Mvondo, S., Eba'a Atyi, R., Humphreys, D. and Maryudi, A. (2010) Forests and sustainability, in Rayner, J., Buck, A. and Katila, P., *Embracing complexity: Meeting the challenges of international forest governance: A global assessment report prepared by the Global Forest Expert Panel on the International Forest Regime* (pp. 75–92), Vienna: International Union of Forest Research Organizations (IUFRO).
77 http://www.cpfweb.org/73035/en/, accessed 6 May 2012.
78 http://web.worldbank.org/WBSITE/EXTERNAL/TOPICS/EXTSDNET/0,,menuPK:64885113~pagePK:7278667~piPK:64911824~theSitePK:5929282~contentMDK:23128447,00.html, accessed 5 May 2012.
79 Tarasofsky, R. (1999) Assessing the international forest regime: Gaps, overlaps, uncertainties and opportunities, in Tarasofsky, R., *Assessing the international forest regime* (pp. 3–12), Geneva: IUCN.
80 Hajjar and Innes (2009) op. cit.
81 Hermosilla, A.C. and Simula, M. (2007) *Review of implementation of the World Bank forest strategy*, Washington DC: World Bank.
82 Thomas *et al*. (2010) op. cit.
83 Diaz, D., Hamilton, K. and Johnson, E. (2011) *State of the forest carbon markets 2011: From canopy to currency*, Ecosystem Marketplace.
84 Zhu, S., Buongiorno, J. and Brooks, D.J. (2001) Effects of accelerated tariff liberalization on the forest products sector: a global modeling approach, *Forest Policy and Economics* 2 (1), 57–78.
85 Bernstein, S. and Hannah, E. (2012) The WTO and institutional (in)coherence and (un)accountability in global economic governance, in Narlika, A., Daunton, M. and Stern, R., *Oxford Handbook on the World Trade Organization*, Oxford: Oxford University Press.
86 Auld *et al*. (2008) op. cit.
87 CBD (2012) *Decision X/2, Strategic plan for biodiversity 2011–2020*.
88 ITTO (2009) *Draft ITTO Handbook*, Yokohama: ITTO, p.6.
89 NLBI (2007) op. cit., Art. 5.
90 McDermott *et al*. (2007) op. cit.
91 Jordan, A., Wurzel, R. and Zito, A.R. (2003) *New instruments of environmental governance? National experiences and prospects*, London: Routledge.
92 Brown, K. (2001) Cut and run? Evolving institutions for global forest governance, *Journal of International Development* 13 (7), 893–905.
93 Humphreys (2006) op. cit., p.13.
94 Steiner, M. (2001) After a decade of global forest negotiations, where are we now? *Review of European Community and International Environmental Law* 10 (1), 98–105.

95 Glück *et al.* (1997) op. cit.; Tarasofsky (1999) op. cit.; Tarasofsky, R. (1995) *The international forests regime: legal and policy issues*, Gland: WWF/IUCN; Humphreys, D. (2001) Forest negotiations at the United Nations: explaining cooperation and discord, *Forest Policy and Economics* 3 (3–4), 125–35.

96 Keohane, R.O. and Victor, D.G. (2011) The regime complex for climate change, *Perspectives on Politics* 9 (1), 7–23; Rayner, J., Buck, A. and Katila, P. (2010) *Embracing complexity: Meeting the challenges of international forest governance: A global assessment report prepared by the Global Forest Expert Panel on the International Forest Regime*, Vienna: International Union of Forest Research Organizations (IUFRO).

4 The emergence of REDD on the global policy agenda

Constanze Haug and Joyeeta Gupta[1]

4.1 Introduction

Having discussed the broader international forest governance architecture in Chapter 3, this chapter discusses the role of forests in the global climate change regime. The climate regime has addressed forest-related issues since its early days, but only to a limited extent. From 2006, the forest issue gained new prominence in the climate arena after the Stern review[2], and the 2008 Eliasch review on the linkage between forests and climate change[3] emphasized cutting emissions from tropical deforestation as a potentially very cost-effective strategy to mitigate climate change. This led to the emergence of the REDD (Reducing Emissions from Deforestation and Forest Degradation) concept. The basic premise of REDD is simple: incentivizing developing countries to reduce their deforestation rates, and in the event they succeed, to compensate them financially.[4] This concept appeals to both developed and developing countries. The former are attracted by its proclaimed cost-effectiveness and its potential to help them meet their emission reduction targets; the latter are drawn by the prospects of funding and view REDD as their opportunity to participate meaningfully in the global climate change regime.[5]

In line with the analytical framework (see 1.5), this chapter examines the evolving role of forests in the climate negotiations (see 4.2), with a special focus on REDD. It traces the evolution of REDD inside and outside of the context of the United Nations Framework Convention on Climate Change (UNFCCC) (see 4.3), examines key issues in REDD design and implementation (see 4.4), the interest of regional organizations in REDD and the broader impact that REDD has produced so far (see 4.5), before drawing some inferences (see 4.6).

4.2 Forests under the UN Climate Convention and the Kyoto Protocol

4.2.1 Early days: the forest-climate pre-Kyoto debate

We distinguish between four phases of 'forest governance' in the history of the climate change regime. In the first pre-1990 phase – 'defining a forest target'— the connection between forests and climate change was discussed at the Ministerial Conference on Atmospheric Pollution and Climate Change held in Noordwijk, the Netherlands, in 1989. Among a series of objectives to deal with climate change, the Conference also adopted a provisional target for net forest growth.[6] If implemented, it would have had a much greater impact on developing than on developed countries.[7] The idea at the time was to link up with the ongoing global discussions on forest governance.

In the second phase (c.1990–97) – 'elaborating a forest policy commitment' – discussions on forests progressed along different strands in the run-up to the 1992 United Nations Conference on Environment and Development in Rio. The role of forests was discussed in the negotiations on the Climate Convention (UNFCCC), but the issue of possible targets on forests was relegated to the parallel negotiations on a forest convention. This was also supported by declarations following the Noordwijk negotiations that expressed the sentiment that the most urgent goal was to first focus on industrial CO_2 emissions. Hence, the UNFCCC recognizes forests as both sources and sinks of greenhouse gases, commits Parties to promote the conservation and enhancement of these and other carbon stocks, and to report on relevant measures in their National Communications. Article 4.2a and b of the Convention elaborate on developed country quantitative commitments – and include an ambiguous target to stabilize their anthropogenic emissions of GHGs.

4.2.2 Forests in the Kyoto Protocol – 'integrating forest commitments into the climate regime'

In the third phase (c.1997–2005) – 'integrating forest commitments into the climate regime'– two key decisions were taken: forest sinks could be used by developed country parties to meet part of their emission limitation and reduction commitments under the Kyoto Protocol; and afforestation and reforestation projects were included in the scope of the Clean Development Mechanism (CDM).

The question of carbon sinks was a controversial issue at Kyoto and subsequent Conferences of the Parties (COPs). Initially, the EU and developing countries opposed the inclusion of sinks into developed country commitments in the Kyoto Protocol, while the United States, Canada, Australia and New Zealand were in favour.[8] Opponents feared the watering down of the quantitative commitments of developed countries and cited concerns over data quality and permanence of carbon storage. At the time, there was relatively little scientific knowledge on this issue, with large uncertainties remaining as to how to account for emissions from land use, land-use change and forestry (LULUCF).[9] Proponents argued that including carbon sinks would make commitments more comprehensive and provide countries with flexibility in implementing them.

Eventually, it was agreed that developed countries would jointly reduce their net emissions of six gases by 5.2 per cent in the period 2008–12 in relation to 1990 emission levels. The term 'net emissions' allowed them to account for afforestation, reforestation and deforestation and other agreed land use, land-use change and forestry activities since 1990 in their emission inventories. However, absorption by soils and the agricultural sector were not included.[10] Definitions and rules for accounting for carbon stocks and fluxes that were adopted at Kyoto and in its aftermath contained some simplifications and shortcomings (due to political pressure by some potential beneficiaries, such as Australia, but also because of scientific uncertainties). This resulted in windfall profits for some countries (such as Russia and Australia) in terms of emission reductions and allowed them to game the process by including or excluding carbon stocks based on national interest, thus diluting the main Kyoto target.[11] LULUCF accounting continues to be a cause of concern for any post-2012 quantitative commitments of developed countries. Negotiations currently centre on the determination of national reference levels in a proposed baseline-and-credit system for emissions from forest management. Baselines that are set too high could again lead to significant windfall profits.[12]

A second agenda item under which forests were discussed was the inclusion of LULUCF projects as eligible activities under the CDM. The CDM allows developed country parties to invest in emission reduction projects in developing countries, provided that they also contribute to sustainable development. The credits generated from such activities can be used by developed countries to meet their commitments under the Kyoto Protocol.[13] The modalities of implementing CDM projects were elaborated at COP-7 in Marrakesh, where it was also decided that the scope for LULUCF activities would be restricted to afforestation and reforestation (A/R) projects. The set-up of A/R projects is generally the same as for other CDM projects, except that the credits they generate carry an expiration date. A/R credits can be used to a maximum of 1 per cent of the Party's base year emissions for each year of the first commitment period.

On the whole, the success of forest projects under the CDM has been very limited. As of 1 May 2012, less than 1 per cent of all CDM projects fall into this category, accounting for only 0.7 per cent of all CERs to be issued until the end of 2012.[14] The reasons lie in the complex rules governing this project type and resulting difficulties in developing robust methodologies and monitoring plans, low buyer interest due to the temporary nature of the credits, demand-side liability in case of their expiration, and the fact that A/R credits are currently excluded from the EU's emissions trading system.[15]

A critical difference between the use of sinks in implementing the Kyoto Protocol's provisions on quantitative commitments and in the CDM is that the latter excludes incentives for reducing deforestation and forest degradation, again due to concerns over permanence, accounting and the question of additionality. These questions proved highly controversial in the CDM negotiations, dividing the Group of 77 developing countries, but also environmental organizations.[16] Opponents, including the EU, the Association of Small Island States (AOSIS) and Brazil, cited the large uncertainties surrounding the crediting of 'avoided deforestation' and argued that this would shift the focus even further away from the need for fossil fuel abatement. Other Parties, such as Costa Rica, Mexico and Bolivia, supported incentives for forest maintenance through crediting.[17] This set the stage for the discussions in the post-2005 stage.

4.3 The emergence of REDD

In the fourth phase (2005–12) – 'tackling avoided deforestation/REDD in developing countries', the focus has turned to finding ways to halt deforestation and forest degradation in the South by rewarding developing countries for their efforts to maintain their forests.

Although the idea of REDD was first floated in the context of the global climate change regime, action on REDD has since increasingly moved outside of it. Given slow progress in the climate regime, multilateral and bilateral initiatives have sprung up to maintain momentum on REDD and support capacity-building. This section traces the evolution of REDD within the UNFCCC regime, before turning to the broader global governance landscape on REDD.

4.3.1 *REDD in the UNFCCC negotiations*

REDD was introduced into the climate negotiations at COP-11 in Montreal in 2005 under the label of 'avoided deforestation'. On behalf of the 'Coalition of Rainforest Nations', Papua New Guinea and Costa Rica submitted a proposal to reduce emissions from tropical deforestation (RED) through a market-based approach. Their submission highlighted the failure of the UNFCCC and the Kyoto Protocol to address the large share of global emissions from deforestation and argued that, as developing nations, the group was 'prepared to stand accountable for our contributions toward global climate stability, provided international frameworks are appropriately modified (...) through fair and equitable access to carbon emission markets.'[18] The proposal constituted a departure from discussions on avoided deforestation at Kyoto as it suggested the use of national reference levels instead of project-based accounting to measure reductions in forest loss, thus avoiding the problem of domestic displacement of emissions. It led to the establishment of a dedicated contact group and a two-year dialogue to further develop the concept. This process stimulated submissions from Parties and observer organizations on various aspects of the design and requirements for a possible REDD mechanism. At COP-13 in 2007, REDD was included in the Bali Road Map as one of the key building blocks of a future post-2012 climate change regime. A decision at the same summit[19] encouraged Parties to *r*educe *e*missions from *d*eforestation and forest *d*egradation (REDD) on a voluntary basis. The scope of REDD was thus expanded to include forest degradation, a move which had been strongly supported by the Central African Forest Commission (COMIFAC).[20] In 2008, in Poznan, its scope was further expanded to include 'the role of conservation, sustainable management of forests and enhancement of carbon stocks in developing countries'.[21]

REDD was one area where substantive progress was made during the Copenhagen climate negotiations in 2009. The global community agreed on the 'immediate establishment of a mechanism'[22] and a 'Green Climate Fund', which among other things should also finance REDD-related activities. At COP-16 in Cancun, the Parties adopted a more comprehensive decision focusing on the broader social and environmental dimensions of REDD, sometimes referred to as REDD+. Elsewhere, the '+' in REDD+ is said to denote an even broader scope for REDD, which encompasses not only deforestation and forest degradation, but also carbon enhancement from sinks.[23] More recently, some have coined the term 'REDD++' or AFOLU (agriculture, forestry and other land uses), calling for the inclusion of emissions from agriculture and other land-use changes into REDD. However, this position has so far not commanded much support in the negotiations. This book uses REDD to refer to all REDD discussion (see 1.1). COP-17 in Durban in 2011 brought moderate progress on the question of determining reference levels from which reductions should be measured (acknowledging that subnational baselines might be developed as an interim measure),[24] and for the first time included (very general) language on the potential role of market-based approaches in REDD.[25]

Whereas many of the design features of REDD are yet to be decided, the greatest challenge at this point in time relates to the question of sources and mechanisms of financing. REDD is strongly interlinked with the broader post-2012 financial architecture for climate change, and also with the concept of nationally appropriate mitigation actions (NAMAs) by developing countries and with emission reduction commitments by developed countries. The emergence of a global framework for REDD therefore primarily depends on advances made in the international climate negotiations as a whole. The failure of the Copenhagen summit and the slow progress since have demonstrated the enormous difficulties associated with reaching a post-2012 climate agreement. Yet whereas negotiations on REDD are well advanced,[26] agreement is so far proving elusive on many other parts of the complex negotiation matrix. This has led proponents of REDD to resort to initiatives outside of the UNFCCC regime in an attempt to move the issue forward while discussions on a global mechanism remain deadlocked. The next section discusses these activities and their key actors in more detail.

4.3.2 REDD developments outside the UNFCCC

When the idea of REDD emerged on the international stage, it generated substantial interest from many public and private actors. At COP-13 in 2007 in Bali, UNEP, UNDP and FAO jointly launched the UN-REDD initiative, which aims at preparing developing countries for REDD implementation. Simultaneously, the World Bank established the Forest Carbon Partnership Facility (FCPF) to bolster its role in the area of carbon finance and forest carbon conservation and sequestration (see Box 4.1). The UN-REDD programme and the FCPF now support almost 50 countries in improving data on historical land use and forest cover, developing reference levels and monitoring strategies, as well as national REDD implementation strategies. As of November 2011, UN-REDD had raised USD 118 million[27] and FCPF USD 232 million.[28] In addition, the FCPF includes the Carbon Fund with USD 212 million in committed finance (as of 2011), for supporting implementation of policies and measures in national REDD strategy documents. While UN-REDD focuses primarily on monitoring, reporting and verification (MRV) strategies and local capacity-building, building on FAO's extensive forest data, access to funds, and networks in 194 member countries, the FCPF concentrates on developing REDD tools and strategies and is seen as a credible standard-setter.[29] However, the latter was also criticized for long delays in disbursing funds.[30] Given the proximity of their mandates, there is a risk of duplication and functional overlap between FCPF and UN-REDD. However, both organizations emphasize their commitment to achieve coordination and coherence through joint publications, workshops, and guidelines for REDD implementation.

Box 4.1 **The World Bank and carbon financing**

The World Bank has been engaged in carbon finance since 2000, when it established its first carbon fund, the 'Prototype Carbon Fund' (PCF). To date, its portfolio spans 12 funds which manage more than 200 active projects and funds totalling USD 2.4 billion.[31] The Bank sees itself as a facilitator and a catalyst for the global carbon market, through knowledge-building and methodology development, by providing an opportunity for 'learning by doing' to governments and the private sector, by generating trust and additional public and private finance for offset projects. While some of its carbon funds are specifically geared towards pioneering new project types or demonstrating high sustainable development benefits, the majority are more commercial in outlook, focusing on helping developed countries meet their offsetting needs for reaching their Kyoto targets.

In 2004, the World Bank established the BioCarbon Fund to gain experience and develop methodologies for terrestrial carbon sequestration projects. The fund focuses on afforestation and reforestation projects. Over two tranches, it has raised USD 90.4 million in capital. Twenty-five per cent of the fund's first tranche was invested in Africa – a much higher proportion than for the CDM as a whole.[32] While the BioCarbon Fund has been instrumental for advancing methodology development for A/R projects, many of its projects have underdelivered in terms of emission reductions, thus demonstrating the difficulties of implementing sequestration projects in practice.[33]

> After the emergence of REDD in the climate regime, the World Bank in 2008 established the Forest Carbon Partnership Facility (FCPF). Its Readiness Fund focuses on institution- and capacity-building for country-wide implementation in 37 countries, while the Carbon Fund pioneers payments for verified emission reductions in countries that are further advanced in terms of REDD readiness.
>
> The World Bank has received recognition, but also severe criticism for its carbon finance activities. An internal evaluation of the World Bank found that the Bank had indeed acted as a catalyst in the carbon market, but had failed to reduce its engagement as a buyer once private investments in the carbon market were taking off.[34] Critics like Friends of the Earth International, Carbon Trade Watch or the Brettonwoods Project further argue that offset projects divert attention from real reduction projects in the West; that such projects (e.g. under the BioCarbon Fund) had failed to deliver expected emission reductions;[35] that the Bank funds many 'dirty' projects, which reduces its legitimacy as a progressive green actor; that its procedures are not adequately transparent and accountable;[36] and that it is affected by a conflict of interest through its role as fund manager, project manager and CDM consultant.[37]

The general stalemate at the Copenhagen COP-15 led 50 Parties to establish the REDD+ Partnership as a global platform to maintain momentum on REDD in 2010. Donors pledged around USD 4 billion for the period 2010–12, yet it is unclear how much of this will actually be 'new' money in addition to existing commitments in official development assistance.[38] The REDD+ Partnership sees itself as a forum for knowledge-sharing and learning on REDD, but has drawn criticism for its limited involvement of civil society representatives and NGOs in the deliberations.[39] As the FCPF and UN-REDD, the REDD+ partnership emphasizes its interim character and its complementarity to REDD negotiations under the UNFCCC and expects 'to be replaced by, or folded into, a UNFCCC mechanism including REDD+ once established and agreed upon by the Parties'.[40] The activities of all three initiatives have made a substantial contribution to the first phase in REDD implementation, often referred to as REDD readiness.

Apart from these multilateral initiatives, individual donor governments have also been active on REDD. Norway, which was instrumental in initiating the REDD+ partnership, is the largest donor on REDD. It has bilateral agreements with Brazil, Guyana, Indonesia and Tanzania, making large REDD payments dependent on demonstrated reductions of deforestation from agreed reference levels. Together with the FCPF's Carbon Fund, the country is thus pioneering the second phase of REDD implementation – performance-based disbursement from a fund-based instrument. The third and final phase would consist of a GHG-based instrument that 'rewards performance on the basis of quantified forest emissions and removals against agreed reference levels'.[41]

In addition, conservation and development NGOs and the private sector have started implementing REDD pilot and demonstration activities on the ground. While much activity concentrates on Indonesia and Brazil, 105 of the 140 non-Annex-I countries have thus far not participated in REDD initiatives.[42]

Yet moving REDD from project-based pilots to implementation at a much wider scale presents formidable challenges. The following sections review key issues related to the design of a global REDD mechanism and the question of financing, and then turn to the impact of REDD to date and the challenges that its implementation poses on the ground in developing countries.

4.4 Key challenges for REDD at the international level: designing an effective, robust mechanism

In the early days of REDD, much of the discussion in international policy circles as well as in the academic and the grey literature focused on how to design a global REDD mechanism under the UNFCCC regime.[43] A few years on, many of the questions (such as the scale of REDD, reference levels, financing, monitoring, permanence and leakage concerns, as well as safeguards) are still relevant in a more fragmented global REDD landscape.

4.4.1 The right scale for REDD

In REDD parlance, 'scale' refers to the level of accounting for REDD activities.[44] The 2005 proposal on REDD by the Coalition of Rainforest Nations had suggested the use of national reference levels to measure reductions in forest loss instead of project-based accounting, with a view to avoiding domestic leakage. While, as of 2012, REDD pilots have mostly been implemented as individual projects, overall there is an expectation that REDD will eventually evolve into a vertically integrated, 'nested' set-up.[45] This would imply a national or at least subnational framework for monitoring and accounting, which includes smaller-scale and spatially explicit project areas where specific REDD activities are being implemented.[46] Such an architecture would ensure that incentives and benefit-sharing extend also to lower governance levels, while avoiding the double-counting of emissions reductions.

4.4.2 Reference levels

The choice and determination of reference levels from which performance is measured and efforts are rewarded is key to both project-based REDD activities and global REDD mechanism design. Both require baselines that define a business-as-usual scenario above which efforts are compensated. In the early phases of REDD, performance will likely be – and in some instances already is – based on a mix of predefined indicators, such as the adoption of certain policies or measures, or the establishment of new institutional structures. The next step would then be the use of proxy indicators for achieved emission reductions, before conditions are in place for more comprehensive carbon accounting. The choice of the reference level has an international equity dimension since it determines how much different countries at different stages of the 'forest transition curve'[47] (see 2.2) are likely to benefit from the planned mechanism. If a historical baseline is chosen (i.e. future reductions in deforestation are rewarded based on the average of a country's past deforestation rates over a certain time frame) those Parties that have deforested rapidly during this period stand to gain more than those who protected their forests. One way of compensating for this effect would be to factor the global average deforestation rate into the calculation of national reference levels.[48] This approach was used in the Norway–Guyana agreement, which relied on an equal weighting of Guyana's (extremely low) deforestation rate and the mean deforestation rate in developing countries with deforestation.[49] Alternatively, historical baselines could be adjusted with a 'development factor' which might include variables such as a country's GDP/capita or its forest cover.[50]

4.4.3 Financing REDD

The question of REDD financing is a practical and a policy design challenge. An effective REDD will require substantial, long-term, stable and predictable funding. The Eliasch Review puts the cost of halving emissions from the forest sector by 2030 at USD 17–33 billion per year, assuming that carbon credits generated from REDD are integrated into a global carbon trading scheme, in addition to an estimated USD 4 billion for capacity-building for REDD in 40 developing countries.[51]

According to the Informal Working Group on Interim Finance for REDD+ (IWG-IFR), achieving an annual 25 per cent cut in global deforestation rates by 2015 could cost 15–25 billion euros between 2010 and 2015.[52] While much REDD-related expenditure to date has been public money (Figure 4.1), leveraging funding from the private sector will be indispensable to make REDD viable in the long term.[53] Yet achieving this is difficult and is often reduced to a simple market versus fund dichotomy.[54] While some argue that only a market-based approach with 'REDD credits' being sold to global carbon markets can raise the necessary funding in the long term,[55] others warn of the risk of cheap forest credits flooding the carbon market[56] and therefore advocate an international fund for REDD or a more decentralized registry approach.[57] Other hybrid public–private financing vehicles such as debt-for-nature swaps (see Box 2.4), forest bonds or credit enhancement have also been suggested,[58] yet to what extent they will materialize and prove able to leverage significant levels of funding remains to be seen.

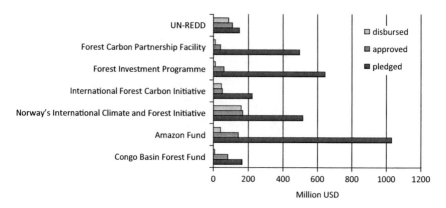

Figure 4.1 Pledged, approved and disbursed finance of key multilateral and bilateral REDD funds

(NB status February 2012. Source: Climate Funds Update, www.climatefundsupdate.org, accessed 27 March 2012)

4.4.4 *Monitoring, reporting and verification*

Payments for emission reductions require reliable systems for monitoring, reporting and verification (MRV), yet developing relevant and accurate country-level data is costly and requires capacity.[59] Concerns regarding MRV had been among the main reasons for Parties to exclude avoided deforestation from the CDM.[60] Estimating net emissions from forest changes requires data on changes in forest area (derived from remote sensing data in combination with ground measurements) as well as on carbon density and carbon fluxes.[61] Still, monitoring deforestation is less difficult than monitoring forest degradation, which primarily depends on costlier ground measurements.[62] Community-based monitoring as part of broader national schemes could involve and reward local peoples, lowering implementation costs.[63]

4.4.5 *Permanence, additionality and leakage*

Permanence, additionality and leakage, all well-known challenges for CDM forest projects, are also relevant for REDD. Permanence refers to the time factor – the risk that the effect of carbon sequestration may be cancelled out later by increased deforestation or forest degradation, leading to net emissions in the future. However, even with permanence concerns in the medium term, proponents argue that REDD may be able to 'buy time' for mitigation until other abatement options are adopted.[64] (Non-)permanence can be partially addressed through mechanism design, like using temporary crediting as under the CDM.[65] Ensuring permanence requires clear liability rules that determine who is held responsible for non-compliance. Given the weak regulatory mechanisms in most developing countries, dedicated national or international insurance facilities or buffer accounts could provide solutions in that regard.[66]

Additionality – ensuring that REDD payments are made in return for actual carbon saved – is closely linked to the question of reference levels. At the national and at the project level, additionality depends on determining realistic business-as-usual scenarios as the basis for national reference levels, beyond which any carbon sequestration can credibly be considered as additional. Whereas lack of additionality under a fund-based approach to REDD would merely result in funds being disbursed without service in return, its consequences would be more severe under a market-based approach, where they would effectively introduce 'hot air' into the system, thus reducing the environmental effectiveness of the cap.

Leakage is the risk that as deforestation or forest degradation is reduced in one place, these activities and thus emissions shift elsewhere. National-level baselines for REDD resolve the issue of project-level leakage at least for international reporting purposes, thereby addressing a key problem of CDM forest projects. However, leakage will continue to be an issue for national REDD registries coordinating and funding individual REDD project activities under the national baseline. Moreover, the risk of the displacement of deforestation abroad remains pertinent.[67] Eventually, broad participation by the largest possible number of tropical countries in future REDD is required to effectively alleviate leakage concerns (see also Chapter 10).[68]

4.4.6 Safeguards

A final theme is the quest to ensure that carbon-focused REDD efforts do not impact negatively on the other ecosystem services (see Table 1.1) provided by forests. This can be achieved through governance safeguards (transparency, accountability and anti-corruption),[69] social safeguards that account for land use rights, knowledge, management systems and resource arrangements, participation and the right of information and informed consent of local communities and indigenous peoples (see Box 4.2) in REDD;[70] and environmental safeguards seeking to conserve biodiversity, and avoid the conversion of natural forests into monocultures.[71] Arguably, the effective preservation of safeguards requires nested, multi-level governance arrangements.[72] This would imply that broad principles are defined at the global level and subsequently elaborated nationally and locally in a bottom-up process, involving a range of stakeholders including the most vulnerable groups. Nonetheless, there remains an inherent tension as to how universal vs. contextual safeguards for REDD should be. It is vital that those who are most affected by REDD implementation can shape their content according to the specific local circumstances. At the same time, cross-scale interactions (including the option of international review of safeguard implementation) are also important for forest peoples to be able to seek recourse to a higher level if they feel that their rights are violated by national or local governments.

The COP-16 Cancun meeting adopted an annex on safeguards in REDD, which arguably represents a 'significant shift in the type of language included in UNFCCC documents to date'.[73] Its inclusion has been attributed to the continuous, collective efforts of many non-state actors.[74] However, the Cancun decision is silent on the design of national reporting systems that REDD countries are required to develop, and whether these will be subject to international review.

4.5 Key challenges for REDD at the domestic level: implementation and benefit-sharing

4.5.1 *The impact of REDD to date*

REDD has taken off in a rather short time frame and has brought new momentum to the forests issue on the global policy agenda. The prospect of REDD has drawn the attention of developing country governments to the value of keeping forests standing; forests have become 'assets to be protected'.[75] REDD+ policies and strategies are now being developed in more than 40 countries.[76] Indonesia, for instance, has an ambitious goal to reduce GHGs by 26 per cent compared to business-as-usual by 2020, and has imposed a two-year moratorium to government-issued licences to convert forest land into other uses, following a bilateral agreement with Norway (see Chapter 6). REDD promises a large-scale influx of funds. According to one estimate, reducing emissions from deforestation in protected areas in the tropics could be valued at USD 6.2–7.4 billion. This corresponds to 1.5 times the estimated current conservation budgets in these regions, which currently fall far short of what is needed.[77] There is, at present, also a general willingness to take on board lessons from previous failed initiatives and to look at the forests issue from a broader, cross-sectoral perspective. Many tropical countries have extensive forest strategies and policies on paper, so what is needed is a strengthening of the implementation and enforcement of these measures.[78]

The advent of REDD has also stimulated major improvements in the availability of forest data and monitoring technology over the last five years. Technological innovations in remote sensing and monitoring increase buyers' confidence in the quality of emission reductions.[79] Some tropical countries (e.g. Brazil and India) now directly survey deforestation with satellite imagery. The new LiDAR (light detection and ranging) technology, for instance, a remote sensing technique which makes use of laser pulses from overhead planes, produces accurate imagery even from tropical forests with consistent cloud cover – a problem that is especially relevant for the case study country of Cameroon (see Chapter 7). At the same time, the issue of MRV raises an inherent tension. Investors will undoubtedly require a high level of scrutiny to ensure that they are buying real emission reductions, yet REDD could easily come to be regarded by developing countries as a 'new form of imperialism' if that level of scrutiny is too high.[80]

4.5.2 Challenges of good governance, tenure and internal benefit-sharing

We have seen above the dilemmas and trade-offs associated with REDD design and the uncertainties surrounding its future under the global climate regime. Yet the greatest challenges may not lie in the governance and design of the global REDD mechanism, but in its domestic implementation in developing countries.[81] Given that the drivers of deforestation in the tropics extend far beyond the forest sector (see 2.3 and Chapters 5–9), effective REDD would require an approach that involves horizontal and vertical coordination across ministries (environment, mining, agriculture, etc.) and policy sectors, looking also outside of REDD focal areas. The success of REDD in the Amazon basin for instance strongly depends on effective poverty alleviation in the Andes (see Chapter 8). Finally, specific instruments within a broader REDD strategy need to be able to target deforestation drivers and match the needs of actors in order to appropriately change their behaviour (see Chapter 2). This may be challenging, however, as local communities may have different interests from private sector actors investing in REDD.[82]

Once REDD moves beyond the pilot stage, implementation in developing countries will likely be led by national governments and subnational agencies in conjunction with local public or private actors.[83] A key issue here is how to create a value chain that identifies the right beneficiaries and reliably distributes benefits down to the local level. This relates to the question of tenure, the 'systems of rights, rules, institutions and processes regulating resource access and use'.[84] With forests becoming an asset under REDD, who has tenure determines carbon rights and thus ultimately financial compensation.[85] It is increasingly evident that entitlements of the state to the ownership of forests are contested domestically by the customary rights of forest dwellers and forest resource users (see Chapters 5–9, 9.4). International pressure to recognize this especially in relation to indigenous peoples has led to many legal declarations (see Box 4.2). However, while indigenous peoples have been able to organize themselves and apply pressure, local and often poor communities who are also dependent on forests continue to remain marginalized in these discussions.

Box 4.2 **Indigenous peoples and REDD**

REDD can improve livelihoods of indigenous peoples, reward them for their conservation activities,[86] help them with their land rights claim,[87] and indigenous organizations are providing local people with information and training.[88] However, REDD, by bringing resources, could make the competition between the state, new REDD actors and the local community more intense leading to unrecognized customary and ancestral rights, poor participatory opportunities,[89] social, economic[90] and even violent conflict. Many countries only tolerate indigenous peoples and may use REDD as an excuse to move these marginalized people.[91] If REDD changes their living circumstances, it may force them to import survival commodities from elsewhere making indigenous people worse off.[92]

Some indigenous people want to participate in REDD, others question it on ethical and cultural grounds,[93] still other 'peoples in voluntary isolation' avoid any contact with outsiders.

In September 2007, the General Assembly adopted the UN Declaration on the Rights of Indigenous Peoples and its principle of free, prior and informed consent (FPIC). This principle is based on the right of indigenous peoples to give or withhold their free, prior and informed consent to actions that affect their lands, territories and natural resources, based on customary laws and practices, free from external interference, manipulation or coercion, following full disclosure of intent and scope of the activity, in a language which is understandable to them.[94] Indigenous groups at COP-13 in Bali influenced the COP text to include language on their needs (but not rights). They promoted discussions on their issues in the subsidiary bodies.[95] A proposal for a formal advisory group on indigenous issues was rejected,[96] but there was a Global Indigenous Peoples Consultation[97] which called for consulting mechanisms for indigenous peoples based on FPIC, an evaluation of indigenous territories, decentralization mechanisms and obligatory monitoring, reporting and anti-corruption measures, recommendations on direct access to REDD Funds (and not via the state).[98] In 2009, the global summit of indigenous peoples on climate change in Anchorage, Alaska adopted a declaration calling for a direct fund for Indigenous Peoples, both to participate in the REDD process and to have access to resources for adaptation, mitigation and monitoring.[99] In 2009, the International Alliance of Indigenous and Tribal Peoples of the Tropical Forests (IIPFCC) sent a submission to the COP stressing in addition to their earlier positions that the 'commercialization and commoditization' of forests conflicts with the principle of self-determination and their value system.[100] Nine Amazonian indigenous organizations stressed that they have preserved 104 million hectares of Amazonian ecosystems with traditional knowledge (and a population of 2,800,000), and their fear that REDD might benefit those who deforest more than those who protect.[101] These groups are now lobbying heavily, but have been disappointed by the response thus far.

Responding to the requirements of REDD, land reform processes have been initiated in a number of developing countries. However, these attempts at harmonizing customary law and statutory tenure have produced mixed outcomes; in Uganda for instance, only few forest dwellers have sought – and have succeeded in – having their customary tenure rights recognized by the state.[102] At the same time, and as the case studies reveal, the recognition of these local rights has been undertaken in very dubious ways (see section 9.3). The clash between statutory and customary tenure may exacerbate the competition between the state and forest dwellers as to who is the real owner; and between the NGOs and private companies and local communities as land grabbing (see Box 9.1) becomes a serious issue in many parts of the world.[103] Unclear tenure also risks deterring the demand side of REDD as it poses an investment risk for private sector investors.[104] Ultimately, what may be required is a better understanding of the bundle of rights within tenure systems and its integration within a broader effort of empowering these local communities.[105] It remains to be seen whether REDD will help address tenure issues, or whether it will perpetuate existing inequalities (see 9.4).

This raises the very legitimate concern that 'the significant financial resources that could become available under REDD+ might exacerbate, rather than address, institutional and social factors that contribute to forest loss and degradation, such as elite capture of benefits and corrupt behaviours'.[106] This is especially likely as many potential REDD countries are 'poor governance hotspots'. The chances for countries like Liberia, the Democratic Republic of Congo and Myanmar, who might stand to gain most from these revenues in relative terms, to access REDD funding may be severely impeded by weak governance.[107] Effective REDD will require far-reaching reforms of public policy and institutions, far beyond the forest sector. Only this can prevent cases like the 'carbon cowboys' in Papua New Guinea, where corrupt government officials persuaded remote forest communities to sell carbon rights, before any regulation to this effect had been adopted.[108]

A key instrument of local empowerment and democratization is decentralization (see 2.4.4). There is a risk that REDD could prejudice or compromise ongoing efforts to decentralize forest governance in the tropics. Previous decentralization efforts had been largely motivated by funding shortages at the national level. This may change however with the influx of REDD finance and may potentially anoint the benefits that decentralized forest management has brought for forest protection.[109]

4.5.3 Risks and implications of commodifying forest carbon

Forests were once an open resource for all to enjoy. Yet some worry that REDD may lead to a commodification of forest carbon, turning forests into club or even private goods.[110] Forest carbon as a key source of income renders forests 'vulnerable to conversion to other more profitable land-uses, changes in carbon prices or international politics'[111] and may lead to a disregard of other, less easily measured forest services. This is salient with regard to biodiversity conservation, especially if the definition of what counts as a forest under REDD does not clearly differentiate between natural and planted forests.[112] Finally, the focus on carbon hardly ties in with local communities' perception of the value of forests: as Boyd found for the CDM, 'local values appear to be closely associated with development and land tenure, jobs, autonomy and political leverage, while administrators and scientists lay claims to carbon and conservation'.[113]

4.6 Inferences

REDD has proven one of the most dynamic areas of international climate policy over the past few years. It has raised great expectations and has been promoted by some as a panacea for simultaneously addressing climate change, deforestation, and alleviating poverty. Agreement on a unified global mechanism for REDD is to some extent deadlocked, however, by the stalemate of the UN climate negotiations as a whole. Meanwhile, a host of initiatives have developed outside of the UNFCCC realm, making it seem likely that progress on REDD will 'occur through complex fragmented pathways of international assistance, bilateral and multilateral agreements, and civil society and market-based processes'.[114]

At a practical level, REDD design poses challenges that will likely require trade-offs in terms of efficiency, effectiveness and the equity of the mechanism. These include questions regarding the stringency of monitoring requirements, the scope of the mechanism, equity considerations in setting baselines, and determining compensation levels and beneficiaries.

The big-picture challenge related to REDD, however, is whether it will succeed in attracting sufficient predictable, long-term funding to compete with other land uses in the context of increasing global pressures on land and rising consumer demand worldwide for forest and non-forest products. On the domestic level, REDD crucially depends on much-needed systemic changes in developing countries that allow for effective multi-level governance all the way down to local communities. There may thus be a contradiction between perceptions in the climate community and the realities of protecting tropical forests on the ground. While some climate experts consider REDD a 'quick fix' before the technologies needed for a low-carbon transition become available, it is highly unlikely that REDD will produce effects quickly enough, given the complexities of its implementation.

Notes

1 The authors would like to thank Fariborz Zelli and Michael Huettner for their thoughtful comments on an earlier draft.
2 Stern, N. (2006) *Stern review: the economics of climate change*, London: HM Treasury.
3 Eliasch, J. (2008) *Climate change: financing global forests*, London: Office of Climate Change.
4 Humphreys, D. (2008) The politics of 'avoided deforestation': historical context and contemporary issues, *International Forestry Review* 10, 433–42.
5 Haug, C. and Pattberg, P. (2008) Der REDD-Hype in den Klimaverhandlungen. Internationale Waldschutzdebatte nach Bali, *Informationsbrief Weltwirtschaft und Entwicklung (W&E)* 19 (9); Agrawal, A., Nepstad, D. and Chhatre, A. (2011) Reducing emissions from deforestation and forest degradation, *Annual Review of Environmental Resources* 36, 11.1–24.
6 The Noordwijk Declaration on Climate Change, Article 21. The target was net forest growth of 12 million hectares annually in the twenty-first century.

7. Bodansky, D. (1993) The United Nations framework convention on climate change: a commentary, *Yale Journal of International Law* 18, 451.
8. Gupta, J. and Ringius, L. (2001) The EU's climate leadership: Reconciling ambition and reality, *International Environmental Agreements: Politics, Law and Economics* 1 (2), 281–99; Oberthür, S. and Ott, H. (1999) *The Kyoto Protocol international climate policy for the 21st century*, Berlin: Springer.
9. Streck, C. and Scholz, S.M. (2006) The role of forests in global climate change: whence we come and where we go, *International Affairs* 82 (5), 861–79; Gupta and Ringius (2001) op. cit.; Boyd, E., Corbera, E. and Estrada, M. (2008) UNFCCC negotiations (pre-Kyoto to COP-9): What the process says about the politics of CDM-sinks, *International Environmental Agreements: Politics, Law and Economics* 8 (2), 95–112.
10. While forests have gained more prominence under REDD, agriculture has until this day remained a stepchild in climate negotiations. One could argue that this omission to integrate agriculture more comprehensively into the climate regime also stands in the way of a more integrated approach to REDD in the current negotiations.
11. Grassi, G., Federici, S. and Pilli, R. (2010) What happened to forests in Copenhagen? *iForest – Biogeosciences and Forestry* 3 (1), 30–32; Macintosh, A.K. (2011) LULUCF in the post-2012 regime: Fixing the problems of the past?, *Climate Policy*, 1–15.
12. Ibid.
13. For a comprehensive history of the CDM, see van Asselt, H. and Gupta, J. (2009) Stretching too far – Developing countries and the role of flexibility mechanisms beyond Kyoto, *Stanford Environmental Law Journal* 28, 311.
14. UNEP Risoe CDM/JI Pipeline Analysis and Database, 1 May 2012. Last accessed 10 May 2012.
15. Dutschke, M. and Angelsen, A. (2008) *How do we ensure permanence and assign liability?* Bogor: Center for International Forestry Research.
16. Fry, I. (2002) Twists and turns in the jungle: Exploring the evolution of land use, land-use change and forestry decisions within the Kyoto Protocol, *Review of European Community and International Environmental Law* 11 (2), 159–68.
17. Fearnside, P.M. (2001) Saving tropical forests as a global warming countermeasure: An issue that divides the environmental movement, *Ecological Economics* 39 (2), 167–84.
18. UNFCCC (2005) *Reducing emissions from deforestation: approaches to stimulate action. Submissions from Parties. FCCC/CP/2005/MISC.1*.
19. COP-13, Bali Action Plan (Decision 1/CP.13) 'Policy approaches and positive incentives on issues relating to reducing emissions from deforestation and forest degradation in developing countries; and the role of conservation, sustainable management of forests and enhancement of forest carbon stocks in developing countries.'
20. Grubb, M. (2011) Durban: the darkest hour? *Climate Policy* 11 (6), 1269–71.
21. UNFCCC/SBSTA/2008/L.23, p.1.
22. Article 6 of the Copenhagen Accord (Decision 2/CP.15).
23. Lawlor, K., Jenkins, A., Olander, L.P. and Murray, B.C. (2010) *Expanding the scope of international terrestrial carbon options: Implications of REDD+ and beyond*, Durham, NC: Nicholas Institute for Environmental Policy Solutions, Duke University.
24. UNFCCC Decision 12/CP.17.

25　UNFCCC Decision 2/CP.17.
26　Grubb, M. (2011) op. cit.
27　http://mdtf.undp.org/factsheet/fund/CCF00, accessed 15 November 2011.
28　FCPF (2011) *Forest Carbon Partnership Facility 2011 Annual Report*, Washington DC: The World Bank.
29　Hardcastle, P., Davenport, D., Cowling, P. and Watson, C. (2011) *Discussion of effectiveness of multilateral REDD+ initiatives*, Bristol: the IDL group.
30　Hardcastle *et al.* (2011) op. cit.
31　http://go.worldbank.org/51X7CH8VN0, accessed 11 May 2012.
32　Independent Evaluation Group (2010) *Phase II: The challenge of low-carbon development. Climate change and the World Bank Group*, Washington DC: The World Bank.
33　Independent Evaluation Group (2010) op. cit.
34　Independent Evaluation Group (2010) op. cit.
35　Horton, A. and Fry, T. (2011) *The role of the World Bank in carbon finance*, London: Brettonwoods Project.
36　Woods, N. (2000) The challenge of good governance for the IMF and the World Bank themselves, *World Development* 28 (5), 823–41.
37　Michaelowa, A. and Michaelowa, K. (2011) Climate business for poverty reduction? The role of the World Bank, *The Review of International Organizations* 6 (3), 259–86.
38　Agrawal *et al.* (2011) op. cit.
39　Silva-Chavez, G. (2010) *Enviro groups criticize REDD+ Partnership's 'unacceptable' exclusion of civil society121*, http://blogs.edf.org/climatetalks/2010/07/16/enviro-groups-criticize-redd-partnership%E2%80%99s-unacceptable-exclusion-of-civil-society/.
40　http://reddpluspartnership.org/65226/en/, accessed 15 November 2011.
41　Angelsen, A., Brown, S., Loisel, C., Peskett, P., Streck, C. and Zarin, D. (2009) *Reducing emissions from deforestation and forest degradation (REDD): An options assessment report*, Meridian Institute.
42　Cerbu, G.A., Swallow, B.M. and Thompson, D.Y. (2011) Locating REDD: A global survey and analysis of REDD readiness and demonstration activities, *Environmental Science & Policy* 14 (2), 168–80.
43　Santilli, M., Moutinho, P., Schwartzman, S., Nepstad, D., Curran, L. and Nobre, C. (2005) Tropical deforestation and the Kyoto Protocol, *Climatic Change* 71 (3), 267–76; Mollicone, D., Achard, F., Federici, S., Eva, H., Grassi, G., Belward, A., Raes, F., Seufert, G., Stibig, H.J., Matteucci, G. and Schulze, E.D. (2007) An incentive mechanism for reducing emissions from conversion of intact and non-intact forests, *Climatic Change* 83 (4), 477–93; Ogonowski, H., Helme, N., Movius, D. and Schmidt, J. (2007) *Reducing emissions from deforestation and degradation: the dual markets approach*, Washington DC: Center for Clean Air Policy; Strassburg, B., Kerry Turner, R., Fisher, B., Schaeffer, R. and Lovett, A. (2009) Reducing emissions from deforestation – the 'combined incentives' mechanism and empirical simulations, *Global Environmental Change* 19, 265–78; Combes Motel, P., Pirard, R. and Combes, J.L. (2009) A methodology to estimate impacts of domestic policies on deforestation: Compensated successful efforts for 'avoided deforestation' (REDD), *Ecological Economics* 68 (3), 680–91.
44　Strassburg *et al.* (2009) op. cit.

45 Cortez, R., Saines, R., Griscom, B., Martin, M., De Deo, D., Fishbein, G., Kerkering, J. and Marsh, D. (2010) *A nested approach to REDD+: Structuring effective and transparent incentive mechanisms for REDD+ implementation at multiple scales*, Washington DC: Nature Conservancy; De Gryze, S. and Durschinger, L. (2010) *An integrated REDD offset program (IREDD) for nesting projects under jurisdictional accounting*, San Francisco: Terra Global Capital LLC; Pedroni, L., Dutschke, M., Streck, C. and Porrúa, M.E. (2009) Creating incentives for avoiding further deforestation: the nested approach, *Climate Policy* 9 (2), 207–20.
46 De Gryze and Durschinger (2010) op. cit.
47 Mather, A.S. (1992) The forest transition, *Area* 24 (4), 367–79.
48 Strassburg *et al.* (2009) op. cit.
49 http://www.regjeringen.no/upload/MD/2011/vedlegg/klima/klima_skogprosjektet/Guyana/JointConceptNote_31mars2011.pdf , accessed 29 November 2011.
50 Parker, C., Mitchell, A., Trivedi, M. and Mardas, N. (2009) *The little REDD+ book: An updated guide to governmental and non-governmental proposals for reducing emissions from deforestation and degradation*, Oxford, UK: Global Canopy Programme.
51 Eliasch (2008) op. cit.
52 IWG-IFR (2009) *Report of the informal working group on interim finance for REDD+*.
53 Eliasch (2008) op. cit.; Agrawal *et al.* (2011) op. cit.; IWG-IFR (2009) op. cit.
54 Stockwell, C., Hare, W. and Macey, K. (2009) Designing a REDD mechanism: the TDERM triptych, in Richardson, B.J.L.B.Y., McLeod-Kilmurray, H. and Wood, S., *Climate law and developing countries: Legal and policy challenges for the world economy* (pp.151–77), Cheltenham: Edward Elgar Publishing Ltd.
55 Eliasch (2008) op. cit.
56 Dixon, A., Anger, N., Holden, R. and Livengood, E. (2008) *Integration of REDD into the international carbon market: Implications for future commitments and market regulation*, Wellington: M-co Consulting and Centre for European Economic Research (ZEW).
57 Reed, D. (2010) *A registry approach for REDD+. The REDD desk*, Climate Registry Option; Obersteiner, M., Huettner, M., Kraxner, F., McCallum, I., Aoki, K., Bottcher, H., Fritz, S., Gusti, M., Havlik, P. and Kindermann, G. (2009) On fair, effective and efficient REDD mechanism design, *Carbon Balance and Management* 4 (11).
58 IWG-IFR (2009) op. cit.
59 Grainger, A. and Obersteiner, M. (2011) A framework for structuring the global forest monitoring landscape in the REDD+ era, *Environmental Science & Policy* 14 (2), 127–39.
60 Boyd, E. (2009) Governing the Clean Development Mechanism: Global rhetoric versus local realities in carbon sequestration projects, *Environment and Planning A* 41 (10), 2380–95.
61 Wertz-Kanounnikoff, S., Verchot, L.V., Kanninen, M., Murdiyarso, D. and Angelsen, A. (2008) *How can we monitor, report and verify carbon emissions from forests?* Bogor: Center for International Forestry Research.
62 De Fries, R., Achard, F., Brown, S., Herold, M., Murdiyarso, D., Schlamadinger, B. and de Souza, C., Jr (2007) Earth observations for estimating greenhouse gas emissions from deforestation in developing countries, *Environmental Science & Policy* 10 (4), 385–94.

63 Danielsen, F., Skutsch, M., Burgess, N.D., Jensen, P.M., Andrianandrasana, H., Karky, B., Lewis, R., Lovett, J.C., Massao, J., Ngaga, Y., Phartiyal, P., Poulsen, M.K., Singh, S.P., Solis, S., Sorensen, M., Tewari, A., Young, R. and Zahabu, E. (2011) At the heart of REDD+: a role for local people in monitoring forests?, *Conservation Letters* 4 (2), 158–67; Palmer Fry, B. (2011) Community forest monitoring in REDD+: the 'M' in MRV?, *Environmental Science & Policy* 14 (2), 181–7; Skutsch, M.M., van Laake, P.E., Zahabu, E.M., Karky, B.S. and Phartiyal, P. (2009) Community monitoring in REDD, in Angelsen, A., *Realising REDD+, National strategy and policy options* (pp.101–12), Bogor, Indonesia: CIFOR.

64 Lubkowski, R. (2008) What are the costs and potentials of REDD?, in Angelsen, A., *Moving ahead with REDD: Issues, options, and implications* (pp.23–30), Bogor: Center for International Forestry Research; Stern, N. (2006) *Stern review: the economics of climate change*, London: HM Treasury.

65 Dutschke and Angelsen (2008) op. cit.

66 Dutschke and Angelsen (2008) op. cit.

67 Meyfroidt, P. and Lambin, E.F. (2009) Forest transition in Vietnam and displacement of deforestation abroad, *Proceedings of the National Academy of Sciences* 106 (38), 16139–44.

68 Dutschke and Angelsen (2008) op. cit.

69 Kanowski, P.J., McDermott, C.L. and Cashore, B.W. (2011) Implementing REDD+: lessons from analysis of forest governance, *Environmental Science & Policy* 14 (2), 111–17.

70 Lyster, R. (2011) REDD+, transparency, participation and resource rights: the role of law, *Environmental Science & Policy* 14 (2), 118–26; Sikor, T., Stahl, J., Enters, T., Ribot, J.C., Singh, N., Sunderlin, W.D. and Wollenberg, L. (2010) REDD-plus, forest people's rights and nested climate governance, *Global Environmental Change* 20 (3), 423–5.

71 Pistorius, T., Schmitt, C.B., Benick, D. and Entenmann, S. (2010) *Greening REDD+: Challenges and opportunities for forest biodiversity conservation*, Policy paper, Freiburg: University of Freiburg; Putz, F.E. and Redford, K.H. (2009) Dangers of carbon-based conservation, *Global Environmental Change* 19 (4), 400–401.

72 Sikor, T., Stahl, J., Enters, T., Ribot, J.C., Singh, N., Sunderlin, W.D. and Wollenberg, L. (2010) REDD-plus, forest people's rights and nested climate governance, *Global Environmental Change* 20 (3), 423–5.

73 http://www.wri.org/stories/2010/12/redd-decision-cancun, accessed 29 November 2011.

74 Agrawal, A., Nepstad, D. and Chhatre, A. (2011) Reducing emissions from deforestation and forest degradation, *Annual Review of Environment and Resources* 36 (1), 373–96.

75 Sunderlin, W.D. and Atmadja, S. (2009) Is REDD+ an idea whose time has come, or gone?, in Angelsen, A., *Realizing REDD+. National strategies and policy options* (pp.47–53), Bogor, Indonesia: Centre for International Forestry Research.

76 Angelsen *et al.* (2009) op. cit.

77 Scharlemann, J.P.W., Kapos, V., Campbell, A., Lysenko, I., Burgess, N.D., Hansen, M.C., Gibbs, H.K., Dickson, B. and Miles, L. (2010) Securing tropical forest carbon: the contribution of protected areas to REDD, *Oryx* 44 (3), 352–7; Clements, T. (2010) Reduced expectations: The political and institutional challenges of REDD+, *Oryx* 44 (3), 309–10.

78 Kanowski, P.J., McDermott, C.L. and Cashore, B.W. (2011) Implementing REDD+: lessons from analysis of forest governance, *Environmental Science & Policy* 14 (2), 111–17.
79 Agrawal *et al*. (2011) op. cit.
80 Clements (2010) op. cit., p.310
81 Corbera, E. and Schroeder, H. (2011) Governing and implementing REDD+, *Environmental Science & Policy* 14 (2), 89–99; Corbera, E., Estrada, M. and Brown, K. (2010) Reducing greenhouse gas emissions from deforestation and forest degradation in developing countries: revisiting the assumptions, *Climatic Change* 100 (3), 355–88; Huettner, M. (2012) Risks and opportunities of REDD+ implementation for environmental integrity and socio-economic compatibility, *Environmental Science & Policy* 15 (1), 4–12.
82 Blom, B., Sunderland, T. and Murdiyarso, D. (2010) Getting REDD to work locally: lessons learned from integrated conservation and development projects, *Environmental Science & Policy* 13 (2), 164–72.
83 Corbera and Schroeder (2011) op. cit.
84 Cotula, L. and Mayers, J. (2009) *Tenure in REDD: Start-point or afterthought?*, London: International Institute for Environment and Development.
85 Doherty, E. and Schroeder, H. (2011) Forest tenure and multi-level governance in avoiding deforestation under REDD+, *Global Environmental Politics* 11 (4), 66–88.
86 Johns, T., Merry, F., Stickler, C., Nepstad, D., Laporte, N. and Goetz, S. (2008) A three-fund approach to incorporating government, public and private forest stewards into a REDD funding mechanism, *International Forestry Review* 10 (3), 458–64.
87 IUCN (2008) *Global indigenous peoples' communication on reduced emissions from deforestation and degradation (REDD)*, Baguio City, Philippines: IUCN Forest Conservation Programme.
88 Griffiths, T. (2008) *Seeing 'REDD'? Forests, climate change mitigation and the rights of indigenous peoples. Update for Poznan (COP 14)*, Forest Peoples Programme.
89 IUCN (2010) *Indigenous peoples and REDD-plus. Challenges and opportunities for the engagement of indigenous peoples and local communities in REDD-plus,* Washington DC: International Union for the Conservation of Nature.
90 Griffiths (2008) op. cit.
91 Gregersen, H., El Lakany, H., Karsenty, A. and White, A. (2010) Does the opportunity cost approach indicate the real cost of REDD+?, *Rights and Resources Initiative*.
92 Karsenty, A. and Ongolo, S. (2008) Can 'fragile states' decide to reduce their deforestation? The inappropriate use of the theory of incentives with respect to the REDD mechanism, *Forest Policy and Economics* 18, 38–45, p.454.
93 Griffiths (2008) op. cit.
94 Forest Peoples Programme (2007) *Making FPIC – Free, Prior and Informed Consent – Work: Challenges and Prospects for Indigenous Peoples*, Moreton-in-Marsh: Forest Peoples Programme.
95 Griffiths (2008) op. cit.
96 Brown, H.C.P., Nkem, J.N., Sonwa, D.J. and Bele, Y. (2009) *Institutional dynamics and climate change in the Congo Basin forest of Cameroon, West Africa*, Copenhagen, Denmark: IARU International Scientific Congress on Climate Change: Global Risks, Challenges and Decisions, 10–12 March 2009.

97 UN-REDD Programme (2009) *Year in review.*
98 IUCN (2008) op. cit.
99 Indigenous Peoples' Global Summit on Climate Change (2009) *The Anchorage Declaration*, Anchorage, Alaska.
100 The International Alliance of Indigenous and Tribal Peoples of the Tropical Forests on behalf of International Indigenous Peoples' Forum on Climate Change (2009) *Submission to Subsidiary Body for Scientific and Technological Advice for Parties (SBSTA) on item 11 of FCCC/SBSTA/2008/L.23, draft conclusions proposed by Chair.*
101 The International Alliance of Indigenous and Tribal Peoples of the Tropical Forests on behalf of International Indigenous Peoples' Forum on Climate Change (2009) op. cit.
102 Doherty and Schroeder (2011) op. cit.
103 Peskett, L., Huberman, D., Bowen-Jones, E., Edwards, G. and Brown, J. (2008) *Making REDD work for the poor*, London: Overseas Development Institute.
104 Corbera, E. and Schroeder, H. (2011) op. cit.
105 Doherty and Schroeder (2011) op. cit.
106 Kanowski, P.J., McDermott, C.L. and Cashore, B.W. (2011) Implementing REDD+: lessons from analysis of forest governance, *Environmental Science & Policy* 14 (2), 111–17.
107 Ebeling, J. and Yasué, M. (2008) Generating carbon finance through avoided deforestation and its potential to create climatic, conservation and human development benefits, *Philosophical Transactions of the Royal Society B: Biological Sciences* 363 (1498), 1917–24.
108 Melick, D. (2010) Credibility of REDD and experiences from Papua New Guinea, *Conservation Biology* 24 (2), 359–61.
109 Phelps, J., Webb, E.L. and Agrawal, A. (2010) Does REDD+ threaten to recentralize forest governance? *Science* 328 (5976), 312–13.
110 McDermott, C.L., Levin, K. and Cashore, B. (2011) Building the forest-climate bandwagon: REDD+ and the logic of problem amelioration, *Global Environmental Politics* 11 (3), 85–103.
111 Clements (2010) op. cit.
112 Sasaki, N. and Putz, F.E. (2009) Critical need for new definitions of 'forest' and 'forest degradation' in global climate change agreements, *Conservation Letters* 2 (5), 226–32.
113 Boyd, E. (2006) Scales of governance in carbon sinks. Global priorities and local realities, in Reid, W.V., Berkes, F., Wilbanks, T. and Capistrano, D., *Bridging scales and knowledge systems: Concepts and applications in ecosystem assessment* (pp.105–26), Washington DC: Island Press.
114 Agrawal *et al.* (2011) op. cit.

5 Case study
Vietnam

*Léa Bigot, Nicolien van der Grijp,
Joyeeta Gupta, and Vu Tan Phuong[1]*

5.1 Introduction

Vietnam lies in the rainforest basin of South East Asia. The country is home to 87 million people of 54 ethnic groups, with the Kinh as by far the largest group.[2] Twenty-eight per cent of the population lives in urban and 72 per cent in rural areas. Nearly half of the Vietnamese people work in the agricultural sector. However, the service sector has been growing fast at the expense of the agricultural sector. In 2008, the per capita GDP was USD 2,787.[3] The forestry sector's contribution to GDP was 2.4 per cent in 2006.

In 1986, the socialist government adopted a '*doi moi*' policy of economic development, liberalization and reform.[4] This economic renewal campaign led to Vietnam's accession to the World Trade Organization (WTO) in 2007, and the development of an export-led and market-based economy.

About 44 per cent of Vietnam is covered by forests of which 28 per cent belongs to the category of dense humid forest, 3 per cent to that of dry forest and 69 per cent to that of mosaics.[5] Seventy-five per cent are natural forests (1 per cent is designated as 'primary forest', 74 per cent as 'other naturally regenerated forest' and 25 per cent as 'planted forests'). Between 1990 and 2000, forest cover in Vietnam increased by 2.28 per cent annually, and between 2000 and 2010 by 1.64 per cent.[6] Since the early 1990s, a forest transition has taken place from net deforestation to net afforestation.[7] However, primary forest degradation[8] and local deforestation in the Central Highlands, the South-East region and the Mekong Delta[9] remain key challenges. Between 1999 and 2005, the area of natural forest classified as rich forest decreased by 10.2 per cent and medium forest decreased by 13.4 per cent.[10] The remaining primary forests are estimated to have declined by 51 per cent.[11] Rich forests can only be found in remote and mountainous areas.[12] Thus, although forest cover has increased during the past two decades, there has been serious forest degradation and a decline of forest density.[13]

This chapter focuses on Vietnam and especially on Dak Lak province in the Central Highlands (see Figure 5.1). Most of the remaining forest with high biomass and biodiversity value is located in this area.[14] Dak Lak is home to four protected areas, with three thousand plant species, 197 bird species and

Figure 5.1 Map of Vietnam and the case study area of Dak Lak province

93 animal species.[15] The Dak Lak Plateau has soils suitable for agriculture, especially for coffee plantations.[16] Within Dak Lak province, the focus of the study was on Buon Ma Thuot city, Krong Ana and Krong Pak districts.

This chapter is structured in accordance with the analytical framework (see 1.5), and discusses driving forces for deforestation and forest degradation (5.2), the forest policy context (5.3), effectiveness and fairness of policy instruments (5.4), implications for REDD (5.5), and draws inferences (5.6).

5.2 Driving forces of deforestation and forest degradation

As shown in Table 5.1, the main drivers of deforestation and forest degradation in Vietnam are: agricultural expansion into perennial cash crop plantations of coffee, pepper, rubber and cashews and the conversion of more than half of the mangroves to shrimp farming;[17] infrastructure extension including several hydropower projects in the Highland Plateau and Central Provinces; overharvesting,[18] illegal[19] and legal wood extraction for the timber and paper industry.[20] Dam construction causes direct deforestation in the uplands and indirect deforestation through the building of roads, power lines and community resettlement.[21] In 2007, 550,000 affected people were resettled on one hectare per household leading to a total conversion of about 80,000 hectares of land to agricultural purposes.[22] Other important drivers include shifting cultivation,[23] and slash-and-burn techniques used by ethnic minorities.[24] These farmers have been pushed into forest lands by a combination of poor policy and the reallocation of land to the private sector.[25]

Table 5.1 Drivers of deforestation and forest degradation in Vietnam and Dak Lak province

Proximate drivers (PD)	Underlying drivers (UD)
Local to national	Local
Agriculture: *industrial crop plantations (e.g. rubber, coffee, pepper, cashew)*; mangroves to shrimp farming; shifting cultivation	Economic: *poverty*
	Local to national
Extraction: *commercial logging, overharvesting, fuelwood, 'illegal' logging*	Cultural: conflict between *forest protection and development – socialist market culture, traditional values under stress*
Infrastructure: *hydropower, roads*	
Bio-physical: forest fires	National
	Demographic: *density, North to Centre migration*, urbanization
	Economic: development goals, industrialization, export-driven market economy
	Technological: *lack of capacity*
	Non-forest policies: *govt. support for forest land conversion for rubber, coffee, pepper, cashew, energy crops; dams and roads*
	Other institutional factors: *challenges in policy implementation, corruption; lack of civil society organizations*
	Global
	Demand/trade/consumption at global level for food, fibre, timber, and forest products, biofuel; climate change

NB The drivers which apply to the specific context of Dak Lak province are marked in italics

Logging is an important driver because of poor management practices of commercial actors, timber harvesting by rural households, and the conversion of forest land for agricultural purposes.[26] Commercial logging is relatively limited, due to logging restrictions and industrial timber imports.[27] However, the average area logged each year from 2002 to 2009 reached 33,824 hectares, of which 72.6 per cent was licensed and 27.4 per cent was unlicensed.[28] Domestic timber originates from forest plantations as well as natural forests.[29] Fuelwood collection, estimated to represent 7 to 25 per cent of the country's energy supply,[30] is another driver. This can be attributed to the dependence of 25 million people on the forests[31] who also benefit from deforestation.[32]

Forest fires result in an average forest loss of 2.353 hectares a year,[33] often caused by slash-and-burn agriculture (63 per cent), careless use of fire (6 per cent) and intentionally burned forest (9 per cent).[34]

The underlying drivers of deforestation and forest degradation in Vietnam include demographic factors (high population density; migration of about 6 million Kinh peasants from the North into the Central Plateaus after the 1975 reunification);[35] the state's economic growth policy; the state's weak forest management capacity, and the limited funding available for forest protection.[36] State support for agricultural policy has also led to some land-use change. Furthermore, there is poor interagency cooperation, and corruption and mismanagement at national to local levels (poorly paid field-level forest officers), which facilitates illegal logging.[37]

These driving forces impact also on Dak Lak province whose poor, marginalized and displaced ethnic minorities produce commercial agricultural crops and practise shifting cultivation.[38] Dak Lak was a deforestation hotspot during the 1990s, mainly because of the coffee boom.[39] After a temporary decrease of the coffee area, it is again on the rise since 2004.

5.3 The forest policy context

This section discusses the organizational framework, the evolution of forest policy, and influence of international treaties and bodies.

5.3.1 *The organizational framework*

The current forest governance framework[40] was established by the Prime Minister's Decision No. 245/1998. The Ministry of Agriculture and Rural Development (MARD) is responsible, through its Vietnam Administration of Forestry (VNForest) established in 2009, inter alia, for managing forests.[41] VNForest is involved in the process of establishing the tools to implement a National REDD+ programme in Vietnam. VNForest's Science, Technology and International Cooperation Department coordinates the national REDD+ programme (see Chapter 4). The Ministry of Natural Resources and Environment (MONRE), through its General Department of Land Administration (GDLA), manages all land, including forest land.[42] It elaborates land use master plans, determines land prices, organizes allotment, issues land use certificates, and settles problems related to land. MONRE is the national focal point for implementing the climate treaties[43] and coordinating CDM projects. The Ministry of Planning and Investment (MPI) coordinates Official Development Assistance projects including those on forests. The Ministry of Industry and Trade (MoIT) promotes the export and import of forest products including biofuels. This ministry is critical for addressing regional leakage as it is the focal point in drafting and preparing regional agreements on illegal timber trade and illegal logging.[44]

Coordination platforms include the Forest Sector Support Partnership (FSSP), established in 2001 by the government and donors to support the National Forest Strategy, and a donor-funded Trust Fund for Forests (TFF), established in 2004. In addition, there are specific coordination groups for the implementation of REDD+ (see 5.5). Other important actors include: public and private forest companies; national associations such as the Vietnam Timber and Forest Product Association, and the Handicraft and Wood Industry Association.[45]

Vietnam is a centralized state, with a strong Communist Party, which controls policies via 63 provinces,[46] 671 districts and 11,773 communities. Each governance level has its respective People's Council and People's Committees, the latter being the executive organ of the local state administration. The district president of a People's Committee reports on forests to the provincial president of a People's Committee who reports to the Prime Minister. The provincial MARD controls forest policies, while the provincial MONRE controls land administration. State forest enterprises fall under provincial authorities. The municipal People's Committee manages community forests, monitors forest ownership and forest resources, and coordinates forest protection with forest rangers and police. The governance system has strong vertical linkages,[47] with the People's Committees formulating policies and MARD and MONRE implementing them.

The governance system has overlapping (e.g. land and forestland management) and weak cross-sectoral coordination (e.g. between MONRE and MARD, especially in relation to REDD).[48] This has resulted in two co-existing official land-use classifications.[49] The General Department of Land Administration under MONRE classifies bare land as 'unused land', while the Forest Inventory and Planning Institute under MARD, which conducts five-yearly national forest inventory, monitoring and assessment programmes, classifies bare land as 'forest land without forest',[50] leading to different forest estimates.[51]

The lack of civil society organizations reinforces state hegemonic control.[52] The few civil society organizations are in fact mass organizations belonging to the Communist Party, such as the Vietnam Fatherland and the Vietnam Union of Science or Technology Association (VUSTA)[53] and support rather than question state policy.[54] Recently, new organizations started emerging, such as PAN-Nature and the Vietnam Network for Agroforestry Education.[55]

5.3.2 *The evolution of forest policy*

Pre-colonial policy focused on forest management, colonial policy on harvesting and illegal quotas, and the Indochina war period (1946–54) on nationalizing forests and using forest products to support the independence effort. Following the partition in 1955, the North retained nationalized forests, the South allowed market mechanisms.[56] After the unification of Vietnam in 1975, forests were managed in a centralized, top-down regulatory manner.[57] State forest enterprises, which harvested wood and processed products, played an important role in forest management.[58]

The *doi moi* process which started in 1986 led to several policy changes. First, under its land tenure reform, the 1991 Forest Protection and

Development Law allocated forest resources to individuals and economic entities and the 1993 Land Law recognized long-term and renewable land use titles, as well as five rights of titleholders (rights to exchange, transfer, inherit, mortgage, and lease). The transfer of forest management to other stakeholders – individuals, households, companies, and even communities – was progressively allowed. The Forest Protection and Development Law (2004) recognized distinct categories of forest ownership, with varying responsibilities and rights for forest management.

Second, the reform of the state forest enterprises – which now own about 38 per cent of the forests according to the Forest Protection Department – led to a decline in its numbers.[59] The reform separates business activities (mainly production forestry) from public goods management (protecting watersheds and conserving biodiversity). In 2004, a decree restructured state forest enterprises into state-operating companies and forest management boards, operating as profit-making public service agencies.[60] However, progress is slow as state forest enterprises continue to resist.

Third, the *doi moi* process encouraged reforestation activities. Programme 327, aiming at 'regreening the barren hills', was implemented between 1992 and 1998. It was replaced by the more ambitious Programme 661, aiming at reforestation of 5 million hectares by 2010. Programme 147 (2007–2015) focused on the development of production forests.

Fourth, *Forest Land Reclassification* has been initiated to increase the area of production forest available for private-sector investment, which partly encourages forest production in degraded forest areas.

The government's policies on environmental protection and climate change response[61] include the 2008 *National Target Programme to Respond to Climate Change* (NTP). This programme aims to assess climate change impacts on sectors and regions in specific periods and to develop short- and long-term feasible action plans to effectively respond to climate change, to promote sustainable development of Vietnam and move towards a low-carbon economy. MONRE coordinates its implementation.

Currently, key forest policies include the 2003 Land Law, the 2004 Law on Forest Protection and Development, the National Forestry Development Strategy (2006–2020), the National Forest Protection and Development Plan (2012–2020) and the National REDD+ programme (see 5.5).[62] The National Forestry Development Strategy includes new ambitious targets for plantations, policy reform, as well as further subsidies for protection and plantations and a greater role and responsibility for local communities.[63] The government has set high targets (20–24 million m^3/year) for domestic harvested timber. In addition, VNForest established in 2006 the Community Forestry Management Pilot Programme, underlining the growing interest and shift towards community forestry in Vietnam.

Vietnam's penal code recognizes several forest crimes – illegal logging, hunting and gathering; forest destruction; ignoring forest fire prevention; encroachment on forest land; illegal forest products transport; illegal grazing;

and various other practices (Law on Forest Protection and Development). MARD has the mandate for ensuring compliance with forest protection and the Forest Protection Department is the principal implementing agency.

5.3.3 *The influence of international treaties and bodies*

Key international organizations working on forests (FAO, World Bank, GEF, UNEP, UNDP) support Vietnam's forest policy. The country has ratified most of the relevant global treaties (CBD, UNFCCC, Kyoto Protocol, UNCCD, CITES, Ramsar and WHC). However, it is not a member of the ITTO. Bilateral donors such as the German development cooperation (GIZ), through its Vietnamese–German Forestry Programme, the Japanese cooperation agency (JICA) and international NGOs (i.e. WWF, IUCN, Forest Trends, etc.) are actively working to help implement the Vietnamese forest policy.

At the regional level, Vietnam participates in regional institutions, such as ASEAN (full member since 1995), Asia-Pacific Forestry Commission (APFC), Asia-Pacific Economic Cooperation (APEC), Asia Forest Partnership (AFP), and in the FLEGT Asia (2008) initiative.

5.4 Key forest policy instruments and their analysis

This section presents and analyses a selection of regulatory and market-based instruments[64] (using the numbering system developed in Table 2.3) in terms of their effectiveness and associated fairness issues, given the existing drivers of deforestation and forest degradation, using the [+ + − −] method developed in 1.5.

Decentralization (ii): The Vietnamese decentralization process started with the adoption of the 1992 Constitution and was formalized by a resolution adopted in 2004.[65] Since then, forests policies have been officially decentralized to subnational levels. However, it is decentralization 'by default' as a 'by-product' of the transition to a state-managed market economy.[66] This decentralization has led to deconcentration without democratization through the relocation of central administrative bodies to subnational levels, enlarging the reach of central control.[67] Despite the introduction of so-called grassroots democratization policies, the devolution of resources and powers to lower government level is relatively low. [+ −]

State forest management (iii): The reform of state forest enterprises led to the splitting of the protection and production tasks, and to the decentralization of forest management from central to provincial level.[68] The enterprises with a protection task became administrative entities (i.e. management boards), managing natural forests and protection/special-use forests. The 157 enterprises with a production task are obliged to produce in accordance with business management criteria.[69] Many state forest enterprises have gone bankrupt because they did not operate efficiently and because their annual harvest quota was reduced from 1 million m^3 to 150,000 m^3 after 2000[70] in order to limit deforestation, which led to income loss;[71] other state forest enterprises have survived through continued state support, for example, in the form of soft loans,[72] and based on forestry programmes.[73] Several enterprises only survived thanks to the cash flows received from governmental Programme 661 – without enhancing their viability as stand-alone enterprises.[74] State forest enterprises specializing in forest protection have had better results because of government support, and the land lease fees exemption. [+ –]

Land rights (vii): The land allocation programme launched in 1994 (and including the provisions in the Land Law 2003, and Protection and Development Law of 2004) aims progressively to give state lands to organizations, communities, households and individuals so they can earn a living and help protect and develop forests. Allocations for households and individuals cannot exceed 30 hectares of forestland, for no more than 50 years. Tenure and use rights differ in accordance with the type of forestland: planted forestland owners have five land use rights including to transfer, rent, inherit, give away, mortgage;[75] natural forest owners cannot transfer, give away, rent or mortgage natural forestland;[76] implying that the latter cannot get a bank loan for land investment. As of 2007, 8 million hectares of forestland was allocated to 1 million organizations, households and individuals.[77] The land tenure programme has suffered from a lack of funding and poor coordination between MARD and MONRE. As of 2006, only 55 per cent of land classified as forestland had been allocated to households as compared to 81 per cent of all agricultural land. Allocation success varies from province to province; measurement methods have given rise to disputes over land rights or land types, and slow certification by land reclassification bodies to state forest enterprises.[78] At the same time, property arrangements have had ambiguous effects upon forest land allocation.[79] Where they are inequitable, they may discourage sustainable forest management by forest owners. Where there are overlapping property rights, they lead to legal wrangling. Especially in some mountain areas where customary regulations still provide equal access to local forests, the land allocation programmes have proven difficult to implement. [+ –]

Law enforcement (x): Vietnam supports criminalization of, and FLEG initiatives against, 'illegal' logging,[80] and promotion of good forest governance.[81] It has criminalized several types of forest activities. Vietnam entered into formal negotiations on a FLEGT Voluntary Partnership Agreement in November 2010, which was still ongoing in early 2012 (see 3.2.7).[82] The amendment of the US Lacey Act (see 3.2.7) involving 'illegal' imports is also a strong incentive since the USA is a major export market for Vietnamese furniture.[83] [+ –]

Subsidies (xiv): Programme 661[84] aimed at five million hectares of reforestation through two million hectares of protection forest and three million hectares of production forest,[85] increasing total forest cover to 14.3 million hectares (43 per cent) by 2010.[86] In 2006, the programme was revised to improve the protection of natural forests. MARD implements this at national level. The provincial People's Committees are in charge of local implementation through a decentralized budget based on projects proposed by forest protection units, management boards and forestry companies. Project developers pay the different stakeholders in charge of protecting or planting forest under contractual arrangements. Programme 661 on reforestation, equalling USD 940 million for 2006–10, addresses the proximate drivers of agricultural expansion and illegal logging, and the underlying drivers of local poverty and mismanagement. It has arguably led to an additional 4 million hectares of forests.[87] However, there are problems of attribution,[88] a shortfall of one million hectares (of mostly fruit trees) less than planned,[89] no impact on forest degradation, uneven afforestation performances,[90] plantations that have focused on fast-growing trees for industrial purposes rather than natural forest regeneration;[91] and inadequate linkage between reforestation payments and actual performance (i.e. results have only been measured after one year of planting and not in the medium or long term, and officials do not have an incentive to report project failures, since it would bring cuts in budget allocation).[92] Furthermore, the subsidies were too small (about 5 to 10 USD per hectare)[93] to independently influence households to plant trees, unless they had decided to do so anyway.[94] While local capacities may have increased,[95] poor farmers' participation and influence on project design and planning[96] led to elite capture of local revenues,[97] unfavourable contractual conditions for farmers[98] and displaced subsistence farming on the bare hills thus constituting a bias against the poor.[99] Interviews reveal that these subsidy programmes could be enhanced if the payments to farmers are based on an opportunity cost analysis, are effectively decentralized and include farmer preferences. Furthermore, the subsidy should not only encourage the increase of forest cover but also other forest services such as biodiversity protection and carbon sequestration.[100] [+ + −]

Payment for Ecosystem Services (PES) (xv): Vietnam introduced a PFES scheme through the Forestry Development Strategy 2006–20 and the national 2008 Biodiversity Law. The latter states that 'organizations and individuals using environmental services related to biodiversity shall pay charges to service providers'. The scheme was further elaborated through Decision 380 and Decree 99. Subsequently, German and US development cooperation agencies have channelled funding, respectively through the Vietnam–German Forestry Programme, and, under the Asian Regional Biodiversity Conservation Programme, to two pilot schemes in Son La and Lam Dong provinces. These schemes focused on watershed protection and are based on a mechanism collecting funds from hydro-power plants and water supply companies in order to compensate for forest protection efforts through direct payments to households in Son La province and payments to groups of households in Lam Dong province. Coefficients (K-factors) have been developed to determine the appropriate level of payment. The scheme is operated through a nested structure of management boards at community, district and provincial levels.[101] After piloting the PFES scheme in Son La and Lam Dong provinces, the Government issued a decree in 2010 elaborating the national policy on payments for PFES. These programmes target the proximate drivers of shifting agriculture and 'illegal' logging, and the underlying drivers of unclear forest tenure and mismanagement. Using decentralized governance structures,[102] the programme was successful[103] in collecting and disbursing funds.[104] For example, in Son La province, four hydroelectric power plants and water supply companies collected more than USD 3 million and 52,000 forest owners in the Da River basin received money from the PFES fund.[105] In Lam Dong province, PFES payments reached USD 4.46 million and were allocated to 22 forest management boards, forestry businesses and 9,870 households.[106] However, there have been problems[107] such as inadequate funds for management, land allocation and implementation,[108] low payments to households relative to forest management boards,[109] delayed payments to farmers in Son La province,[110] farmers not understanding what they were being paid for in Lam Dong province,[111] and constant K-factors for all types of forests.[112] Payment to groups of households was more successful than payment to individual households as the resources could be used for the community and elite capture could be avoided.[113] Awareness has to be raised about the importance of using K-factors. Finally, clarification of land rights is a prerequisite for the implementation of a wider PFES scheme. [+ + −]

Carbon offset funds (xviii): As of March 2011, seven carbon trading projects were being implemented and another seven were planned.[114] The Cao Phong A/R CDM project in Hoa Binh province is the only one being implemented under the compliance market, although the potential area available for A/R CDM projects has been estimated at 3.2 million hectares of bare land.[115] This small-scale project will realize plantations on 365 hectares of currently degraded grass and shrub land, thus rehabilitating the land, sequestering carbon and increasing local income.[116] It is implemented through the Japanese development cooperation agency (JICA) and sponsored by the Vietnam Honda Company, with the technical assistance of the Vietnam Forestry University and the Research Centre for Forest Ecology and Environment (RCFEE). A/R CDM projects target the drivers of agricultural expansion and local poverty. They may be successful in rehabilitating degraded land and decentralizing the project management structure. Furthermore, local farmers may eventually benefit from carbon credits and wood harvesting revenues.[117] However, these projects have complex governance aspects (i.e. establishment of baselines, proof of additionality, selection of appropriate methodology, leakage accounting, monitoring and validation procedures, etc.).[118] Thus far, they have led to huge investments in A/R CDM readiness[119] and rigorous methodologies.[120] Even small projects of about 300 hectares in Cao Phong District cost USD 17,000 in preparation[121] and average full transaction costs are between USD 50,000 and 200,000.[122] An additional factor is the low price of forest carbon credits (USD 5)[123] which led to the fact that farmers did not see the difference between an A/R CDM project and a traditional tree plantation project[124] and started selling the timber.

High transaction costs and poor local participation are critical factors here. Besides, there is a need for a revision of the A/R CDM instrument at international level especially in terms of flexibility of the mechanism and simplification of the methodology in order to limit transaction costs. [+ – –]

CSR/Certification (xix): The national strategy is to certify 30 per cent of the country's production forests in the next decade.[125] As of June 2011, 16,000 hectares of forests had received FSC certification[126] and 150 companies had Chain of Custody certification.[127] This process is supported by international non-governmental organizations such as the Forest Trust and WWF which coordinates the Vietnam Forest and Trade Network, as part of the Global Forest Trade Network. The presence of international certification bodies in Vietnam, such as SGS Systems and Services Certification, Control Union Vietnam and GFA Consulting Group, is a sign of the dynamism in this sector.[128] However, the progress in certifying forest operations has been slow because there is little experience with market instruments. Many companies are not using FSC or other standards[129] because of limited knowledge of certification procedures and insufficient incentives, the inability of state corporations/companies to take action with short-term rental contracts[130] and the high direct and indirect costs,[131] especially for small producers.[132] The government is promoting FSC certification by developing national guidelines tailored to the national context,[133] but has not succeeded in receiving approval from the FSC headquarters, ostensibly because of a lack of equitable representation in the FSC national committee.[134] Based on the research, certification can be improved by enhancing capacity building of local farmers, increasing the representativeness of the national FSC committee in order to allow it to adapt FSC guidelines to the national context and the standards should be in line with a clear road map for sustainable forest management.[135] Furthermore, the government should ensure that imported wood for export is certified. Finally, the reform of state forest companies should be pursued so that these companies have incentives to adopt voluntary standards. [+ –]

5.5 Implications for REDD

Vietnam is considered a frontrunner in the race to capture international support for its REDD+ readiness activities.[136] In 2008, Vietnam was selected as a participant in the Forest Carbon Partnership Facility (FCPF) implemented by the World Bank and in 2009 it was chosen as the first country to pilot the UN-REDD Programme. The government is currently developing a national REDD+ strategy with the support of the UN-REDD programme. The latter has already allocated 20 per cent of its budget to Vietnam to enable REDD-readiness, which includes the development of a national REDD+ strategy defining the overall goals and plans as well as responsibilities of agencies at provincial and district level, monitoring procedures, guidelines for local benefit distribution, the relevant processes at the central and provincial level, including assessment of all the activities.

Furthermore, Vietnam is a member of the core group of the Interim REDD+ Partnership, a fast-start interim programme to speed up the REDD+ process. Several smaller REDD+ readiness activities are being implemented by international NGOs and aid agencies: Japanese International Cooperation Agency (JICA), Netherlands Development Cooperation (SNV), World Agroforestry Centre (ICRAF), Forest Carbon Partnership Facility of the World Bank (FCPF), the German Development Cooperation (GIZ), the Australian Development Cooperation (AusAID) and the Finnish Government (through FAO and bilaterally). These organizations directly or indirectly work with REDD+ elements at provincial level and deal with practical issues.

Vietnam was allocated USD 4.3 million by UN-REDD to become REDD ready by 2012.[137] However, achieving complete REDD readiness may need a longer time,[138] requiring capacity development at national and subnational levels and greater regional cooperation in the Lower Mekong Basin.[139] Vietnam was also allocated USD 3.6 million from the FCPF to implement its Readiness Preparation Proposal (R-PP). MARD established the Vietnam REDD+ Office in 2009, which coordinates all REDD-related activities, including a national REDD+ network which promotes information sharing between stakeholders and a technical working group with four sub-working groups on REDD+ governance, MRV, financing and benefit distribution and local implementation.[140] Furthermore, Vietnam is seen as a pilot country regarding the development of a Free Prior and Informed Consent (FPIC) mechanism (see Box 4.2).

REDD activities are in an advanced stage of readiness, but challenges include: (i) the establishment of a clear, consistent and enforceable legal framework in the context of good governance[141] which legitimately settles the issue of rights to carbon, land and forests,[142] enhances coordination between MARD and MONRE,[143] addresses forest tenure issues and the rights of local communities[144] and provides political commitment to deal with corruption.[145] (ii) REDD needs to be made flexible to the national context[146] and mainstreamed in national and subnational environmental plans.[147] (iii) Lessons learnt from A/R CDM include that methods to measure GHGs or carbon stocks need to be simple, transaction costs lowered and should take opportunity costs into account.[148] (iv) Sufficient and predictable funding is needed – about USD 3 million to get REDD running in the country[149] but resources provided have been lower.[150]

Related to the lessons learned from experiences with other instruments, our research leads us to a number of implications for REDD in Vietnam (see Figure 5.2). (a) REDD needs a good enabling framework, political will, a country-specific approach, simple methods and financial resources. (b) Its implementation requires making effective links with existing national mechanisms such as Programme 661 and the PFES national strategy.[151] (c) FSC's experience with the development of sustainable forestry standards should be built on when developing guidelines on REDD+ social and environmental safeguards;[152] having complementary approaches with FLEGT implementation processes (i.e. definition of timber legality, etc.) in order to improve

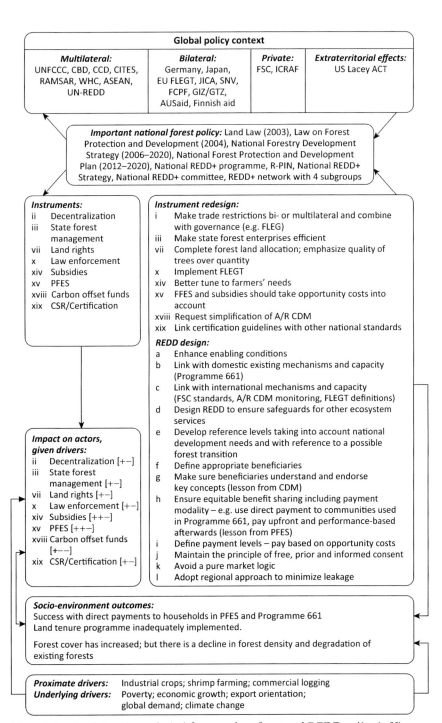

Figure 5.2 Applying the analytical framework to forest and REDD policy in Vietnam

credibility and efficiency of the REDD+ strategy.[153] (d) REDD needs to be designed to ensure safeguards for other ecosystem services. (e) Reference levels should take into account national development needs and with reference to a possible forest transition. (f) Rights and beneficiaries of the schemes should be clarified. (g) The new scheme should be properly explained to stakeholders (lesson from A/R CDM);[154] (h) Equitable benefit sharing including payment modality should be ensured, building on the successful experience of direct payments to groups of households, taking the lesson into account that administratively set payments are less effective than those based on an opportunity cost analysis,[155] introducing coefficients that would allow differentiated benefit distribution according to different ecological, economic and sequestration conditions at each governance level,[156] and recognizing the need for early initial payments and in a later stage performance-based payments, based on the experiences with the PFES pilot projects.[157] (i) Payment levels should be based on opportunity cost. (j) Free prior informed consent is needed to ensure that indigenous communities are effectively and equitably engaged in the discussions (see Box 4.2). (k) Develop REDD as a combination of market and regulatory characteristics. Pure market-based instruments cannot work in a forestry sector that is mainly state-controlled and has limited market freedom. (l) Develop a regional approach to REDD[158] with a focus on Cambodia and Laos.[159] In this process, safeguarding other ecosystem services is also vital.

5.6 Inferences

Vietnam's transition to a market economy and its integration in regional and global markets have clearly contributed to shaping the factors affecting deforestation and forest degradation. However, reforestation efforts have been important and Vietnam is one of the first countries to start a forest transition without having first completely depleted its forests. The government is facing difficult trade-offs between supporting its export industry, heavily dependent on the agricultural and timber sectors, and its desire to serve as a model for biodiversity conservation and climate change action.

The proximate drivers of deforestation are the conversion of forest land for agricultural purposes, the construction of hydropower plants and legal and illegal wood logging while the underlying drivers are the *doi moi* economic policy, mismanagement practices including corruption, unclear forest land tenure, as well as the lack of a strong civil society. Shifting cultivation, illegal logging and the construction of roads are the main drivers of forest degradation.

In terms of policy, Vietnam's forest sector has undergone major evolutions following the *doi moi* renovation process, notably the implementation of a land allocation programme, a reform of state forest enterprises and the issuance of a legal framework allowing the introduction of market-based instruments. International institutions have also shaped national policies and supported the introduction of economic incentives such as PFES, A/R CDM, FSC standards and FLEGT.

Vietnam has well-developed policies in place for forest protection but the practical implementation of these policies has been characterized thus far by numerous failures and weaknesses, such as limited funding, poor coordination, low stakeholder involvement and corruption.[160] The government itself also recognizes that there is room for improvement in legislation, better coordination between ministries, agencies and provinces, improved mainstreaming of climate change into social and economic development programmes, and capacity building for better dealing with climate change at all state levels.

Forestry instruments are to some extent able to address certain drivers of deforestation and forest degradation. However, they do not deal with the underlying drivers, nor are they able to address the opportunity costs of deforestation and forest degradation. Market-based instruments have contributed to the adoption of decentralized management structures and to an increased involvement of local actors in forest management. However, important issues to consider are the high transaction costs and the risk of leakage.

The emphasis in the Vietnamese approach is on subsidies for reforestation efforts, payment for forest ecosystem services and sustainable forest management through the development of a national certification scheme. Improving the legal framework is also important for Vietnam's regional and international image.

The policy instruments to combat deforestation and forest degradation in Vietnam are not fully effective because of institutional weaknesses and their inability to adequately address the full spectrum of drivers. They have a moderately positive impact on equity, providing new opportunities for the poor.

Vietnam has received considerable support for REDD activities and is far in the process of REDD readiness. This chapter has recommended 12 approaches to strengthen REDD implementation which include synergizing a future REDD instrument with existing instruments like FLEGT and PFES policy plans. Thanks to its rich and diverse experience, the country has the opportunity to contribute to the regional and global debate on REDD.

Notes

1 The authors wish to thank Ms Ngan Bui for her research and interview work and Ms Thu Thuy Pham and Mr Patrick Meyfroidt for their critical comments on an earlier version of this chapter.
2 FAO (2011a) *The state of forests in the Amazon Basin, Congo Basin and Southeast Asia*, Rome: Food and Agriculture Organization of the United Nations.
3 World Bank (2010) *World Development Indicators Database*. Data for 2008. Available at http://databank.worldbank.org/data/home.aspx.
4 Riedel, J. and Turley, W.S. (1999) *The politics and economics of transition to an open market economy in Vietnam*, Working Paper 125, Paris, France: OECD Development Center; ADB (2006) *Economic transition in Viet Nam: Doi Moi to WTO*, Manila, Philippines: Asian Development Bank.
5 FAO (2011a) op. cit.
6 FAO (2011a) op. cit.

7 Meyfroidt, P. and Lambin, E.F. (2008a) The causes of reforestation in Vietnam, *Land Use Policy* 25, 182–97; Meyfroidt, P. and Lambin, E.F. (2008b) Forest transition in Vietnam and its environmental impacts, *Global Change Biology* 14, 1319–36; FAO (2010a) *Global forest resources assessment 2010: country report Vietnam*, Rome, Italy: FAO.
8 Meyfroidt, P. and Lambin, E.F. (2009) Forest transition in Vietnam and displacement of deforestation abroad, *Proceedings of the National Academy of Sciences* 106 (38), 16139–44; FAO (2011a) op. cit.
9 Long, V. and Phuong, V.T. (2011) *Review of forest policies in Vietnam*, Hanoi, Vietnam: Research Centre for Forest Ecology and Environment (RCFEE); Hoang, M.H., Do, T.H., van Noordwijk, M., Pham, T.T., Palm, M., To, X.P., Doan, D., Nguyen, T.X. and Hoang, T.V.A. (2010) *An assessment of opportunities for reducing emissions from all land uses. Vietnam preparing for REDD, Final national REALU report*, Nairobi, Kenya: Partnership for the Tropical Forest Margins.
10 FCPF (2008) *Readiness plan idea note (R-PIN) for Vietnam*, Forest Carbon Partnership Facility.
11 EIA and TelePak (2008) *Borderlines: Vietnam's booming furniture industry and timber smuggling in the Mekong region*, Bogor, Indonesia: Environmental Investigation Agency and TelePak.
12 Long and Phuong (2011) op. cit.
13 Pham, T.T., Moeliono, M., Nguyen, H.T., Nguyen, T.H. and Vu, T.H. (2012) The context of REDD+ in Vietnam: Drivers, agents and institutions, *Occasional Paper* 75.
14 Meyfroidt and Lambin (2008a) op. cit.
15 FLITCH (2007) *Feasibility study report-Forestry Development Project to improve life in the Central Highlands {in Vietnamese}*, Hanoi, Vietnam: The Ministry of Agriculture and Rural Development-Project Management Board of Forestry.
16 Berding, F.R., Tan, T.M., Tuyen, T.D., Hue, T.V., Deckers, J. and Langohr, R. (1999) *Soil resources of Dak Lak Province: Correlation with the World Reference Base for Soil Resources*, Leuven, Belgium: K.U. Leuven University.
17 The total mangrove forest area has decreased from 400,000 hectares to 156,500 hectares between 1943 and 2002: Son, N.T. and Tu, N.A. (2008) Determinants of land-use change: a case study from the lower Mekong delta of southern Vietnam, *Electronic Green Journal* 1 (27).
18 Interviews by L. Bigot, 2011: 5, 8.
19 Illegal logging is limited (McElwee, P. (2004) You say illegal, I say legal: the relationship between 'illegal' logging and land tenure, poverty, and forest use rights in Vietnam, *Journal of Sustainable Forestry* 19 (1–3), 97–135), although it doubled between 1987 and 2006 (Meyfroidt and Lambin (2009) op. cit.; Long and Phuong (2011) op. cit.) Its effect is small, and the illegal trade of the larger parties is more important than the illegal logging by the small groups (EIA and TelePak (2008) op. cit.; Meyfroidt and Lambin (2009) op. cit.).
20 Vietnam exports inexpensive hardwood furniture to Europe, North America and Asia (FCPF (2008) op. cit.) mostly made from imported wood (Agroinfo (2009) *80% of wood in Vietnam is imported {in Vietnamese}*). Only 20 per cent of timber used is sourced domestically because of a government annual harvesting quota for natural forests.
21 FCPF (2008) op. cit.
22 Dang, N.A. (2007) Resettlement in hydro-power plants construction in Vietnam [in Vietnamese], *The Communist Magazine* 14 (134).

23 Interviews by L. Bigot, 2011: 4.
24 Sam, D.D. (1994) *Shifting cultivation in Vietnam: Its social, economic and environmental value relative to alternative land use*, Forestry and Land Use Series 3, London: International Institute for Environment and Development (IIED).
25 Interviews by L. Bigot, 2011: 4; see also Hoang *et al.* (2010) op. cit., p.69.
26 Pham *et al.* (2012) op. cit.; McElwee (2004) op. cit.
27 Meyfroidt, P. and Lambin, E.F. (2009) Forest transition in Vietnam and displacement of deforestation abroad, *PNAS* 106 (38), 16139–44; Agroinfo (2009) op. cit.
28 Agroinfo (2009) op. cit.; Pham *et al.* (2012) op. cit.
29 Pham *et al.* (2012) op. cit.
30 Government of Vietnam (2005) *National Report to the Fifth Session of the United Nations Forum on Forests*; Castrén, T. (1999) *Timber trade and wood flow study, Vietnam*, Regional Environmental Technical Assistance 5771, Greater Mekong Subregion (GMS) Watersheds Project.
31 World Bank (2010) *Socialist Republic of Vietnam – Forest law enforcement and governance. East Asia and the Pacific Region*, Washington DC, USA: World Bank; Government of Vietnam (2005) op. cit.; FCPF (2008) op. cit.
32 Interviews by L. Bigot and N.L. Bui, 2011: 1, 2, 3; Sunderlin, W.D. and Huynh, T.B. (2005) *Poverty reduction and forests in Vietnam {in Vietnamese}*, CIFOR; McElwee, P. (2008) Forest environmental income in Vietnam: household socioeconomic factors influencing forest use, *Environmental Conservation* 35 (2), 147–59; McElwee, P. (2010) Resource use among rural agricultural households near protected areas in Vietnam: the social costs of conservation and implications for enforcement, *Environmental Management* 45, 113–31.
33 Long and Phuong (2011) op. cit.
34 Long and Phuong (2011) op. cit.
35 de Konninck, R. (1999) *Deforestation in Vietnam*, Ottawa, Canada: International Development Research Centre; FCPF (2008) op. cit.; Sikor, T. and Nguyen, T.Q. (2007) Why may forest devolution not benefit rural poor? Forest entitlements in Vietnam's central highlands, *World Development* 45 (11), 2010–25.
36 Pham *et al.* (2012) op. cit.
37 Interviews by L. Bigot and N.L. Bui, 2011: 4, 8; Alley, P. (2011) *Corruption: a root cause of deforestation and forest degradation*, Transparency International, Global Corruption Report 2010: Climate Change; McElwee (2004) op. cit.; FCPF (2008) op. cit.
38 Sunderlin and Huynh (2005) op. cit.; Meyfroidt, P., Vu, T.P. and Hoang, V.A. (2012) *Trajectories of deforestation, coffee expansion and displacement of shifting cultivation in the Central Highlands of Vietnam*; UN-REDD (2010a) *UN-REDD Vietnam programme: phase II: operationalising REDD+ in Vietnam*, UN-REDD Programme and Vietnam's Ministry of Agriculture and Rural Development.
39 D'haeze, D., Deckers, J., Raes, D., Phong, T.A. and Loi, H.V. (2005) Environmental and socio-economic impacts of institutional reforms on the agricultural sector of Vietnam: Land suitability assessment for Robusta coffee in the Dak Gan region, *Agriculture, Ecosystems and Environment* 105 (1–2), 59–76.
40 Government of Vietnam (1998a) *Decision No. 245/1998 of the Prime Minister 'on the exercise of state managerial responsibility of various levels concerning forests and forest land'*.
41 Tasks include supervising implementation at provincial level, conducting forest surveys and inventory monitoring, publishing annual forest coverage status,

establishing criteria for forest classification and plantations, carrying out forest protection and development measures, regulating forest allocation, implementing national programme against desertification, directing prevention on forest fires, etc. (GoV, 2008c); Government of Vietnam (2010) *Decision No. 04/2010/QD-TTG of the Prime Minister on 'regulation of functions, tasks, authorities and organisational structure of the Directorate of Forestry'*.

42 Government of Vietnam (2008a) *Decree No. 25/2008/ND-CP defining the functions, tasks, powers and organizational structure of the Ministry of Natural Resources and Environment*.
43 MONRE (2010) *Viet Nam's second national communication to the United Nations Framework Convention on Climate Change*, Hanoi, Vietnam: Ministry of Natural Resources and Environment.
44 This insight is based on the review comments of Ms Thu Thuy Pham.
45 Interviews by L. Bigot, 2011: 10.
46 This includes 59 provinces and four municipalities with provincial status (Hanoi, Ho Chi Minh City, Danang and Can Tho).
47 Vien, T.D. and Quang, N.V. (2004) *Decentralisation in forest management and in three communities in Vietnam's uplands*, Hanoi Agriculture University: Center for Agricultural Research and Environmental Studies (CARES).
48 Interviews by L. Bigot, 2011: 1.
49 Hoang *et al.* (2010) op. cit.
50 Hoang *et al.* (2010) op. cit.
51 For example, the total area of forest land in 2005 and 2007 differed by four million hectares to 2.6 million hectares between the two classification systems (Hoang *et al.*, 2010: 48).
52 Thayer, C.A. (2008) *One-party rule and the challenge of civil society in Vietnam. Presentation to remaking the Vietnamese State: implications for Viet Nam and the Region*, Hong Kong: Viet Nam Workshop, City University of Hong Kong, 21–22 August 2008.
53 Thayer (2008) op. cit.
54 Thayer (2008) op. cit.
55 Interviews by L. Bigot, 2011: 4.
56 FSSP (2003) *Forest sector manual*, Hanoi, Vietnam: Forest Sector Support Partnership.
57 Sam, D.D. and Trung, L.Q. (2001) Forest policy trends in Vietnam, *Policy Trend Report 2001*, 69–73.
58 de Jong, W., Sam, D.D. and van Hung, T. (2006) *Forest rehabilitation in Vietnam: histories, realities and future*, Bogor, Indonesia: Centre for International Forestry Research (CIFOR).
59 de Jong *et al.* (2006) op. cit.
60 Hoang *et al.* (2010) op. cit.
61 MONRE (2010) op. cit.
62 For details, see Long and Phuong (2011) op. cit.
63 Government of Vietnam (2007) *Vietnam Forestry Development Strategy 2006–2020. Promulgated and enclosed with the Decision No. 18/2007/QD-TTG of the Prime Minister*.
64 Other forest policy instruments used in Vietnam, such as community forest management, spatial planning and protected areas are not included in this analysis.
65 Government of Vietnam (2004) *Resolution No. 08/2004/NQ-CP on further stepping up the State management decentralization between the Government and the provincial/ municipal administrations*.

66 Painter, M. (2008) Reforming social services – Asian experiences: From command economy to hollow State? Decentralisation in Vietnam and China, *The Australian Journal of Public Administration* 67 (1), 79–88.
67 Fritzen, S.A. (2006) Probing system limits: Decentralisation and local political accountability in Vietnam, *The Asia-Pacific Journal of Public Administration* 28 (1), 1–23.
68 FSIV (2009) *Vietnam forestry outlook study*, Working Paper No. APFSOS II/WP/2009/09, Bangkok: Food and Agriculture Organization of the United Nations, Regional Office for Asia and the Pacific.
69 Long and Phuong (2011) op. cit.
70 Long and Phuong (2011) op. cit.
71 Long and Phuong (2011) op. cit.
72 Long and Phuong (2011) op. cit.
73 Clement, F. and Amezaga, J.M. (2009) Afforestation and forestry land allocation in northern Vietnam: analysing the gap between policy intentions and outcomes. *Land Use Policy* 26 (2009), 458–70; McElwee, P.D. (2012) Payments for environmental services as neoliberal market-based forest conservation in Vietnam: Panacea or problem?, *Geoforum* 43, 412–26.
74 EASRD (2005). State Forest Enterprise Reform in Vietnam. World Bank Technical Note. Clement and Amezaga (2009) op. cit.; McElwee (2011) op. cit.
75 Land Law 2003: Art. 113.
76 Land Law 2003: Art. 63 (4).
77 MONRE (2010) op. cit.
78 ADB (2006) op. cit.; FCPF (2011) *Forest Carbon Partnership Facility 2011 Annual Report118*, Washington DC: The World Bank.; FCPF (2008) op. cit.
79 Sikor and Nguyen (2007) op. cit.; Geist, H.J. and Lambin, E.F. (2002) Proximate causes and underlying driving forces of tropical deforestation, *Bioscience* 52 (2), 143–50; Castella, J.C., Boissau, S., Nguyen, H.T. and Novosad, P. (2006) Impact of forestland allocation on land use in a mountainous province of Vietnam, *Land Use Policy* 23, 147–60; Meyfroidt and Lambin (2008a) op. cit.
80 EIA and TelePak (2008) op. cit.; Phuc, T.X. and Dressler, W. (2010) *How FLEGT and REDD+ can help address illegal logging? A case from Vietnam*, EU FLEGT Facility Vietnam.
81 ProForest (2009a) *Joint FLEGT Vietnam. Scoping Study – Part 1. Main report.*
82 Http://www.illegal-logging.info/approach.php?a_id=320.
83 Interviews by L. Bigot, 2011: 5.
84 This is a follow-up to the 'Greening the Barren Hills Programme' (Programme 327) implemented from 1992 to 1998 and influenced by UNCED.
85 Government of Vietnam (1998b) *Decision No. 661/QD-TTG of the Prime Minister on Objectives, tasks, policies and implementation arrangements of the 5 Million Hectares Reforestation Project.*
86 de Jong *et al.* (2006) op. cit.
87 GSO (2011) *The official national site for statistics*, General Statistics Office of Vietnam; Meyfroidt and Lambin (2009) op. cit.
88 McElwee, P. (2009) Reforesting 'Bare Hills', in Vietnam: Social and environmental consequences of the 5 Million Hectare Reforestation Program, *Ambio: a Journal of the Human Environment* 38 (6), 325–33.
89 Interviews by L. Bigot, 2011: 9.
90 UN-REDD (2010a) op. cit.
91 McElwee (2009) op. cit.

92 UN-REDD (2010a) op. cit.
93 Interviews by L. Bigot, 2011: 1.
94 Wunder, S., The, B.D. and Ibarra, E. (2005) *Payment is good, control is better: Why payments for forest environmental services in Vietnam have so far remained incipient*, Bogor, Indonesia: Center for International Forestry Research; Sikor, T. (2009) *Finance for household tree plantations in Viet Nam: Current programs and options for the future*, Bogor, Indonesia: Center for International Forestry Research; McElwee (2011) op. cit.
95 UN-REDD (2010a) op. cit.
96 van der Poel, P. (2007) *Towards a program-based approach in the forest sector in Viet Nam?*, Hanoi, Vietnam: GFA Consulting.
97 McElwee (2009) op. cit.
98 Phuc, T.X. (2009) Why did the forest conservation policy fail in the Vietnamese uplands? Forest conflicts in Ba Vi national park in Northern region, *International Journal for Environmental Studies* 66 (1), 59–68.
99 McElwee (2009) op. cit.
100 Interviews by L. Bigot, 2011: 10.
101 UN-REDD (2010b) *Design of a REDD-Compliant benefit distribution system for Vietnam*, UN-REDD Programme and Vietnam's Ministry of Agriculture and Rural Development.
102 UN-REDD (2010a) op. cit.
103 Interviews by L. Bigot and N.L. Bui, 2011: 3, 11, 12.
104 Hess, J. and To, T.H. (2010) *GTZ accompanies Vietnam in development and implementation of policy on payment for environment services (PFES)*.
105 Hess and To (2010) op. cit.
106 Winrock International (2011) *Payment for Forest Environmental Services: a case study on pilot implementation in Lam Dong Province Vietnam 2006–2010*.
107 Interviews by L. Bigot, 2011: 1.
108 MARD (2010) *Mid-term evaluation report on the implementation of the Decision 380/QQD-TTG of the Prime Minister on the policy for PFES*.
109 Interviews by L. Bigot, 2011: 5.
110 Interviews by L. Bigot, 2011: 11.
111 Interviews by L. Bigot, 2011: 1.
112 Interviews by L. Bigot, 2011: 1; UN-REDD (2010b) op. cit.; Winrock International (2011) op. cit.
113 Interviews by L. Bigot, 2011: 1.
114 Phuong, V.T. and Hai, V.D. (2011) *Report on Forest Carbon Trade activities in Vietnam*, Hanoi, Vietnam: Forest Science Institute of Vietnam.
115 Doets, C., Son, N.V. and Tam, L.V. (2006) *The Golden Forest: practical guidelines for AR-CDM project activities in Vietnam*, Hanoi: SNV Vietnam.
116 Phuong and Hai (2011) op. cit.
117 Phuong and Hai (2011) op. cit.
118 Thomas, S., Dargusch, P., Harrison, S. and Herbohn, J. (2010) Why are there so few afforestation and reforestation Clean Development Mechanisms projects?, *Land Use Policy* 28, 880–87.
119 Interviews by L. Bigot, 2011: 9.
120 Chokkalingam, U. and Vanniarachchy, S.A. (2011) *CDM-AR did not fail and is not dead*, Forest Carbon Asia.
121 Interviews by L. Bigot, 2011: 9, 13 (with N.L. Bui).

122 Thomas *et al.* (2010) op. cit.
123 Interviews by L. Bigot and N.L. Bui, 2011: 13.
124 Doets *et al.* (2006) op. cit.
125 Government of Vietnam (2007) op. cit.
126 Sojitz Corporation (2006) *Sojitz acquires FSC certification for its forestry business in Vietnam*; WWF (2010) *'First forest smallholder group achieves FSC certification in Vietnam'*.
127 ProForest (2009b) *Joint FLEGT Vietnam. Scoping Study – Part 1. Annexes*.
128 Interviews by L. Bigot, 2011: 12.`
129 Interviews by L. Bigot and N.L. Bui, 2011: 9, 10, 11, 17.
130 Interviews by L. Bigot, 2011: 8; ProForest (2009b) op. cit.
131 The minimum initial costs of certification are around 1,000 USD too high for small loggers (Durst, P.B., McKenzie, P.J., Brown, C.L. and Appanah, S. (2006) Challenges facing certification and eco-labelling of forest products in developing countries, *International Forestry Review* 8 (2), 193–200). The recurrent costs are estimated at around 5 to 10 per cent of the total logging costs, while the premium on certified products is also estimated at around 5 to 10 per cent (ProForest, 2009b); interviews by L. Bigot, 2011: 4.
132 Interviews by L. Bigot, 2011: 3.
133 Interviews by L. Bigot, 2011: 3.
134 Interviews by L. Bigot, 2011: 9.
135 CISDOMA (2009) *Development of a new sustainable forest management planning methodology: Final Report*, Hanoi: Consultative Institute for socio-economic development of rural and mountainous areas.
136 IGES (2010) *Developing REDD+ systems: progress challenges and ways forward. Indonesia and Vietnam country studies*, Japan: Institute for Global Environmental Strategies; Hoang, M.H., Do, T.H., Pham, M.T., van Noordwijk, M. and Minang, P.A. (2011) Benefit distribution across scales to reduce emissions from deforestation and forest degradation (REDD+) in Vietnam, *Land Use Policy* (in press).
137 The REDD Desk (2011) *Vietnam: an overview from the REDD countries database*.
138 Interviews by L. Bigot and N.L. Bui, 2011: 15.
139 UN-REDD (2009) *UN-REDD Vietnam programme: Revised standard joint programme document*, UN-REDD Programme and Vietnam's Ministry of Agriculture and Rural Development.
140 The REDD Desk (2011) op. cit.
141 Interviews by L. Bigot, 2011: 4 ;UN-REDD (2010a) op. cit.
142 Interviews by L. Bigot, 2011: 1, 5.
143 Interviews by L. Bigot and N.L. Bui, 2011: 16.
144 UN-REDD (2010a) op. cit.; IGES (2011) *Payment for Environmental Services in Vietnam: an analysis of the pilot project in Lam Dong Province. Forest Conservation Project, Occasional Paper 5*, Japan: Institute for Global Environmental Strategies.
145 Kanninen, M., Murdiyarso, D., Seymour, F., Angelsen, A., Wunder, S. and German, L. (2007) *Do trees grow on money? The implications of deforestation research for policies to promote REDD*, Bogor, Indonesia: Center for International Forestry Research (CIFOR).
146 Interviews by L. Bigot, 2011: 3, 5; Kanninen *et al.* (2007) op. cit.
147 Interviews by L. Bigot, 2011: 3.
148 Interviews by L. Bigot and N.L. Bui, 2011: 2, 4, 16; Hoang *et al.* (2010) op. cit.; UN-REDD (2010b) op. cit.

149 Interviews by L. Bigot, 2011: 3, 15 (with N.L. Bui).
150 Interviews by L. Bigot and N. L. Bui, 2011: 13.
151 UN-REDD (2010b) op. cit.
152 FSC (2011) *Recommendations and strategic framework on FSC climate change engagement*, Forest Stewardship Council; Vickers, B. (2010) *FSC and REDD+*, Workshop Decoding REDD; Kanninen *et al.* (2007) op. cit.
153 ProForest (2011) *FLEGT and REDD+ linkages: working together effectively. Briefing note 3*, Kuala Lumpur, Malaysia: ProForest.
154 Interviews by L. Bigot and N.L. Bui, 2011: 13; RECOFTC (2011) *Forests and climate change after Cancun: an Asia-Pacific experience*, Bangkok, Thailand: The Center for People and Forests; UN-REDD (2010b) op. cit.
155 Wertz-Kanounnikoff, S. and Kongphan-Apirak, M. (2008) *Reducing forest emissions in Southeast Asia: A review of drivers of land-use change and how payments for environmental services (PES) schemes can affect them*, Working Paper No. 41, Bogor, Indonesia: Centre for International Forestry Research (CIFOR).
156 UN-REDD (2010b) op. cit.
157 IGES (2011) op. cit.
158 Meyfroidt and Lambin (2009) op. cit.
159 EIA and TelePak (2008) op. cit.
160 Pham *et al.* (2012) op. cit.

6 Case study

Indonesia

Mairon Bastos Lima, Joyeeta Gupta, Nicolien van der Grijp, and Fahmuddin Agus[1]

6.1 Introduction

Indonesia (1,919,317 sq km), with more than 17,000 islands, is home to about 240 million people, including over 300 ethnic groups. The country's per capita income is USD 4,200, with its gross domestic product coming mostly from industry (47 per cent), services (38 per cent) and agriculture (15 per cent).[2] It has a democratic government that is transitioning from centralized policies that lasted until 1998.

Indonesia's territory is officially divided into forest land (71 per cent) and non-forest land (29 per cent).[3] However, that does not necessarily match actual forest cover. Only about 68 per cent of the official forest land is under forest cover, and 15 per cent of what is officially non-forest land is actually under forests.[4]

Forest areas are under significant pressure in Indonesia. Deforestation took place at an average rate of 0.51 per cent between 2000 and 2010, a loss of nearly 500,000 hectares per year – about five million hectares in total.[5] Forest degradation is also at pace, with a continuous transformation of primary (i.e. undisturbed) into secondary (i.e. disturbed or logged-over) forest.[6] Finally, there is a notable increase of tree plantations in Indonesia, from 0.01 per cent of the country's forest cover in 1990[7] to 4.7 per cent in 2009.[8] Still, about 94.4 million hectares of rainforest cover remain, the third largest in the world (after Brazil and the Democratic Republic of Congo), and a number of policies are in place aiming at forest conservation and sustainable management.[9]

This chapter analyses Indonesia's forest policy context utilizing a layered case-study approach based on literature review, analysis of forest policies, and field work focused on Jambi Province in Sumatra (see Figure 6.1) and the districts of Muaro Jambi, Muara Bungo and Tanjung Jabung Barat. Jambi Province offers a good representation of Indonesia's larger context as it has experienced both large forest conversion into agricultural plantations as well as forest conservation projects, such as socially inclusive 'village forest' (*Hutan Desa*) initiatives.

The chapter is structured in accordance with the analytical framework (see 1.5), and discusses driving forces for deforestation and forest degradation

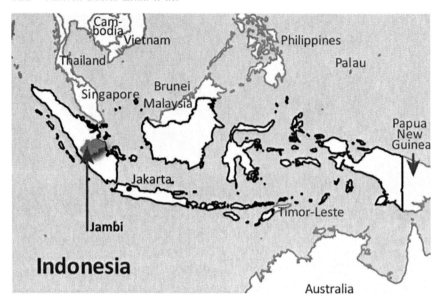

Figure 6.1 Map of Indonesia and case study area of Jambi province

(see 6.2), the forest policy context (see 6.3), describes and assesses effectiveness and fairness of policy instruments (see 6.4), the implications for REDD (see 6.5) and draws inferences (see 6.6).

6.2 Driving forces of deforestation and forest degradation

As shown in Table 6.1, deforestation and forest degradation in Indonesia can be attributed primarily to logging, agriculture and mining. Logging is regulated by the State through long-term concessions to industries. Timber-harvesting concessions (*Hak pengusahaan hutan*) last for 20 years and take place in natural forests (i.e. not on plantations), allowing companies to extract wood following certain rules (e.g. minimum trunk girth) and procedures (e.g. restoration programme following harvesting). However, rule enforcement is weak, and thus much forest has been degraded.[10] On occasion, such forests have been cleared to give place to profitable industrial timber plantations – HTI (*Hutan tanaman industry*) concessions, using fast-growth species such as eucalyptus or acacia.[11]

Illegal logging without forestry concessions occurs at large and small scales. At larger scales, private companies engage in illegal logging by operating without licences or beyond their concession areas and selling the plywood in domestic and international markets, sometimes transforming it into 'legal' wood abroad.[12] At the local level, villagers use timber for subsistence needs such as house-building or income generation. These customary activities have

Table 6.1 Drivers of deforestation and forest degradation in Indonesia and Jambi Province

Proximate Drivers	Underlying Drivers
Local to national	*Local*
Agriculture: oil palm, rubber, coffee	Economic: *poverty*
Extraction: commercial timber plantations (for acacia, eucalytus), violation of SFM rules in concessions, small-scale encroachment	Local to national
	Cultural: e.g. Papua very connected to nature; *in Jambi few indigenous people, generally a utilitarian approach to nature,* perceived need for development
Infrastructure: *forest roads*	
Industry: *coal*/gold *mining*	
Biophysical: *fires*	*National*
	Demographic: growth, density, migration
	Economic: growth, domestic demand for land-based commodities
	Technological: *access to modern services*
	Policies: timber-harvesting concessions; mining and agricultural policy
	Other institutional factors: poor monitoring, lack of synergy among ministries, corruption
	Global
	Demand/trade/consumption at global level for food, fibre, timber, palm oil, biofuel, paper and plywood; climate change

* The drivers which apply to the specific context of Jambi Province are marked in italics

been relabelled as 'illegal' and are contested. Illegal brokers (*cukong*) link these villagers to the existing wood market.[13] Furthermore, villagers or plantation companies may use timber extraction to raise resources for investing in, for example, oil palm plantations.[14]

Another proximate driver is agriculture.[15] Large-scale companies expand primarily oil palm plantations.[16] Many land concessions have been given nominally for oil palm development but without deployment actually taking place, suggesting that commodity agriculture has been also used as a means to speculate on land prices or to simply harvest existing timber. Small-scale farmers frequently encroach on forests to expand subsistence agriculture or cash-crop plantations such as cacao, coffee and rubber.

Mining, too, is an important driver, particularly coal mining in Sumatra and Kalimantan.[17] Given its dwindling oil reserves, Indonesia is turning to coal for meeting its energy needs and for exports, particularly to India and China, which buy about 75 per cent of Indonesia's coal exports.[18]

Natural forest fires (i.e. not human-induced) have also been important drivers over the last decades.[19]

Underlying drivers of deforestation in Indonesia include, first, the drive for economic growth. Government officials argue that many provinces have a higher forest cover than the world average (sometimes reaching 80 per cent of the area) and that they should not be prevented from 'developing'.[20] Frequently these forests are surrounded by poverty,[21] so rural dwellers not only encroach on these forests for their needs but also tend to welcome infrastructure development and plantation companies that reduce their isolation and dependency on donors[22] and enhance their wealth.[23] This reflects the common perception that the forest has less intrinsic value than the services it provides,[24] or that short-term goals supersede long-term issues.[25] In Jambi Province, for instance, most smallholders are migrants with much less of a connection with nature than indigenous peoples tend to have.[26] Meanwhile, the government sees the opportunity costs of not converting the forest land or accessing the mines as unaffordable for a developing country.[27]

Second, the strong global demand for land-based commodities such as timber, paper, coal, edible oil and biofuels and high international commodity prices has encouraged greater investment in resource exploitation in Indonesia.[28] The country has explicitly aimed to provide for the world's growing food and energy needs: in 2010, President Susilo Bambang Yudhoyono announced that Indonesia would help to meet domestic and global food and energy security through projects such as the Merauke Integrated Food and Energy Estate (MIFEE), which has targets of about 1 million hectares of concessional land in West Papua for timber, food and feedstock crop exports.[29]

Third, the increasing international demand for arable land and private land acquisitions in the developing world[30] has led speculators to engage in land-banking by leasing land for plantations, removing the valuable wood, and then letting the land lie fallow until the permit can be sold at a high price,[31] possibly to the pulp and paper industry.[32] For instance, of the 20 million hectares under nominal oil palm concessions, only 8–9 million hectares of oil palm is actually under cultivation.[33] This is clearly related to the global demand for commodities,[34] but it has become in itself a motivation for land-clearing and concession-seeking.

6.3 The forest policy context

This section discusses the organizational framework, the evolution of forest policy, and the influence of international policies and bodies.

6.3.1 *The organizational framework*

In Indonesia all land is officially classified either as forest (*kawasan hutan*) or non-forest land (*areal penggunaan lain*). Official forest lands correspond to 71 per cent of the territory; however, there are forested areas which fall outside of that and there are areas without tree cover which on paper are classified as 'forest'. The Ministry of Forestry regulates all activity on official forest lands and holds the exclusive authority to issue permits for land conversion in those areas. For lands under the 'non-forest land' category, whether forested or not, provincial and district levels have the authority to decide over land uses and land-use change.

6.3.2 The evolution of forest policy

Indonesia has had a centralized forest policy since colonial times under Dutch rule. During that period, customary ownership of land and its resources was accepted, but subordinated to the interest of the colonizers.[35] A similar approach to land and forests was maintained after independence (1949) in the country's first major related policy, the Basic Agrarian Law of 1960. Although it recognized customary land rights, these remained subordinated to national state interests. This principle was reaffirmed for forests in the Forestry Basic Law of 1967, which declared State-ownership of forested areas and prescribed rules for timber-harvesting concessions.[36] This centralized approach prevailed during the 30 years Indonesia was under President Suharto (1967–98). Following Suharto's fall, decentralization and devolution of authority to local governments has been gradually taking place. In 2004 the *Autonomy Law* officially devolved to provincial and district governments the authority to issue land-development permits and conduct semi-autonomous planning, but forest areas have remained under the jurisdiction of the central government through the Ministry of Forestry.[37] Law 41/1999 amended the older Basic Forestry Law but reaffirmed State-ownership of forests and the Ministry of Forestry's exclusive authority to issue permits for any activity on official forest areas. [38]

SFM has been the underlying concept of most Indonesian policies on forestry. This was recently included in Decree No P38/Menhut-II/2009, which contains general guidelines for forest utilization and conservation.

Indonesia's forests are classified into Conservation Forests (national parks, other reserves, and areas devoted to biodiversity conservation – 17 per cent of the actual forests), Protection Forests (forests maintained for hydrological and other such services – 26 per cent) and Production Forests (forests allocated for economic use, mainly timber extraction – 57 per cent). Production forests may be non-convertible or convertible (areas which can be legally cleared for other land-uses – what some government officials refer to as 'planned deforestation' [39]).[40]

6.3.3 The influence of international treaties and bodies

Four types of international institutional influences can be identified: multilateral agreements, bilateral agreements, private certification schemes, and the extraterritorial effects of foreign policies of other governments. With regard to multilateral agreements, Indonesia is a party to all major agreements that relate to forests (CITES, CBD, UNFCCC, KP, UNCCD,

ITTA, Ramsar, WHC and UNPFII), in addition to the non-binding UN Forum on Forests (UNFF) (see Chapter 3).[41] These agreements have limited influence on Indonesia's forest policy and its implementation,[42] but the discourses in these agreements are influential, sometimes shifting the focus from sustainable forest management to climate change.[43] Although Indonesia does not have binding emission reduction commitments under the UNFCCC, it is seen as the world's third largest gross GHG emitter because of high emissions from deforestation and peatland degradation.[44] Concern about its image and the impacts of that on trade have led Indonesia to adopt a voluntary commitment to reduce emissions by 26 per cent by 2020 relative to a business-as-usual trend.[45] Furthermore, the climate change framing has been associated with the promise of financial compensation for forest conservation, such as through REDD (see Chapter 4).

6.4 Key forest policy instruments and their analysis

This section presents and analyses a selection of regulatory and market-based instruments[46] (using the numbering system developed in Table 2.3) in terms of their effectiveness and associated fairness issues, given the existing drivers of deforestation and forest degradation, using the [++ − −] method developed in 1.5.

Trade restrictions (i): Indonesia has a bilateral agreement with the European Union (its biggest export market for timber products) to combat illegal logging. In 2003 the European Commission proposed an Action Plan on Forest Law Enforcement, Governance and Trade (FLEGT), a framework that promotes trade in legal forest products.[47] Upon FLEGT's legal adoption in 2005, the EU launched a *Voluntary Partnership Agreements* (VPAs) scheme where, together with timber-exporting countries, the EU can help develop a licensing system to halt imports of illegal timber.[48] Although voluntary, these partnerships become legally binding once they are agreed on – for five years, with the possibility of renewal. Indonesia joined VPA negotiations in 2007, and it is expected that in January 2013 the agreement will become fully operational.[49] Meanwhile, Indonesia has been receiving technical assistance to develop a *Legality Assurance System* that will allow it to emit legality licences to its timber industry, without which shipments will be refused at EU borders.[50] Indonesia plans to go beyond that and request a legality licence from *any* timber exporter, whether to the EU or not, and to apply the same to domestically consumed timber in the future.[51] Stakeholders consulted see this as a promising agreement despite the expected monitoring and enforcement challenges.[52] It may be promising because it addresses EU demand and promotes capacity building on legal assurance in Indonesia. It may reduce illegal logging and have effects which go beyond the fraction of the timber industry that exports to the EU.

Sustainability criteria in the EU Biofuels Directive exclude Indonesian palm-oil-based biodiesel as it does not reduce GHG emissions sufficiently.[53]

As such, EU countries may import that biofuel but it will not count towards their mandatory EU targets of 10 per cent renewable transport fuel by 2020. Indonesian critics argue that this policy fails to recognize Indonesia's development needs, choosing a zero-conversion policy instead of a mid-way compromise,[54] and some accuse the EU of being protectionist under the guise of sustainability concerns.[55]

The EU biofuel sustainability policy has not yet succeeded in reducing imports, though it may have worked as a disincentive to further investments. Indonesian stakeholders see this as trade protectionism under the guise of sustainability and as ignoring Indonesia's development needs.[56] At least two lessons can be drawn from the above. First, bilateral negotiations (FLEGT) are more constructive than unilateral standards and trade restrictions (Biofuel directive) which create resistance. Second, it could trigger domestic governance improvements. The European restriction on biofuel imports would gain in both receptiveness and effectiveness by engaging on these two points. [EU FLEGT: +; EU Biofuel Standards: – –]

Decentralization (ii): Since its re-democratization in 1998, Indonesian law is relocating power to subnational authorities.[57] District governments have the authority to decide over land-uses in areas which are not official forest land (the latter remains under the Ministry of Forestry). Decentralization relies heavily upon the capacity of subnational governments to produce land-use plans in line with national regulations and development strategies. However, local governments do not always understand the devolutionary system and think they have complete autonomy on decisions regarding land use.[58] They are often eager to clear forest land for development even without permission from the central government.[59] The responsibility for raising part of their budget – instead of only receiving transfers from the central government – has added to that eagerness, possibly creating a race-to-the-bottom in the form of competition among districts to attract private investment.[60] Provincial and district governments need further capacity enhancement to produce land-use plans that are in line with national regulations and development strategies, as a way to improve the effectiveness of conservation strategies and reduce land conflicts. *In situ* verification is necessary to update data on forest land. Local governments may need revenue-generating, sustainable programmes that can match the opportunity costs of deforesting. [+ –]

Binding forest rules (iv): These rules include the classification of forest areas according to their main purpose (biodiversity conservation, ecosystem protection or commercial production) and determine their use (see 6.3.2). Furthermore, Presidential Decree N.32/1990 regulates the use of peatlands and establishes that only areas where the peat is shallower than 3m can be converted, thereby limiting the impacts of land-use changes. The zoning of forest areas and the peatland use policy are viewed positively.[61] However, weak enforcement and some contradictory regulatory incentives have severely limited the effectiveness of these binding rules. Plantations have frequently encroached on protection and conservation forests. Prominent examples include World Heritage Sites such as the Kerinci Seblat, Bukit Barisan Selatan and Gunung Leuser national parks in North Sumatra, which together comprise UNESCO's 2.5 million hectare Tropical Rainforest Heritage of Sumatra,[62] and which since June 2011 figure in UNESCO's 'Danger List' due to the damages of illegal encroachment.[63] Effectiveness is often undermined by lenient authorities who wish to accommodate agricultural and infrastructure development interests. For instance, the Ministry of Forestry approved almost 3 million hectares in development concessions on 31 December 2010, on the eve of the date the moratorium from the agreement with Norway could potentially come into force, releasing more than 40 new concession holders from the stricter controls of that policy.[64] Eleven days after the moratorium decree was signed by the president, the ministry adopted another decree[65] changing the status of more than 1.6 million hectares of forest land in Central Kalimantan (the province selected for pilot work of the Norway–Indonesia partnership). Much of what was considered protection or conservation forest became convertible production forest, and part of the overall forest area had its status changed into non-forest. This rests upon Government Regulation N.10/2010, which allows for status changes in forest areas when their de facto condition no longer supports their classification. Indeed, much of the relabelled area in Central Kalimantan had already been converted to oil palm plantations.[66] This implies that forestry categorization may sometimes not suppress illegal land-use change but rather legitimizes it. If this continues as a trend, it provides a perverse regulatory incentive for illegal land-clearers to convert forests into plantations and expect that sooner or later the land status will be changed. [+ –]

Spatial planning (v): Indonesia's Law 24/1992 regulates spatial planning. However, it was only after Law 26/2007 that provinces and districts gained the right and the responsibility to conduct spatial planning at their respective levels and feed it into a national database. Still, any plan related to official forest areas requires Ministry of Forestry approval and must be in conformity with the binding rules described above.[67] However, between 2007 and 2011 only six out of 33 provinces had submitted their spatial plans to the National Development Planning Agency.[68] Hence, there are often conflicts between departments and across levels in terms of what a map shows, its boundaries between zones, and what is intended for each area. For instance, there are hardly clear boundaries between official forest and non-forest areas; or sometimes there are but maps at the province and at the Ministry of Forestry give different information.[69] This has led to increased social conflicts and environmental degradation often with negative impacts on areas of biodiversity or watershed conservation value.[70] It is not uncommon for district governments to issue land-use permits that go against regulations from the Ministry of Forestry – or even for overlapping permits to be issued by different authorities for the same area.[71]

Spatial planning rules frequently define who wins and who loses out of land-use policy implementation.[72] Finally, it is controversial that the government categorizes timber plantations as forests, although this is in line with FAO and UNFCCC definitions (see Box 1.1). Including plantations is acceptable from a carbon stock perspective[73] but not from a biodiversity and ecosystem service perspective.[74] Critics refer to it as a 'discursive trick' and argue that it masks the loss of natural forests in forest cover statistics.[75] [spatial planning + –; peatland use + +]

Forest logging/concessions (vi): Logging concessions are widely prevalent since all forest land belongs to the State and thus companies cannot buy land but only lease it. These concessions are of two main types: logging concessions on natural forests (Law 5/1967, *Hak pengusahaan hutan*) and concessions for industrial timber plantations (P.3/Menhut-II/2008, *Hutan tanaman industri*). Logging concession on natural forest rules include annual extraction plans, limits on what timber can be harvested, and payments for forest rehabilitation. These concessions have been controversial to the extent that they have granted timber industries exclusive access to large areas, ignoring non-timber forest products, sometimes excluding traditional users and exacerbating land conflicts. In addition, weak enforcement of SFM has led to poor conservation.[76] As a consequence of government dissatisfaction with the performance of concession-holders, natural forest depletion and relatively low profits for industry, there is decreasing interest in such logging concessions on natural forests.[77] They have decreased to about 300 concessions in 2008, from an average of 600 in the 1990s.[78] Meanwhile, there is growing interest in timber plantations, primarily for the pulp and paper industry. These plantations are promoted as 'forest improvement' as they increase carbon stocks, but monocultures of fast-growing species such as acacia and eucalyptus may be replacing areas of lower carbon stocks but of higher conservation value.[79] Moreover, policies have facilitated the expansion of tree plantations: formerly, natural vegetation could not be cleared for timber plantations if it exceeded a certain minimum level of density but this level has been lowered, facilitating land use change. In this case, enforcement is also an issue, as those rules are not implemented rigorously.[80] Forestry rules and management regulations on concessions need stronger enforcement of regulations on sustainability and stricter rules on when a forest can be handed to logging or converted into a tree plantation. [Logging concessions on natural forests + + –; industrial timber plantation concessions + – –]

Land rights (vii): All forest land belongs to the national state. Customary land rights are recognized by law, but subordinated to national interests (see 6.3.2). Lack of land tenure security and overlapping land rights remain a problem making land tenure rights unclear. The issuing of business operation permits on customarily held lands routinely leads to land conflicts.[81] This involves not only traditional communities but also smallholders who squat and settle on State forestland and remain there until the area is licensed for

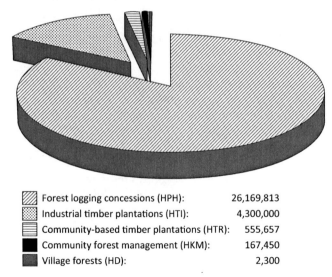

Forest logging concessions (HPH):	26,169,813
Industrial timber plantations (HTI):	4,300,000
Community-based timber plantations (HTR):	555,657
Community forest management (HKM):	167,450
Village forests (HD):	2,300

Figure 6.2 Areas under different land-use-right concessions from the Ministry of Forestry (in hectares)

Sources: Akiefnawati, R., Villamor, G.B., Zulfikar, F., Budisetiawan, I., Mulyotami, E., Ayat, A. and van Noordwijik, M. (2010) Stewardship agreement to Reduce Emissions from Deforestation and Degradation (REDD): case study from Lubuk Beringin's *Hutan Desa*, Jambi Province, Sumatra, Indonesia, *International Forestry Review* 12 (4), 349–60; Ministry of Forestry (2009a) Forestry Statistics of Indonesia: 2008, Jakarta: Ministry of Forestry; Obidzinski, K. and Dermawan, A. (2010) Smallholder timber plantation development in Indonesia: what is preventing progress? *International Forestry Review* 12 (4), 339–48.

business operation, which then works as a trigger for (sometimes violent) conflict.[82] Indirectly, this also exacerbates deforestation. As these communities live mostly on areas considered secondary forest, companies may prefer peatlands or primary forests, where the likelihood of a conflict is lower but the impacts on the environment are larger.[83] There are also serious distributional issues. Because most land belongs to the State, equity in access to land is largely determined by the allocation of concessions given by the Ministry of Forestry granting land-use rights. As of 2008, there were 4.3 million hectares conceded to large-scale companies doing timber plantations, and an additional 9 million hectares are sought by 2016.[84] More than 26 million hectares were under private forest-logging concessions in 2008.[85] In contrast, less than 1 million hectares had been granted to local communities in all community-based forest management programmes combined, and more than half of that consists of timber plantations held under supply contracts with the timber industry (the HTR scheme).[86] As such, there is inequality in the distribution of recognized land-use rights between companies and forest-based local communities (see Figure 6.2). It is critical to improve land tenure security and the de facto recognition of customary land- and forest-use rights. [+ – –]

Reporting (viii), Monitoring (ix) and Enforcement (x): Although many laws have strong reporting, monitoring and enforcement provisions, implementing these is challenging.[87] For instance, timber traceability is problematic and even more so because illegal timber may become 'legal' at some point in the chain.[88] Indonesia may be too large for easy monitoring, and large conservation areas and forest management units require not only satellite mapping but also *in situ* monitoring and evaluation.[89] As a government official puts it:

> It is hard for satellite imagery to trace shifting cultivation or to distinguish it from fallow land, or to tell different types of forests and plantations apart. Agroforestry is even a more difficult category to comprehend. Statistics keep going up and down, so you have to verify it on the ground.[90]

Indonesia also faces a lack of manpower to deal with its forests.[91] Cases of bribery, data manipulation and other forms of corruption are also not uncommon, limiting the effectiveness of the policies in place.[92] It is important to improve timber traceability and *in situ* monitoring and evaluation. This requires investment in the quality and quantity of personnel. [– –]

Debt-for-nature swaps (xvii): Indonesia has engaged twice in debt-for-nature swaps with the US, in 2009 and 2011. First, nearly USD 30 million in debt was forgiven and instead put into the management of conservation forests in Sumatra. The second swap involved USD 28.5 million of pardoned debt for the management of conservation areas in Kalimantan.[93] Although it provides additional financial resources for conservation, it does not attempt to improve domestic policies nor does it address enabling conditions for effective conservation such as unclear land tenure issues. Moreover, the swapped amounts are too little to provide significant debt relief to Indonesia's budget.[94] This instrument can be improved by increasing the amount of swapped debt (to enhance budgetary relief) and tying it to domestic forest policy improvements. [+ – –]

CSR/certification (xix): International certification instruments exist on timber (e.g. FSC certification) and on palm oil (e.g. the RSPO certification). Although these schemes were initiated in Northern countries, many Indonesian stakeholders such as NGOs and producers are active members. In 2011 the government decided to launch its own scheme – the Indonesian Sustainable Palm Oil certification – to become effective in 2014.

While certification schemes have improved forest management through scrutiny of the timber chain of custody and the sustainability of the oil palm sector (reducing forest conversion),[95] they cover a small fraction of the timber and oil palm sectors.[96] As of April 2012, only about 1 million hectares in forest concessions were covered by the FSC[97] from a total of more than 30 million hectares currently under logging concessions. As of October 2011, only 16 of the 435 members of the Indonesian Palm Oil Association were certified by the RSPO. Moreover, the Association itself decided to leave it, arguing that the requirements were too stringent, that it added little market benefits (small take-up and large markets which do not require certification, particularly India and China), and given that the Indonesian government now had its own national certification scheme for palm oil.[98] The certification schemes can be improved by increasing the price premium and market absorption of certified timber and agricultural products. [+ −]

Community-based management (xxiv): Indonesia has multiple concession schemes promoting community-based forest management, partly as a way to resolve land tenancy disputes.[99] However, these concessions remain relatively small in number and area when compared to industrial logging concessions. They include Community-based Timber Plantations (*Hutan Tanaman Rakyat*) and Village Forests (*Hutan Desa*). Community-based Timber Plantation concessions are contracts between smallholders and industries for supplying timber.[100] They have, however, suffered from design problems (e.g. permits cannot be transferred or inherited), low timber prices, poor bargaining power, and faced competition from other land uses.[101] 'If they [concession holders] die, their families starve.'[102] By 2011, about 600,000 hectares were under these concessions, lower than the 2 million hectares targeted by the government,[103] but enough to raise concerns of degraded forests of higher biodiversity being replaced by timber monocultures.[104] The Village Forests scheme, on the other hand, has been more successful in reconciling forest conservation and the maintenance of traditional livelihoods that use forest resources sustainably. It officially recognizes local use rights and provides an assurance that forests will not be converted for other purposes.[105] Villagers, NGOs and government authorities in the Muaro Bungo district of Jambi Province showed great satisfaction at the results of the programme in conserving natural forests and providing legal recognition of community rights over the forest.[106] Some villagers argued, for instance, that it further legitimizes their stewardship and empowers them to stave off encroachers.[107] Moreover, forest ecosystem maintenance has helped conserve watershed resources and generate electricity from micro-hydropower units.[108] The main challenge, however, is protecting an 'island' of forest in a sea of unregulated inhabited land.[109] There has been limited adoption of the Village Forests scheme, so its effectiveness remains limited to small areas. As of May 2011, only six villages had obtained a Village Forests permit, while 23 had started an application procedure.[110] Some analysts feel that district governments show little interest in supporting a village's application to the Ministry of Forestry for such a permit as it does not generate revenues for them.[111] In addition, bureaucratic transaction costs are perceived as too high for local communities to handle on their own if NGOs do not step into the process.[112] Preference should be given to schemes that promote the conservation of natural vegetation rather than the establishment of plantations, which may lead to further land clearing. [Community-based Timber Plantation − −; Village Forests + +]

NGO-based management (xxv): Indonesia utilizes ecosystem restoration concessions, 100-year-long concessions to private entities for forest rehabilitation.[113] Launched in 2004,[114] amended in 2007[115] and 2008[116], these concessions covered 100,000 hectares (2 million hectares under application) by 2011.[117] Ecosystem restoration concessions have gained momentum from the ongoing PES and REDD debates and attracted many Indonesian and foreign NGOs.[118] Their effectiveness as conservation tools remains to be seen. However, the challenges of illegal encroachment and border monitoring persist.[119] In addition, there are dubious benefits for local people (some included,[120] some fenced off[121]), leading some critics to refer to it as a new form of 'green land grabbing'[122]. [+ −]

6.5 Implications for REDD

Indonesia adopted a REDD+ National Strategy in 2010 as part of a bilateral agreement with Norway to reduce emissions from land-use changes. This non-binding agreement includes a Letter of Intent where Norway pledged to provide USD 1 billion as official development assistance funds,[123] for Indonesia to devise and implement a REDD+ strategy including provisions in domestic law such as a two-year moratorium on issuing new permits to convert forest lands. Following this agreement Indonesia launched in 2010 a REDD+ National Strategy and a task force to recognize the challenges and propose guidelines for improvement of forest management and REDD implementation in the country, in addition to building capacity for measurable, reportable and verifiable emissions reductions from land-use change.[124] Following its preparatory stage (2010–11), there has been a transformation stage (2011–13) with pilot work in Central Kalimantan testing land tenure, law enforcement, developing a degraded lands database, and imposing a two-year moratorium on issuing new development concessions on peatland and areas of natural forest nationwide. The third phase (2014 onwards) is to see REDD becoming operational and contributing to verified emission reductions against payments from Norway and, potentially, other donors. Initial assessments of this Letter of Intent and the two-year moratorium indicate that its coverage is limited and enforcement weak. Three-quarters of the primary forest and peatland areas covered by the moratorium were already protected by Indonesian law, suggesting limited additionality.[125] Secondary forests, which tend to be at frontier areas and more vulnerable, have been excluded from the coverage. And, finally, initial evidence reveals more than 100 illegal clearings in forest areas covered by the moratorium during its first three months.[126] Including secondary forests in the moratorium is key to reducing land conversion. More generally, funds for reducing emissions from deforestation (such as REDD+) should also be associated with biodiversity conservation and social goals to ensure more inclusive and environmentally friendlier approaches rather than tree plantations for carbon absorption only.

In 2009 the Ministry of Forestry issued a decree on REDD+ fund-sharing, only for the Ministry of Finance to later declare it illegal and make it invalid.[127] Officials recognize that policies and programmes under different directorates sometimes conflict and generally lack the articulation that could optimize results.[128] For example, a senior forest officer argues that forest rehabilitation funds hardly ever match the rehabilitation planned schedule and the planting season.[129] These governance problems are recognized in the country's National REDD+ Strategy[130] as well as by non-State stakeholders, who complain also of bureaucratic hurdles especially in relation to policies aimed at local communities, who often lack the skills and the resources to navigate the bureaucracy.[131] Ecosystem restoration concessions should have provisions to ensure the inclusion of local communities into their conservation programmes.

The two-year moratorium on issuing new permits to convert forest lands provoked the oil palm sector to react[132] as the deal almost halves oil palm expansion from 350,000 hectares per year to about 200,000 hectares per year.[133] Regardless, in May 2011 the Indonesian president finally signed the moratorium decree,[134] and until May 2013 central and local governments cannot issue new mining, logging or plantation concessions on peatlands or areas of *primary* forest. Still, existing concessions can be renewed, and concessions to sectors considered strategic for national development (oil, gas, rice and sugar cane) are exempted. As of April 2012, Indonesia has also set up a national REDD+ Task-Force[135] and initiated pilot works in the province of Central Kalimantan. Norway, in turn, has paid USD 30 million through a UNDP Trust Fund, the rest being conditional on further implementation.[136]

Indonesian stakeholders generally argue that REDD remains unclear with respect to its operational aspects.[137] In addition, local actors generally have very little understanding of it.[138] It is evident that (a) enabling conditions (such as land tenure, weak governance and conflicts between environmental and development policy) need to be improved prior to REDD implementation.[139] (b) It would be appropriate to link with existing mechanisms and knowledge gained such as with community forests, ecosystem restoration, moratorium, peatland policy and protected areas, as well as (c) with international mechanisms (FLEGT, FSC). (d) REDD needs to be designed to ensure that ecosystem services (see Table 1.1) and equity challenges are safeguarded. REDD and equity issues (e.g. fear of land grabbing;[140] the lack of recognition of tenure rights[141]) need to go in tandem: '*No rights, no REDD; no REDD, no rights.*'[142] For instance, the recognition of collective customary ownership of local people may allow for REDD benefit sharing with forest dwellers.[143]

Successful REDD implementation will also require improving Indonesia's domestic governance, particularly in terms of better vertical and horizontal coordination, harmonized land-use planning, and strong institutions to prevent corruption.[144] REDD has already provided a forum for different sectors and actors to come together and dialogue on land-use and emissions issues, but synergetic action must follow. For this it will be useful to (e) develop baselines based on development aspirations and the implications for its forest transition and to clearly (f), equitably define the beneficiaries (especially necessary due to unclear tenure), (g) explain to them the key concepts involved in REDD (a highly abstract concept), and (h) develop modalities of benefit sharing. Where payments are made, they should be made partially upfront and then become performance-based. (i) REDD funds must prove to be at least as significant as the deforestation alternative – in other words, they must match the opportunity costs. Still, this alone may not suffice, as halting deforestation may raise the price of commodities, making deforestation even more profitable.[145] This could lead to a race to offer the higher payments – a race in which REDD could pour in large resources and still lose. (j) The principle of free prior informed consent should be implemented to ensure that indigenous communities are effectively and equitably engaged in the discussions (see Box 4.2). Furthermore, REDD policy (k) should not adopt a pure market logic and utilize incentive-based instruments only; rather, it needs to be tailored to the necessary implementation reforms which would allow for the regulatory instruments in place to also become effective. Finally, it seems useful to (l) adopt a regional approach in order to minimize leakage.

A question remains as to whether REDD policies should be project-based or landscape-based. Project-based supporters see this as simple and manageable, as test cases that can be later scaled up.[146] However, this may lead to problems similar to that of CDM, such as exclusion of smallholders and leakage. A landscape-based approach, instead, could deliver concrete benefits to local people (improvement in access to energy services, health and educational facilities, etc.) in addition to public investment in development and poverty-alleviation programmes.[147] Whichever the option, REDD+ needs to be aligned with development policies.

6.6 Inferences

The steady pace of deforestation in Indonesia is not accidental. Logging, mining and agriculture have been very strong drivers of land-use change stimulated by an agenda of development and economic growth and a global demand for land-based commodities such as coal, minerals, wood and palm oil. Indonesia's forests have suffered both from the cumulative effects of small-scale encroachment by poor local villagers and from large-scale land-conversion endeavours of governmental programmes (e.g. Merauke Integrated Food and Energy Estate) or private investors aiming at the growing national and international markets. Notably, there has been also a demand for arable land, with speculative interests mindful of the growing need for land-based commodities and increasing prices.

Against this background, Indonesia's policy instruments for forest conservation have been ill-suited and largely ineffective to counter the deforestation drivers at work. There has been a strong reliance on regulatory instruments, such as zoning and SFM rules for logging concessions, but coupled with inadequate monitoring and weak enforcement. Additionally, these regulations have not met Indonesia's development needs, thus conflicting with the

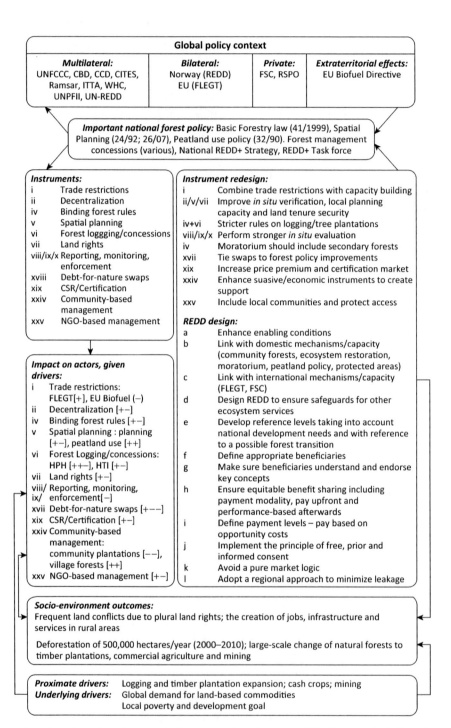

Figure 6.3 Applying the analytical framework to forest and REDD policy in Indonesia (HPH = Hak pengusahaan hutan; HTI = Hutan tanaman industry)

economic growth agenda. Therefore, it becomes of prime importance for forest conservation policies to compensate for the opportunity costs of not expanding land-based economic activities. Moreover, market mechanisms and economic instruments that link conservation to trade, such as the partnership with the EU on combating illegal wood exports, seem to hold great promise to the extent that they target the *underlying* drivers of deforestation.

Lastly, there may be a need to construe viable, alternative sustainable and socially inclusive paths of local economic development. The current agenda has led to large losses of forests and increased land conflicts due to a focus on short-term economic gains, corporate-controlled resource extraction, and forest policies that aim at 'improving carbon stocks' through tree plantations rather than at the rehabilitation of natural habitats or the integration between conservation and socio-economic development. For one, the emphasis solely on carbon and REDD's dissociation from biodiversity and ecosystem services risks legitimizing environmentally damaging paths as 'sustainable' – and rewarding them as such – simply because they have climate benefits. Furthermore, rural communities and local governments are eager to welcome large-scale investment in oil palm (and other cash crops) mostly because this has been the only option offered – the other being stagnation. It is uncertain whether environmental degradation (of soil, water resources, etc.) will not compromise the very viability of these businesses in the mid and long terms. Moreover, the fact that they accrue some benefits to the local people should not hide the fact that the major beneficiaries are still those trading the wood and other land-based commodities – a matter of equity. Therefore, this should not signify that alternative development paths cannot or should not be conceived.

Notes

1. The authors thank Meine van Noordwijk for extensive comments on previous versions of this chapter.
2. http://www.indexmundi.com/indonesia/economy_profile.html
3. Ministry of Forestry (2009a) *Forestry Statistics of Indonesia: 2008*, Jakarta: Ministry of Forestry.
4. Ministry of Forestry (2009a) op. cit.
5. FAO (2011) *The state of forests in the Amazon Basin, Congo Basin and Southeast Asia*, Rome: Food and Agriculture Organization of the United Nations.
6. Ekadinata, A., Widayati, A., Dewi, S., Rahman, S. and van Noordwijk, M. (2011) *Indonesia's land-use and land-cover changes and their trajectories (1990, 2000, 2005)*, Bogor, Indonesia: World Agroforestry Center (ICRAF) Southeast Asia Regional Program.
7. Ekadinata *et al.* (2011) op. cit.
8. Ministry of Forestry (2009b) *Eksekutif: Data Strategis Kehutanan 2009 (Executive: Strategic Data Forestry 2009)*, Jakarta: Departemen Kehutanan (Department of Forestry).
9. FAO (2011) op. cit.
10. Interviews by M. Bastos Lima, 2011: 6, 7, 8, 52.

11 Interviews by M. Bastos Lima, 2011: 6, 7, 8, 19, 52.
12 Interviews by M. Bastos Lima, 2011: 1, 6, 14, 52.
13 Interviews by M. Bastos Lima, 2011: 1, 4.
14 Interviews by M. Bastos Lima, 2011: 6, 8, 19, 52.
15 Interviews by M. Bastos Lima, 2011: 2, 4, 5, 6, 7, 8, 9, 11, 14, 18, 19, 21, 48.
16 Interviews by M. Bastos Lima, 2011: 1, 2, 4, 5, 6, 7, 8, 9, 10, 11, 13, 14, 15, 16, 17, 18, 19, 22, 50; Koh, L.P. and Wilcove, D.S. (2008) Is oil palm agriculture really destroying tropical biodiversity? *Conservation Letters* 1 (2), 60–64; Koh, L.P., Miettinen, J., Lieuw, S.C. and Ghazoul, J. (2011) Remotely sensed evidence of tropical peatland conversion to oil palm, *Proceedings of the National Academy of Sciences* 108 (12), 5127–32.
17 Interviews by M. Bastos Lima, 2011: 9, 11, 16, 19.
18 Ewart, D.L. and Vaughn, R. (2009) *Indonesian Coal*, US: Marston & Marston Inc.
19 Miettinen, J., Shi, C. and Lieuw, S.C. (2012) Two decades of destruction in Southeast Asia's peat swamp forests, *Frontiers in Ecology and the Environment* 10 (3), 124–8.
20 Interviews by M. Bastos Lima, 2011: 2, 5, 6, 22.
21 Interviews by M. Bastos Lima, 2011: 22.
22 Interviews by M. Bastos Lima, 2011: 6.
23 Interviews by M. Bastos Lima, 2011: 2, 5, 6, 15.
24 Interviews by M. Bastos Lima, 2011: 6, 11, 12, 14, 16, 19, 22, 48.
25 Interviews by M. Bastos Lima, 2011: 11, 16, 19.
26 Feintrenie, L., Chong, W.K. and Levang, P. (2010) Why do farmers prefer oil palm? Lessons learnt from Bungo District, Indonesia, *Small-scale forestry* 9, 379–96.
27 Interviews by M. Bastos Lima, 2011: 6, 11, 12, 16, 17, 19, 48.
28 Interviews by M. Bastos Lima, 2011: 1, 2, 6, 8, 9, 11, 16, 17, 22.
29 Interviews by M. Bastos Lima, 2011: 14, 16, 52; Maulia, E. (2010) Indonesia pledges to 'feed the world', *The Jakarta Post*.
30 Cotula, L., Vermeulen, S., Leonard, R. and Keeley, J. (2009) *Land grab or development opportunity? Agricultural investments and international land deals in Africa*, London/Rome: IIED/FAO/IFAD; Von Braun, J. and Meinzen-Dick, R. (2009) *'Land grabbing' by foreign investors in developing countries: risks and opportunities*, IFPRI Policy Brief 13, Washington DC: International Food Policy Research Institute.
31 Interviews by M. Bastos Lima, 2011: 7, 13, 21, 22.
32 Interviews by M. Bastos Lima, 2011: 7 and 21.
33 Colchester, M., Jiwan, N., Andiko, Sirait, M., Firdaus, A.Y., Surambo, A. and Pane, H. (2006) *Promised Land: Palm oil and land acquisition in Indonesia – Implications for local communities and indigenous peoples*, Moreton-in-Marsh, England and Bogor, Indonesia: Forest Peoples Programme, Perkumpulan Sawit Watch, HuMA, and the World Agroforestry Centre; interviews by M. Bastos Lima: 7 and 13.
34 Von Braun and Meinzen-Dick (2009) op. cit.
35 Contreras-Hermosilla, A. and Fay, C. (2005) *Strengthening forest management in Indonesia through land tenure reform: Issues and framework for action*, Washington DC: Forests Trends and Nairobi: World Agraforestry Centre.
36 Interviews by M. Bastos Lima, 2011: 9 and 16.
37 See Law 32/2004.
38 Government of Indonesia (1999) *Undang-Undang Republik Indonesia, Nomor 41 Tahung 1999 tentang Kehutanan (The Government Law No 41/1999: The Basic Forestry Law, issued in September 1999)*.
39 Interviews by M. Bastos Lima, 2011: 2, 5, 12, 16.

40 Ministry of Forestry (2009a) op. cit.
41 FAO (2010) Global forest resources assessment 2010 – main report, *FAO Forestry Paper* 163, Rome: Food and Agriculture Organization.
42 When asked about the relevance of international forestry obligations for Indonesia's policy-making on forestry, none of the interviewees cited one of those agreements. Rather, they stressed – positively or negatively – the bilateral agreements recently signed. Some interviewees explicitly elaborated on how those multilateral agreements mean little to Indonesia in practice: interviews by M. Bastos Lima: 5, 7, 18.
43 Interviews by M. Bastos Lima, 2011: 12.
44 Hooijer, A., Silvius, M., Wösten, H. and Page, S. (2006) *PEAT-CO2, Assessment of CO2 emissions from drained peatlands in SE Asia*, Delft Hydraulics report Q3943.
45 Interviews by M. Bastos Lima, 2011: 1, 3, 6, 7; Simamora, A.P. (2010) Govt denies RI is world's 3rd largest emitter in new report, *The Jakarta Post*.
46 Other forest policy instruments used in Vietnam, such as community forest management, spatial planning and protected areas are not included in this analysis.
47 European Commission (2003) *Forest Law Enforcement, Governance and Trade (FLEGT): Proposal for an EU action plan*, COM (2003) 251, Brussels.
48 European Union (2005) *Council Regulation (EC) No 2173/2005 of 20 December 2005 on the establishment of a FLEGT licensing scheme for imports of timber into the European Community*; European Union (2008) *Commission Regulation (EC) No 1024/2008 of 17 October 2008 laying down detailed measures for the implementation of Council Regulation (EC) No 2173/2005 on the establishment of a FLEGT licensing scheme for imports of timber into the European Community*.
49 European Union (2011) *FLEGT Voluntary Partnership Agreement between Indonesia and the European Union*, VPA Briefing Note – May 2011.
50 European Union (2011) op. cit.
51 European Union (2011) op. cit.
52 Interviews by M. Bastos Lima, 2011: 9, 15, 16, 17, 18.
53 Bastos Lima, M.G. (2009) Biofuel governance and international legal principles: is it equitable and sustainable?, *Melbourne Journal of International Law* 10 (2), 479–92.
54 Personal interviews 2, 6, 13.
55 Personal interview 6.
56 Interviews by M. Bastos Lima, 2011: 2, 6, 13.
57 Autonomy Law 26/2007
58 Interviews by M. Bastos Lima, 2011: 2, 6, 16.
59 Interviews by M. Bastos Lima, 2011: 6, 13, 14, 15, 16, 19, 22, 48.
60 Interviews by M. Bastos Lima, 2011: 14.
61 Interviews by M. Bastos Lima, 2011: 1, 2, 4, 6, 11, 19.
62 Interviews by M. Bastos Lima, 2011: 6, 13, 14, 22, 25
63 UNESCOPRESS (2011) *Danger listing for Indonesia's tropical rainforest heritage of Sumatra*, World Heritage Committee, United Nations Educational, Scientific and Cultural Organization.
64 Ministry of Forestry decree P.50/Menhut-II/2010; Lang, C. (2011) *On the eve of the logging moratorium, Indonesia's Ministry of Forestry issued almost three million hectares of concessions*, REDD-Monitor.
65 Ministry of Forestry decree SK.292/Menhut-II/2011.
66 Effendi, E. (2011) *Minister of Forestry proves Indonesia's moratorium is toothless*, Greenomics Indonesia.

67 Personal interview 16.
68 Interviews by M. Bastos Lima, 2011: 1, 2, 5, 6, 7, 9, 11, 14, 16, 18, 19, 51, 52.
69 Interviews by M. Bastos Lima, 2011: 6, 11.
70 Interviews by M. Bastos Lima, 2011: 14, 19.
71 Interviews by M. Bastos Lima, 2011: 6, 8, 9, 11, 18, 21; Barr, C., Resosudarmo, I.A.P., Dermawan, A. and McCarthy, J. (2006) *Decentralization of forest administration in Indonesia: Implications for forest sustainability, economic development and community livelihoods*, Bogor, Indonesia: CIFOR (Center for International Forestry Research).
72 Interviews by M. Bastos Lima, 2011: 51.
73 Interviews by M. Bastos Lima, 2011: 2, 3, 5.
74 Fitzherbert, E.B., Struebig, M.J., Morel, A., Danielsen, F., Bruhl, C.A., Donald, P.F. and Phalan, B. (2008) How will oil palm expansion affect biodiversity? *Trends in Ecology & Evolution* 23 (10), 538–45; interviews by M. Bastos Lima, 2011: 7, 8, 10, 16, 21.
75 Interviews by M. Bastos Lima, 2011: 7.
76 Interviews by M. Bastos Lima, 2011: 5, 6, 14, 16.
77 Barr, C. (2001) *Banking on sustainability: structural adjustment and forestry reform in post-Suharto Indonesia*, Washington DC, USA and Bogor, Indonesia: WWF Macroeconomics Program Office and CIFOR (Center for International Forestry Research); interviews by M. Bastos Lima, 2011: 5, 6, 14, 16.
78 Barr (2001) op. cit.; Ministry of Forestry (2009a) op. cit.
79 Fitzherbert *et al.* (2008) op. cit.
80 Interviews by M. Bastos Lima, 2011: 6, 14.
81 Interviews by M. Bastos Lima, 2011: 5, 6, 7, 8, 12, 14, 15, 16, 51; Colchester *et al.* (2006) op. cit.
82 Interviews by M. Bastos Lima, 2011: 7, 8, 13, 16.
83 Interviews by M. Bastos Lima, 2011: 14, 52.
84 Obidzinski, K. and Dermawan, A. (2010) Smallholder timber plantation development in Indonesia: what is preventing progress? *International Forestry Review* 12 (4), 339–48.
85 Ministry of Forestry (2009a) op. cit.
86 Obidzinski and Dermawan (2010) op.cit; interviews by M. Bastos Lima, 2011: 7, 52.
87 Interviews by M. Bastos Lima, 2011: 1, 2, 3, 4, 6, 12, 14, 23.
88 Interviews by M. Bastos Lima, 2011: 5, 6, 18.
89 Interviews by M. Bastos Lima, 2011: 1, 2, 11.
90 Interviews by M. Bastos Lima, 2011: 2.
91 Interviews by M. Bastos Lima, 2011: 1, 2, 4, 5, 6, 11, 49.
92 Interviews by M. Bastos Lima, 2011: 1, 2, 4, 5, 6, 11, 14, 49.
93 http://www.state.gov/r/pa/prs/ps/2011/09/174803.htm
94 Cassimon, D., Prowse, M. and Essers, D. (2011) The pitfalls and potential of debt-for-nature swaps: A US-Indonesian case study, *Global Environmental Change* 21 (1), 93–102.
95 Interviews by M. Bastos Lima, 2011: 2, 13, 14, 18.
96 Interviews by M. Bastos Lima, 2011: 6, 14, 18.
97 FSC (2012) *Global FSC certificates: type and distribution*, Bonn, Germany: FSC, AC.
98 Baskoro, F.M. and Al Azhari, M. (2011) Gapki rejects world standards, says Indonesia must set pace, *Jakarta Globe*; interviews by M. Bastos Lima, 2011: 2, 51.
99 Hall, D., Hirsch, P. and Li, T. (2011) *Powers of exclusion: land dilemmas in Southeast Asia*, Singapore: NUS Press; interviews by M. Bastos Lima, 2011: 7, 11, 16, 52;

Ministry of Forestry (2007) *Tata cara permohonan izin usaha pemanfaatan hasil hutan kayu pada hutan tanaman rakyat dalam hutan tanaman (Business license application procedures for utilization of timber in the forests of the people in plantation crops)*, Ministerial regulation P.23/Menhut-II/2007.

100 Schneck, J. (2009) *Assessing the viability of HTR – Indonesia's community-based forest plantation program*, Master's thesis, Duke University.
101 Ministry of Forestry (2007) op. cit., art. 15; Obidzinski and Dermawan (2010) op. cit.
102 Interviews by M. Bastos Lima, 2011: 52; This is said to be different in Vietnam, where similar permits exist but are transferable, and this apparently has led to higher take-up and success of the policy there.
103 Obidzinski and Dermawan (2010) op. cit.; personal interview 52.
104 Obidzinski and Dermawan (2010) op. cit.
105 Akiefnawati, R., Villamor, G.B., Zulfikar, F., Budisetiawan, I., Mulyoutami, E., Ayat, A. and van Noordwijk, M. (2010) Stewardship agreement to Reduce Emissions from Deforestation and Degradation (REDD): case study from Lubuk Beringin's *Hutan Desa*, Jambi Province, Sumatra, Indonesia, *International Forestry Review* 12 (4), 349–60; interviews by M. Bastos Lima, 2011: 16, 21, 23, 36, 45.
106 Akiefnawati *et al.* (2010) op. cit.; interviews by M. Bastos Lima, 2011: 7, 11, 16, 21, 23, 36, 37, 45, 46, 47.
107 Interviews by M. Bastos Lima, 2011: 45, 46.
108 Akiefnawati *et al.* (2010) op. cit.
109 In particular, the community finds it hard to repel encroachers who are kin or friends from neighbouring villages, and they are normally unwilling to create enmity in the region; interviews by M. Bastos Lima, 2011: 35, 45, 46.
110 Interviews by M. Bastos Lima, 2011: 21.
111 Interviews by M. Bastos Lima, 2011: 5, 21, 23, 45, 46.
112 Interviews by M. Bastos Lima, 2011: 21, 45, 46.
113 Ministry of Forestry regulation SK. 159/Menhut-II/2004.
114 Ministry of Forestry (2004) *Restorasi ekosistem di kawasan hutan produksi (Restoration of forest ecosystems in production)*, Ministerial regulation SK. 159/Menhut-II/2004.
115 Government Regulation No.6/2007.
116 Government Regulation No.3/2008.
117 BirdLife International (2011) *Restoration model set to transform Indonesia's forest sector*.
118 BirdLife International (2011) op. cit.; interviews by Mairon Bastos Lima, 2011: 14, 16.
119 BirdLife International (2011) op. cit.; interviews by Mairon Bastos Lima, 2011: 14, 18.
120 BirdLife International (2011) op. cit.
121 Hall *et al.* (2011) op. cit.; personal interviews 13 and 21.
122 Interviews by M. Bastos Lima, 2011: 13.
123 Caldecott, J., Indrawan, M., Rinne, P. and Halonen, M. (2011) *Indonesia–Norway REDD+ Partnership: first evaluation of deliverables*, Helsinki: GAIA Consulting Ltd in association with Creatura Consulting Ltd.
124 Kingdom of Norway and Republic of Indonesia (2010) *Letter of Intent between the Government of the Kingdom of Norway and the Government of the Republic of Indonesia on 'Cooperation on reducing greenhouse gas emissions from deforestation and forest degradation'*.

125 Austin, K., Sheppard, S. and Stolle, F. (2012) *Indonesia's moratorium on new forest concessions: key findings and next steps*, WRI Working Paper, Washington DC, USA: World Resources Institute.
126 Austin *et al.* (2012) op. cit.
127 Interviews by M. Bastos Lima, 2011: 5, 21; Simamora, A.P. (2010) No decision yet no REDD fund sharing mechanism, *The Jakarta Post*.
128 Interviews by M. Bastos Lima, 2011: 9, 11.
129 Interviews by M. Bastos Lima, 2011: 9.
130 Wardoyo, W., Masripatin, N., Sugadirman, R., Ginting, N., Syam, M., Widiaryanto, P., Royana, R., Situmorang, A.W., Khatarina, J. and Bimmo, A. (2010) *REDD+ national strategy, Draft 1 revised*.
131 Interviews by M. Bastos Lima, 2011: 18, 21, 36, 45, 46, 51, 52.
132 Interviews by M. Bastos Lima, 2011: 8, 13, 14, 52.
133 Slette, J. and Wiyono, I.E. (2011) *Indonesia Forest Moratorium 2011, GAIN Report ID 1127*, Jakarta, Indonesia: USDA Foreign Agricultural Service.
134 Presidential Decree No. 10/2011.
135 Presidential Decree No. 19/2010.
136 Caldecott *et al.* (2011) op. cit.
137 How carbon will be counted, how benefits will be shared, or how problems such as leakage or permanence will be tackled. It is also unclear who owns the carbon and who is responsible for keeping it.
138 Interviews by M. Bastos Lima, 2011: 1, 2, 5, 6, 14, 15, 16, 19.
139 Interviews by M. Bastos Lima, 2011: 2, 5, 7, 8, 9, 12, 13, 22.
140 Interviews by M. Bastos Lima, 2011: 7.
141 Interviews by M. Bastos Lima, 2011: 5, 8; Hall *et al.* (2011) op. cit.
142 Interviews by M. Bastos Lima, 2011: 13.
143 Interviews by M. Bastos Lima, 2011: 8.
144 Interviews by M. Bastos Lima, 2011: 2, 7, 8, 10, 11, 16.
145 Persson, U.M. and Azar, C. (2010) Preserving the world's tropical forests – A price on carbon may not do, *Environmental Science and Technology* 44, 210–15.
146 Interviews by M. Bastos Lima, 2011: 12, 17, 21.
147 Interviews by M. Bastos Lima, 2011: 13, 18, 45.

7 Case study

Cameroon

*Jonathan Y.B. Kuiper, Nicolien van der Grijp,
Joyeeta Gupta, and Channah Betgen*[1]

7.1 Introduction

Around 17 per cent of global forests are in Africa, of which 37 per cent are to be found in the Congo Basin.[2] It is the second largest tropical forest after the Amazon, holding 46 megatonnes of carbon.[3] Cameroon lies in the North-Eastern part of the Congo Basin. In 2010, 42 per cent of its land area was covered by forests, of which 16 per cent was primary forest area and 83 per cent other naturally regenerated forest area.[4] The forest provides food, resources and livelihoods for rural communities, raw materials for industries,[5] and ecosystem services such as watershed management, soil and biodiversity conservation and carbon sequestration.[6]

With 19 million people from many different ethnic groups,[7] Cameroon has the highest population density in the region. Fifty-seven per cent of the population lives in urban and 43 per cent in rural areas. In 2008, the per capita GDP amounted to USD 2,195 and its annual growth rate to 3.9 per cent. Most people are active in the agricultural and livestock sector (44 per cent of GDP), services (40 per cent) and industry (16 per cent).[8] The forest sector contributed 1.9 per cent to GDP in 2006.[9] Oil accounts for 49.9 per cent of exports, followed by cocoa, cotton and wood with 6.5 per cent of exports. Cameroon is classified as a highly indebted poor country, and is supported by the International Monetary Fund which leveraged governance reform as a component for eligibility.[10] Forty per cent of the population still lives below the national poverty line.[11]

Cameroon has the second highest deforestation rate of the Congo basin. Between 1990 and 2000, forest cover diminished by 0.94 per cent per year, and between 2000 and 2010 by 1.04 per cent per year.[12] Forest degradation occurred at an annual rate of 0.01 per cent between 1990 and 2000.[13] However, although data collection is quite well established, there are inconsistencies, and variations in the collected data and continuous cloud cover limits high-quality remote sensing images.[14]

This case study focuses on Cameroon and its South-West region (see Figure 7.1) which hosts dense tropical humid forests. The northern part of the country has different forest types mixed with savannah and desert.

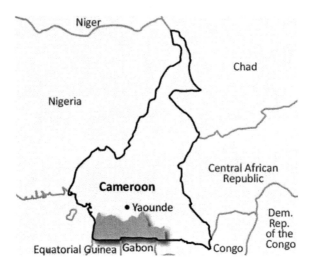

Figure 7.1 Map of Cameroon and the case study area of the South-West region

In accordance with the analytical framework, this chapter discusses the driving forces for deforestation and forest degradation (7.2), the forest policy context (7.3), effectiveness and fairness of policy instruments (7.4), implications for REDD (7.5), and draws inferences (7.6).

7.2 Driving forces of deforestation and forest degradation

As shown in Table 7.1, agriculture is a major proximate driving factor for deforestation and forest degradation in Cameroon.[15] The larger-scale agro-industry produces industrial quantities of bananas, rubber, sugar cane and oil palm, and has replaced 30 per cent of original plant cover of inland coastal zones. Small-scale agroforestry, producing cocoa and coffee in the Southern region, utilizes 914,000 hectares of forest land.[16] Two contested agricultural drivers are slash-and-burn and reduced fallow periods.[17] Forest degradation is caused by small-scale chainsaw milling (artisanal logging, mostly without authorization) which provides 75 per cent of domestic timber.[18] Industrial logging takes place within forest concessions and wood and wood products are primarily exported to the European Union and Asia. Degradation occurs more in Northern forests and around urban areas in response to demands for fuel wood of 1.3 million urban households.[19] Roads, mining and hydropower projects and the proximity to ports have also had negative impacts on forests.[20] Planned mining in the South-East is expected to affect forest cover in the future.

The underlying driving forces include population growth (2.3 per cent per year), urbanization, demand for cash crops,[21] and migration.[22] Economic drivers

Table 7.1 Drivers of deforestation and forest degradation in Cameroon and the South-West region*

Proximate drivers	Underlying drivers
Local to national	*Local*
Agriculture: *cash crop plantations (cocoa, coffee in South-West humid zone*; bananas, oil palm, sugar cane, rubber in inland coastal zones), *reduction of fallow periods, shifting cultivation*	Economic: poverty
	Local to national
	Cultural: reproduction and forest tenure customs, conflict between economic growth and conservation ethic
	National
	Demographic: growth, urbanization (demand for cash crops)
	Economic: growth, low price for wood and forest products
	Technological: poor technologies available
Extraction: industrial wood, fuel wood, chainsaw milling, illegal logging	Non-forest policies: agricultural and mining policy
	Other institutional factors: poor law enforcement; lack of coordination between ministries, corruption
Infrastructure: forest roads, hydropower	*Global*
Industry: mining	Demand/trade/consumption at global level for food, fibre, timber and other forest products, biofuel; debt and structural adjustment programmes; climate change
Biophysical: poor soil	

* The drivers which apply to the specific context of the South-West region are marked in italics

include the national growth and employment strategy which aims at enhancing transport infrastructure to promote mining and agriculture.[23]

Many people living in rural areas are dependent on the rainforests for wood, fuel wood, bush meat, medicinal plants and food. In the 1970s and 1980s, economic, agricultural, monetary and commodity exports policies (oil, coffee and cocoa) led to agricultural expansion until the collapse of these export markets. Subsequent debt problems were solved with loans from the World Bank accompanied by structural adjustment programmes. These programmes led to devaluation of the Central African franc (CAF) by 50 per cent, reduced subsidies for fertilizers and pesticides, a shift towards market-oriented food crops, high unemployment, urban/rural re-migration, stronger relations between rural production and urban demands and increased demand for fuel wood as household fuel prices increased.[24] The crisis also led to slashing of wages of ministry employees, increasing corruption to compensate for lost wages,[25] indirectly allowing illegal logging and the non-collection of taxes. It is estimated that officials of the Ministry of Flora and Fauna (MINFOF) collected €9.1 million in informal payments between 2008 and 2009.[26] However, corruption is said to occur at all levels.[27]

7.3 Key policies and instruments

This section discusses the organizational framework, the evolution of forest policy, and the influence of international treaties and bodies.

7.3.1 *The organizational framework*

In 1992, the Ministry of Environment and Forests was established to promote sustainable and community-based management. In 2004, this Ministry was split into the Ministry of Flora and Fauna and the Ministry of Environment and Nature Protection. The former is responsible for the sustainable management of forests and wildlife in all forest estates; marking physical boundaries of community forests; issuing and establishing logging titles for all harvested areas; approving management plans for concessions and community forests; regeneration, reforestation and updating the forest inventory; and forest law enforcement (conducted by the National Forest Law Enforcement Brigade, in partnership with independent observers based on the National Strategy for Forest and Wildlife Law Enforcement, which was established in 2005).[28]

The Ministry of Environment and Nature Protection tracks environmental impacts from all ministries, including impacts on deforestation and forestland degradation through, inter alia, environmental impact assessments of logging projects.[29] It formulates, monitors and evaluates the implementation of the sectoral master plans for environmental protection. Furthermore, it develops, distributes and tracks indicators of sustainable development, promotes the implementation of Agenda 21, and coordinates climate policy including REDD. However, it lacks the authority, leadership, capacity and institutional authority to actually exert control.[30]

The Ministry of Economy and Finance is involved in the chain of custody of timber products and controls the collection of the annual area fee generated by forestry activities in concessions and community and council forests[31] and subsequently transfers it to the treasury, local governing bodies and local communities. The Ministry of Territorial Administration and Decentralization is responsible for the demarcation of official boundaries of the council forests. The National Forest Development Agency encourages financially and environmentally viable plantations and develops alternative revenue streams for their establishment; however, its progress has been slow.[32]

Activities to protect the forests are often countered by the Ministry of Agriculture and Rural Development, which promotes food security and revenue through food, cash crops and biofuel production for international markets, and also by the Ministry of Planning and Regional Development, which promotes local development.[33]

In Cameroon, decision-makers seem to be aware of climate change issues and agree on its potential impacts. They also recognize the importance of forests as a carbon sink.[34] However, while there is communication between institutions on a national scale on issues like climate change, there is no indication of such exchange with lower levels of government or communities. Some NGOs indicated that climate change has become integrated into their work, but also felt that there was no communication between governmental institutions responsible for biodiversity conservation and those responsible for climate change.[35]

7.3.2 The evolution of forest policy

The 1982 Forest Code, focusing on short-term leases, quick extraction and export duty based taxation, was repealed in 1993, when a radically different forest policy was adopted. This new policy was institutionalized in the 1994 Forest Law[36] (which took four years to become active),[37] the Constitution, and the 1996 Framework Law on Environmental Management.[38] The 1994 Forest Law classified the entire dense humid forest area into zones, and allocated long-term forest harvesting rights through a public auctioning process based on technical and financial criteria. It furthermore reorganized the government to improve forest governance, while productive activities were transferred to local councils and production forest operations. It granted private firms long-term concessions to operate in permanent commercial forests and obligated them to develop and implement forest management plans to be monitored by the forest administration. In addition, the Forest Law granted communities and local councils rights to manage forests based on a contract, including provisions on access, customary rights, forest area, title allocation, sustainable logging practices, and the management of flora and fauna.[39]

7.3.3 The influence of international treaties and bodies

Cameroon is party to the relevant global agreements (UNFCCC, CBD, CCD, RAMSAR, CITES, ILO, NLBI, UNPFII, see Chapter 3). As a party to these agreements, Cameroon is explicitly searching for revenue streams emerging from these international processes. Although it is also a party to the Central Africa Forests Commission (COMIFAC) and the Forest Law Enforcement and Governance process (FLEG), these processes appear to have less influence on the domestic policy process. In recent years, there have been efforts to engage Cameroon in the implementation of REDD (see Chapter 4).

As mentioned above, structural adjustment has been very influential in Cameroon. Following its financial crisis of 1987, Cameroon borrowed from the World Bank, a process accompanied by three consecutive structural adjustment programmes in the 1990s. The first and second programmes did not influence the forest sector, but the third programme introduced forest sector conditionalities, including the design and enforcement of a new forest taxation system, improved transparency and governance, and addressing corruption in the forest sector.[40] The forest sector was included within additional International Monetary Fund initiatives including the Poverty Reduction and Growth Facility and activities surrounding the Heavily Indebted Poor Countries initiative. This process was subsequently criticized for not focusing on sustainable forest management and later led to a code of conduct for signatories[41] to utilize the Forest and Environment Sector Programme as a common framework for aligning efforts to further develop the economic reforms and sustainable development.[42]

International NGOs, such as the World Resource Institute, Global Witness, and Resource Extraction Monitoring, have participated in Cameroon as independent observers whose public reports on irregularities in logging activities, auctions and harvesting practices[43] forced the Ministry of Flora and Fauna to take action.[44] Furthermore, Global Forest Watch, a branch of the World Resource Institute, is helping the ministry to prepare a forest database, and World Wildlife Fund, the Wildlife Conservation Society and the International Union for Conservation of Nature promote biodiversity conservation and conservation management.[45]

7.4 Key forest policy instruments and their analysis

This section presents and analyses a selection of regulatory and market-based instruments[46] (using the numbering system developed in Table 2.3) in terms of their effectiveness and associated fairness issues, given the existing drivers of deforestation and forest degradation, using the [+ + – –?] method developed in 1.5.

Decentralization (ii): Since 2004, Cameroon's formerly highly centralized, top-down government has started to decentralize policy authority to ten regions, including 339 municipalities. Furthermore, it has delegated powers to 58 divisions, with 275 subdivisions and 53 districts; and to technical entities.[47] The decentralization process in Cameroon is still incomplete, due to lack of political will, competing interests between ministries, and bureaucracy.[48] However, in comparison with other sectors, the forest sector is the most decentralized. Yet this decentralization still implies top-down policy formulation which effectively disenfranchises local populations, allows elite capture and is accompanied by poor management structures.[49] The Ministry of Flora and Fauna controls policies to a large extent. In 2010 a new joint order was passed to regulate all revenues derived from forest and wildlife activities to be redistributed to riparian communities and councils. Several measures were included to ensure accountability and transparency in the management of revenues such as capping operating costs to ensure adequate investment in the community.[50] To strengthen the potential of the joint order, technical support is needed. REDD will need to be implemented within the context of effective decentralization.[51] Discussions about REDD are still very much centralized in the national government and there is little engagement of local people concerned. [+ –]

Spatial planning (v): Zoning and gazetting implies classifying forest land and legalizing boundaries, ownership and land-use designations. The process clarifies access rights and authority of control.[52] The 1984 Forest Law distinguishes between: (i) the permanent forest estate, which can only be used for forestry or as a wildlife habitat, and (ii) the non-permanent forests estate, consisting of forested land which can be converted to non-forest uses.[53] The permanent forest domain includes state forests and council forests. State forests include protected areas (National Parks, wildlife reservations, community-managed hunting zones, state-owned game ranches, state-owned zoos, wildlife sanctuaries and buffer zones)[54] and forest reserves (ecological reservations, production forests, protected forests, recreational forests, forests for learning and research, plant sanctuaries, botanical gardens and reforestation areas). Council forests, including several communities, are managed by a municipal board, and have a management plan approved by the forest administration. The non-permanent forest domain includes community forests which fall under a 25-year management arrangement between the community and the forest services, and other forests which could be privately owned. In 2009, 80 per cent of the forestlands were designated as permanent and the remaining 20 per cent as non-permanent forest domain.[55]

Starting in the 1990s, protected areas have been established in Cameroon to conserve natural heritage and biological diversity.[56] The establishment of the permanent forest domain is an important step towards protecting flora and fauna. In protected areas, timber exploitation and agricultural activity is prohibited, with the exception of particular protected areas that grant user rights to indigenous peoples to maintain traditional lifestyles. Protected areas are quite effective and are supported by regulations. They require a management plan as outlined by the National Directive on the Management of Protected Areas of the Ministry of Flora and Fauna in partnership with World Wide Fund for Nature.[57] The directive requires management plans to provide data on biological, socio-economic, infrastructure, and management structure. It contains provisions about financing, zoning, biodiversity protection, and stakeholder participation. Moreover, the effectiveness of management has increased over the years in several parks (Bakossi Kupe, Benoué, Boumba Bek, Nki, Campo Ma'an, Korup, M'Bam and Djerem, Ndongoré, and Waza).[58]

Although 8 million hectares have been allocated as forest management units, council forests and protected areas, gazetting, which involves public consultation, is yet to be completed. Currently, 4.5 million hectares of council forests and protected areas and 8.6 million hectares of forest management units still await gazetting.[59] The designation of national parks, game reserves, forest reserves and botanical gardens has arguably reduced forest access for indigenous peoples.[60] In Campo Ma'an, locals were not consulted or compensated until the park had already been created.[61] The same is allegedly the case for Koli/Nkongabok, in the Ottotomo Forest Reserve, where communities lost their access rights. In a study of the communities of Dja, Korup, Boumba Bek and Lobéké, only Korup villagers were asked to move voluntarily with compensation (189 people/23 households received USD 22,000 per household), while Baka Pygmies of the Dja Reserve reported a 'perceived loss'.[62] In the case of Boumba Bek national park, efforts were made to secure the rights of the Bakas in relation to their agricultural practices. The surrounding communities of Lobéké national park were consulted on the establishment of the protected area and provisions were made for the conversion of traditional practices and agriculture.[63]

Furthermore, at national level, protected areas cannot compete with other economic interests such as mining.[64] Mining permits have been issued within the boundaries of Lobéké and Boumba Bek national parks, part of the Sangha Tri national park area, and the Douala Edea wildlife reserve.[65] At local level, shifting cultivation probably occurs in densely populated protected areas, such as Campo Ma'an and Korup,[66] as well as encroachment by agroforestry and agricultural activities in border areas.[67] 'Illegal' logging may occur where parks are near populations,[68] and occasionally informal extractive activities are tolerated within park boundaries.[69] If protected areas do not provide local people with alternative income, their dependence on the forests will lead to encroachment.[70] Moreover, if protected areas do not account for cultural dynamics such as the slash-and-burn techniques of the Baka in the South-West region by consulting and compensating local people, their policies will be ineffective.[71] Although the central government and NGOs have allocated funding for forest management,[72] this is too little and local populations are frequently not consulted.[73] At the same time, the central government is losing revenues due to reduced available lands for timber and mining exploitation. Efforts to create a coherent and strategic zoning plan need to be aligned between the ministries involved. [+ + −]

Forest logging/concessions (vi): A concession relates to an area of forested land, zoned for industrial timber production. Each concession is typically broken up into several forest management units. Concessions are state owned and auctioned to private operators. Concession units produce industrial timber, implement sustainable forest management, generate revenues for the state and help organizing production forests. The 1990 economic reforms modernized logging titles that had previously been administered at random to serve a non-transparent patronage system.[74] Since 1990, within the permanent forest domain, logging titles are distributed to forest management units (to maintain biodiversity and industrial logging) and council forests (for logging).

Operators of forest management units may exploit forests under a three-year provisional licence during which they need to develop a forest management plan to guarantee the unit's long-term economic value by selecting and managing key commercial species, and managing regeneration rates to regulate the rate of recovery of a given species from one harvesting cycle to the next. Forest management plans require minimum harvesting cycles of 30 years. This licence is succeeded by a *Convention Définitive* that grants full harvesting rights for 15 years.[75]

Forest management plans use topographical, biological and socio-economic data of the concession to achieve conservation, agroforestry and forest production objectives.[76] These plans must ensure that timber harvesting does not jeopardize the sustainability of forest goods and services, the intrinsic values and future productivity of harvested areas or create environmental and social externalities.[77] They should focus on attaining silvicultural parameters such as a list of key species on which yield calculation is to be based, and the determination of the minimum diameter at which these species can be cut. These parameters directly influence the annual allowable cut, which can correspond either to the maximum theoretical volume of timber or the maximum area that can be harvested annually.[78] Forest management units utilize selective harvesting techniques by selecting key commercial species per hectare on 30-year harvesting cycles to ensure that enough commercial species remain per hectare to regenerate and ensure availability for the next harvesting cycle. The plans should constrain the harvesting of high-value species to encourage the harvesting of 'secondary species'.[79]

Permits for the non-permanent forest domain are allocated as *Ventes du Coup* and *Petite Titres*. *Ventes du Coup* are allocated under the same auctioning system as forest management units for three years. They cannot exceed 2,500 hectares and do not require a management plan. *Petites Titres* include forest product exploitation permits, timber recuperation permits, timber removal permits and personal logging permits.[80] These are reserved for citizens.

Forest concessions with a forest management plan have improved the organization and management of production forests by balancing environmental, social and economic aspects. However, these plans have not ensured de facto sustainable forest production, due to flaws in the legal framework and political lethargy in regard to updating the provisions of forest management plans.[81] Concession holders are obliged to reduce harvesting rates to 'sustainable' limits (1/30th of the concession area annually) and most adhere to these.[82] The high cost of bidding has brought in larger international companies, leaving the smaller national companies to engage in illegal informal logging.[83] Concessions expand the frontier through road development and subsequently provide space for small-scale agricultural practices and artisanal logging.[84] The use of often poorly paid imported rather than local labour[85] leads employees to engage in agricultural cultivation and timber exploitation in and around concessions[86] and local unemployed people to engage in slash-and-burn cultivation.[87] Further, pressure from illegal logging by neighbouring countries, such as Nigeria, in border areas is impossible for timber companies to monitor.[88] These companies are often unable or unwilling to implement regeneration of problematic species or enrichment planting; this implies that natural regeneration is the only means of regeneration for harvested species.[89]

Similarly as for protected areas, forest management plans for forest management units are assumed to promote sustainability.[90] However, the Ministry of Flora and Fauna has been slow to administer *Conventions Définitives* sometimes because of competing economic interests, such as biofuels and mining,[91] resulting in operators continuing to exploit their allocated forest management unit illegally. Second, forest management units do not always implement minimum precautionary safeguards for valuable species, and may harvest in disregard of the forest management plan.[92] Third, although the ministry must review and update forest management plans every five years, it has not initiated such reviews. Fourth, guidelines prescribe 30-year harvesting cycles while operating licences have 15-year cycles without a guarantee of contract renewal, discouraging operators from responsible behaviour.[93] Fifth, these regulations and their associated costs have forced small operators out of the forest sector into the informal sector. The above weaknesses need to be dealt with through improved policy design. [+ −]

Land rights (vii): In 1974, all unclaimed forested lands (which were almost all lands) were nationalized and ancestral rights were no longer recognized. People had to register if they wanted land ownership.[94] The 1976 Land Tenure Ordinance listed conditions for obtaining land titles and required nationals to obtain a 'land certificate' to be entitled to land regardless of ancestral[95] or customary rights.[96] The Land Tenure Law[97] of 1979, which has already been debated for decades, provides a highly uncertain legal framework for all stakeholders who utilize forest resources.[98] The 1994 Forest Law recognizes customary use rights allowing state, local councils, village communities and private individuals to make use of forest resources, but excludes land ownership thereby denying indigenous forest dwellers, such as the Bantu and Pygmy, their cultural, spiritual and ancestral rights. This has created land insecurity and indirectly excludes stakeholders from benefits and there is increasing need to rectify this situation. [− −]

Law enforcement (x): In recent years, 'illegal' activity has declined in forest concession areas and in logging outside of permitted areas, but has not disappeared completely.[99] Poor law enforcement continues and results from an inadequate monitoring and reporting system, out-of-court settlements, insufficient financial resources, failing disbursement procedures, bureaucracy, and diversion of funds to the single treasury.[100] While in general foreign companies perform better than local ones,[101] corruption pervades the system as poor company performance is most likely influenced by auction irregularities where logging titles are granted to unqualified companies.[102] Furthermore, several ministries are competing to enhance control over forest resources and do not effectively coordinate their policies.[103] Additional reasons for weak law enforcement are a lack of integrated knowledge and research. [− −]

Forest taxes (xiii): The two current forest taxes create predictable revenue streams, promote efficiency and reduce dependence on export taxes. One is a levy on the concession area, the other is a sawmill entry tax. Both encourage efficiency of timber operations. The taxes are collected through the Enhanced Forest Revenue Programme (Programme de Sécurisation des Recettes Forestières, PSRF) established in 1999 by the Ministries of Flora and Fauna, and Environment and Nature Protection.[104]

Furthermore, an annual forest fee is levied on forest management units and redistributed to the state (50 per cent), local councils (40 per cent) and communities (10 per cent) according to the 1998 Finance Law. This fee empowers approximately 50 councils[105] and communities financially and creates 'ownership' and 'stewardship' of local resources.[106] However, it is unclear whether the collected funds have found their proper destination. The instrument of the annual area fee should be made more accountable and transparent to enhance its effectiveness and prevent elite capture. [+]

CSR/Certification (xix): Certifying community forests in Cameroon provides economic benefits (access to international markets and price security for forest products), social benefits (international recognition, credibility and participation in training) and environmental benefits (improved conservation).[107] By mid 2009, ten forest management units had been certified by the Forest Stewardship Council, accounting for 70 per cent of Cameroon's exports of forest products.[108] Certification has improved industrial logging by decreasing annual harvested volumes per concession, on average by 7 to 16 per cent,[109] requires higher management standards than the government, and brings stability to producers through contracts and better prices thereby arguably reducing the incentive to overproduce to receive a reasonable rate of return. It has increased accountability towards local populations, respect for the rights of adjacent communities and ensured revenue distribution contributing to the welfare of communities. However, irregularities have occurred,[110] and small companies who cannot afford certification are marginalized.[111] Adopting certification is difficult without external assistance.[112] If sustainable forest management is to be achieved through certification, then the government should consider providing technical assistance to community forests, promoting public support through information and education, and synchronizing its standards with international certification standards. [+ −]

Community-based management (xxiv): The current Constitution[113] allows for community-based management. The 1994 policy reforms established community and council forests[114] to include and empower communities in forest management while improving rural livelihoods.[115] This scheme is embedded in the Forest and Environment Sector Programme. In 2010, this programme developed a four-year strategic plan for the South-West region based on stakeholder participation. Forest management plans for council forests are implemented by elected municipal bodies, which further decentralized decision making. However, the preparation of forest management plans is expensive, requires expertise and often leads the council forest to cooperate with a private company, which is granted logging rights and puts the council forest in a dependency relationship. Local livelihoods have often not improved because of reduced opportunities for agriculture leading to food imports and increased food prices,[116] and elite capture of revenues by the mayors of council forests,[117] creating local mistrust.[118] This affects the willingness of the population to comply with regulations set out by the council forest,[119] especially as mayors have close links with military officials and there is an informal network of collusion and clientelism. These problems may be solved by the 2010 accountability law, for the collection and redistribution of forest and wildlife revenues, which aims to ensure accountability and transparency, and caps operating costs to ensure adequate investment in the community.[120] According to Brown and Lassoie,[121] community-based forest management has potential, but imposition of inappropriate institutions, which do not reflect local systems of accountability in resource management, has limited its success. Community forests need to become more accountable, transparent, involve stakeholders in decision making and revenue distribution,[122] and have enhanced management capacity. A national directive guaranteeing the right of local communities, including indigenous people, to benefit from revenues generated by council forests would provide support in the political process.[123] Allowing council forests to exploit timber, on the same terms as forest management units, may help generate the needed revenue to develop a forest management plan.[124] [+ + −]

7.5 Implications for REDD

Cameroon and other COMIFAC countries support the national implementation of REDD, as it is seen as an economic development opportunity,[125] that can generate funds to support poverty reduction, the national development goals, biodiversity conservation, and mitigation of and the adaptation to climate change.[126] REDD is also viewed as an opportunity to bring together different stakeholders such as government, civil society, indigenous groups, and the private sector.[127] However, the details on how the benefits from REDD can be secured are not clear.[128]

Cameroon has taken a proactive position by protecting over 18.24 per cent of forested lands under various forms of conservation regimes. This means that there is ample opportunity for REDD to support the conservation of protected areas in Cameroon.[129] Cameroon is currently preparing a national REDD strategy and implementation plan. However, reports indicate that progress is insufficient.[130] One of the reasons for lack of progress is that the R-PIN does not contain any specific plans for consultation of indigenous peoples and forest-dependent communities.[131] Several REDD pilot projects are currently being tested by national and international organizations.[132]

Despite international discussions of the opportunities presented by various carbon trading systems, questions remain as to whether they can be effective solutions to multiple challenges when implemented in countries, such as Cameroon, with weak governance systems and monitoring mechanisms.[133] It is suggested that REDD is trying to be too many things to too many different groups and that, ultimately, it would end up so complicated that it would not accomplish anything.[134]

REDD clearly needs to build on the existing forest experiences in Cameroon. One could argue that if REDD is to be successful, there must be improved (a) enabling conditions. This includes action to build adaptive capacity through communicating climate change information more broadly and effectively, and building awareness of its potential impacts and strategies to adapt at all levels.[135] Furthermore, public and policy awareness of the role of forests in national development is vital.[136] Given Cameroon's institutional weaknesses, one could ask whether support should be given to NGOs instead.

However, this could antagonize the government.[137] Increasing the capacity of society to understand and take advantage of REDD is also important.[138] (b) REDD could effectively link up with existing national mechanisms such as protected areas and help create compensatory and participatory mechanisms for those who rely on protected areas and buffer zones for their survival. REDD could use council forests to push for genuine political representation of all stakeholders and ensure that local populations derive benefits from having restricted access to forest exploitation, implying transparency and accountability of the distribution of council forest and REDD revenues. Where concessions operate under a forest management plan based on sustainable forest management principles, these principles could be utilized for REDD implementation. REDD may have to provide substantial incentives to the Ministry of Flora and Fauna to increase measures of enforcement and management. REDD can only be effective if monitoring, reporting and verification are arranged. It is important to find an appropriate partner for these tasks. The existing ministries will need considerable support if they are to become equal to the task. Some suggest that such tasks can be entrusted to the FSC.[139] (c) REDD may need to build on and reinforce existing international instruments and processes such as COMIFAC, FLEG and international certification processes. (d) REDD may need to ensure that local land, food and work security is enhanced, and develop livelihood adaptation strategies on a framework of forests food and services.[140] The government should be supported in efforts to formalize small-scale illegal logging and to bring sustainable practices to a completely unregulated market. Simply banning small-scale illegal logging has not worked and does not address the demand of the domestic timber market.

(e) Reference levels for REDD may need to take the possible state of the country in terms of development and the forest transition into account. (f) For REDD to be effective, it is essential to understand who are the potential direct and indirect beneficiaries in the scheme whose interests need to be taken into account. (g) REDD is only likely to be successful if all direct and indirect beneficiaries to the scheme understand what it is about; this is a major lesson that emerges from experiences with respect to the CDM. (h) The modality of payment needs to be understood. One option is to build on the AAF system, and pay partially upfront to subsidize the implementation process, applying the decentralized benefit sharing model of council forests for REDD and modifying it to include stakeholder participation and better enforcement of recently established laws aimed at elite capture. (i) A key question is what the payment levels should be. If the levels are too low, this may not be effective. The extent to which opportunity costs need to be taken into account has to be deliberated. For example, REDD may not be able to deal with the opportunity costs associated with mining and commercial agriculture.

(j) International discussions on prior free and informed consent (PFIC) from indigenous peoples need to be taken into account (see Box 4.2). (k) In

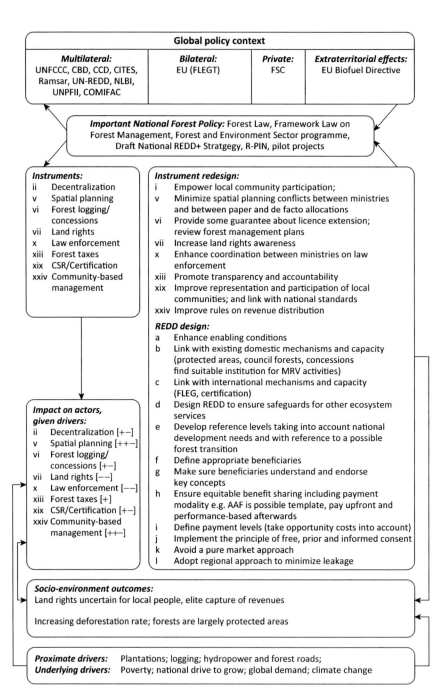

Figure 7.2 Applying the analytical framework to forest and REDD policy in Cameroon

Cameroon, with incomplete and often deficient markets, REDD should not be developed as a pure market-based instrument. (l) Finally, the potential for leakage implies that it may be effective to develop REDD as a regional instrument. Figure 7.2 illustrates the application of the analytical framework (see 1.5) to Cameroon based on the information in this chapter.

7.6 Inferences

Cameroon has a very large forest area which is under threat of increased deforestation and forest degradation. The proximate drivers are shifting agriculture, cash crops, mining, infrastructure and illegal logging. The underlying drivers are population growth and migration, and at local level poverty.

Cameroon's sophisticated model of forestry governance is developed on principles of sustainability. Cameroon forestry policy has taken progressive action by establishing permanent forest estates in the form of protected areas, implementing sustainable forest management practices, conservation through protected areas and decentralized models of community-based forestry. However, the analysis shows that there is a disconnect between drivers and policies because of a lack of resources, capacity and corruption. The decentralization process has begun but has still a long way to go. Furthermore, current instruments do not address illegal logging.

This will have implications as to how a REDD mechanism could be effective taking into account the cultural, economic and political complexities. Although Cameroon is seen as a probable candidate for the implementation of a REDD mechanism, preparations have still not been initiated nor has REDD been clarified at the international level causing difficulty in determining what form REDD may take in the context of Cameroon. The key question is: can the international community provide the support to enable Cameroon to develop and its people to achieve food and security, while still protecting the forests?

Notes

1 The authors would like to thank Dr Carolyn Peach Brown for kindly providing us with suggestions for relevant literature.
2 The Congo Basin includes Angola, Burundi, Cameroon, Central African Republic, Congo, Equatorial Guinea, Gabon, Rwanda, and Sao Tomé/Principe; FAO (2011a) *State of the world's forests 2011*, Rome: FAO.
3 FAO (2011b) *The state of forests in the Amazon Basin, Congo Basin and Southeast Asia*, Rome: Food and Agriculture Organization of the United Nations.
4 FAO (2011b) op. cit.
5 Somorin, O.A., Brown, H.C.P., Visseren-Hamakers, I.J., Sonwa, D., Arts, B. and Nkem, J.N. (2012) The Congo Basin forests in a changing climate: policy discourses on adaptation and mitigation (REDD+), *Global Environmental Change* 22, 288–98.

6 Brown, H.C.P., Smit, B., Sonwa, D., Somorin, O.A. and Nkem, J.N. (2011) Institutional perceptions of opportunities and challenges of REDD+ in the Congo Basin, *Journal of Environment & Development* 20 (4); Somorin *et al*. (2012) op. cit.
7 FAO (2011b) op. cit.
8 Cerutti, P.O., Ingram, V. and Sonwa, D. (2008a) *The Forests of Cameroon in 2008*, Food and Agricultural Organization; Topa, G., Karsenty, A., Megevand, C. and Laurent, D. (2009) *The rainforests of Cameroon: experience and evidence from a decade of reform*, Washington, DC: The World Bank.
9 FAO (2011b) op. cit.
10 Cerutti *et al*. (2008a) op. cit.; IMF (2010) *Cameroon: poverty reduction strategy paper – joint staff adivsory note*, Washington: International Monetary Fund.
11 World Bank (2011) *Cameroon at a glance*, Washington, DC: World Bank.
12 FAO (2011a) op. cit.
13 Robiglio, V., Ngendakumana, S., Gockowski, J., Yemefack, M., Tchienkoua, M., Mbile, P., Tchawa, P., Thoundjeu, Z. and Bolognesi, M. (2010) *Reducing emissions from all land uses in Cameroon: Final national report*, Nairobi: Partnership for the Tropical Forest Margins.
14 Dkamela, G.P. (2011) *The context of REDD+ in Cameroon: drivers, agents, institutions*, Bogor Barat: Center for International Forestry Research.
15 Robiglio *et al*. (2010) op. cit.; Cerutti *et al*. (2008a) op. cit.
16 Dkamela (2011) op. cit.
17 Robiglio *et al*. (2010) op. cit.; Ickowitz, A. (2006) Shifting cultivation and deforestation in tropical Africa: critical reflections, *Development and Change* 37 (3), 599–626; Freudenthal, E., Nnah, S. and Kenrick, J. (2011) *REDD and rights in Cameroon: a review of the treatment of indigenous peoples and local communities in policies and projects*, Moreton-in-Marsh, UK: Forest Peoples Programme.
18 Cerutti, P.O., Lescuyer, G., Assembe Mvondo, S. and Tacconi, L. (2010) *Les défis de la redistribution des bénéfices monétaires tirés de la foret pour les administrations locales (The challenges of the redistribution of monetary benefits from the forest for local governments)*, Bogar Barat: Center for International Forestry Research; Dkamela (2011) op. cit.
19 FAO (2011a) op. cit.; Topa *et al*. (2009) op. cit.
20 Dkamela (2011) op. cit.
21 FAO (2011b) op. cit.
22 Robiglio *et al*. (2010) op. cit.
23 Dkamela (2011) op. cit.; Cerutti *et al*. (2010) op. cit.
24 Dkamela (2011) op. cit.
25 Assembe Mvondo, S. (2009) State failure and governance in vulnerable states: an assessment of forest law compliance and enforcement in Cameroon, *Africa Today* 55 (3), 85–102.
26 Cerutti *et al*. (2010) op. cit.
27 Dkamela (2011) op. cit.
28 REM (2010) *IM-FLEG Cameroon: Progress in tackling illegal logging in Cameroon*, Cambridge, UK and Yaoundé, Cameroon: Resource Extraction Monitoring.
29 Decree No. 2005/0577 /PM of 23 February 2005, Arête No. 0070/MINEP of 22 April 2005, Arête No. 0001/ MINEP of 3 February 2007, all on EIA.
30 Dkamela (2011) op. cit.
31 The 1994 Forest Law permits a local council to create its own forest 'estate' (Domaine privé de la Commune) in the permanent forest domain, on the condition that it prepares a management plan that is approved by the Forest Administration.

32 Dkamela (2011) op. cit.
33 Dkamela (2011) op. cit.
34 Brown, H.C.P., Nkem, J.N., Sonwa, D. and Bele, Y. (2010) Institutional adaptive capacity and climate change response in the Congo Basin forests of Cameroon, *Mitigation and Adaptation Strategies for Global Change* 15, 263–82.
35 Brown *et al.* (2010) op. cit.
36 Law No. 94/01 of 20 January 1994 on Forestry, Wildlife and Fisheries.
37 Topa *et al.* (2009) op. cit.
38 Law No. 96/12 of 5 August 1996 on Environmental Protection.
39 Cerutti *et al.* (2008a) op. cit.; Dkamela (2011) op. cit.; Topa *et al.* (2009) op. cit.
40 Topa *et al.* (2009) op. cit.
41 Signatories included France, Canada, Germany, the Netherlands, the United Kingdom, the European Union, the African Development Bank, the World Bank, the Food and Agriculture Organization, the United Nations Development Programme, World Wildlife Fund, the International Union for the Conservation of Nature, the Netherlands Development Organization.
42 Topa *et al.* (2009) op. cit.
43 Topa *et al.* (2009) op. cit.
44 Dkamela (2011) op. cit.
45 Topa *et al.* (2009) op. cit.
46 Other forest policy instruments used in Vietnam, such as community forest management, spatial planning, and protected areas are not included in this analysis.
47 Dkamela (2011) op. cit.
48 Interviews conducted by J. Kuiper, 2011: 2; Cerutti *et al.* (2008a) op. cit.; Singer, B. (2008) *Cameroonian forest-related policies: A multisectoral overview of public policies in Cameroon's forests since 1960*, Institut d'EtudesPolitiques.
49 Dkamela (2011) op. cit.
50 Dkamela (2011) op. cit.
51 Brown *et al.* (2011) op. cit.
52 Topa *et al.* (2009) op. cit.
53 Topa *et al.* (2009) op. cit.; Brown, H.C.P. and Lassoie, J.P. (2010) Institutional choice and local legitimacy in community-based forest management: lessons from Cameroon, *Environmental Conservation* 37 (3), 1–10.
54 Cerutti *et al.* (2008a) op. cit.
55 Dkamela (2011) op. cit.
56 National directive for management of protected areas.
57 Topa *et al.* (2009) op. cit.
58 As noted by Topa *et al.* (2009): 'WWF used a metric known as "protected area management effectiveness" (PAME) and scored each area using the PAME Tracking Tool, which is based on 30 indicators that provide a comprehensive assessment of management effectiveness. The tool, developed jointly by WWF and the World Bank, is consistent with management effectiveness recommendations of the World Commission on Protected Areas and with the Global Environment Facility's monitoring and evaluation policies.'
59 Topa *et al.* (2009) op. cit.; Dkamela (2011) op. cit.
60 Oyono, R., Kouna, C. and Mala, W. (2003) Benefits of forests in Cameroon. Global Structure, issues involving access and decision-making hiccoughs, *Forest Policy and Economics* 7 (2005), 357–68.
61 Singer (2008) op. cit.

62 Curran, B.E. (2009) Are Central Africa's protected areas displacing hundreds of thousands of rural poor?, *Conservation and Society* 7 (1), 30–45.
63 Curran (2009) op. cit.
64 Interviews conducted by J. Kuiper, 2011: 9.
65 Dkamela (2011) op. cit.
66 Interviews conducted by J. Kuiper, 2011: 9, 12.
67 Singer (2008) op. cit.
68 Singer (2008) op. cit.
69 Curran (2009) op. cit.
70 Interviews conducted by J. Kuiper, 2011: 10; Singer (2008) op. cit.
71 Singer (2008) op. cit.
72 Interviews conducted by J. Kuiper, 2011: 10; Dkamela (2011) op. cit.; Topa *et al.* (2009) op. cit.
73 Interviews conducted by J. Kuiper, 2011: 10; Singer (2008) op. cit.; Oyono *et al.* (2003) op. cit.
74 Topa *et al.* (2009) op. cit.
75 Interviews conducted by J. Kuiper, 2011: 8, 10.
76 Topa *et al.* (2009) op. cit.
77 Topa *et al.* (2009) op. cit.
78 Cerutti *et al.* (2008a) op. cit.
79 Topa *et al.* (2009) op. cit.
80 Cerutti *et al.* (2008a) op. cit.; Dkamela (2011) op. cit.; Mertens, B., Steil, M., Nsoyuni, L.A., Shu, G.N. and Minnemeyer, S. (2007) *Interactive Forestry Atlas of Cameroon. Version 2.0*, Washington: World Resources Institute.
81 Cerutti, P.O., Nasi, R. and Tacconi, L. (2008b) Sustainable forest management in Cameroon needs more than approved forest management plans, *Ecology and Society* 13 (2).
82 REM (2010) op. cit.; Topa *et al.* (2009) op. cit.
83 Interviews conducted by J. Kuiper, 2011: 18.
84 Dkamela (2011) op. cit.; Topa *et al.* (2009) op. cit.
85 Interviews conducted by J. Kuiper, 2011: 1.
86 Interviews conducted by J. Kuiper, 2011: 8.
87 Interviews conducted by J. Kuiper, 2011: 10.
88 Interviews conducted by J. Kuiper, 2011: 10; Singer (2008) op. cit.
89 Interviews conducted by J. Kuiper, 2011: 2, 15.
90 REM (2010) op. cit.
91 Interviews conducted by J. Kuiper, 2011: 9.
92 Cerutti *et al.* (2008b) op. cit.
93 Topa *et al.* (2009) op. cit.
94 Ordinance No. 74/1 of 6 July, 1974; Dkamela (2011) op. cit.
95 Article 9 of Decree No. 76/165 of 27 April 1976; Costenbader, J. (2009) *Legal Frameworks for REDD: Design and implementation at the national level*, Gland, Switzerland: International Union for the Conservation of Nature; Nguiffo, S., Etienne, P. and Mballa, N. (2009) *Land rights and the forest peoples of Africa*, Moreton-in-Marsh, UK: Forest Peoples Programme.
96 Costenbader (2009) op. cit.
97 Law No. 79/05 of 29 June 1979 pertaining to the land and estate regime.
98 Cerutti *et al.* (2008a) op. cit.; Costenbader (2009) op. cit.
99 REM (2010) op. cit.

100 REM (2010) op. cit.; Dkamela (2011) op. cit.
101 Interviews conducted by J. Kuiper, 2011: 2.
102 Cerutti *et al.* (2010) op. cit.
103 Dkamela (2011) op. cit.
104 Topa *et al.* (2009) op. cit.
105 Cerutti *et al.* (2008a) op. cit.
106 Dkamela (2011) op. cit.
107 Alemagi, D., Hajjar, R., David, S. and Kozak, R.A. (2011) Benefits and barriers to certification of community-based forest operations in Cameroon: an exploratory assessment, *Small-scale forestry* (in press).
108 Cerutti, P.O. and Lescuyer, G. (2011a) *The domestic market for small-scale chainsaw milling in Cameroon*, Bogar Barat: Center for International Forestry Research.
109 Cerutti, P.O., Tacconi, L., Nasi, R. and Lescuyer, G. (2011b) *Legal vs. certified timber: preliminary impacts of forest certification in Cameroon*, Bogar Barat: Center for International Forestry Research.
110 Singer (2008) op. cit.
111 Interviews conducted by J. Kuiper, 2011: 10.
112 Alemagi *et al.* (2011) op. cit.
113 Law No. 96/06 of 18 January 1996 to amend the Constitution of 2 June 1972.
114 Currently, only six local council forests are gazetted (Dimako, Djoum, Gari Gombo, Messondo, Mouloundou and Yokadouma), totalling about 153,500 hectares, of which 121,000 hectares are managed according to an approved plan. Another 11 forests (nearly 247,000 hectares) are at various stages of gazetting, and a further four (nearly 92,000 hectares) are preparing to initiate gazetting (MINFOF (2011) Le progrès de la foresterie communal au Cameroun. Yaounde, Cameroon: Ministry of Flora and Fauna).
115 Topa *et al.* (2009) op. cit.
116 Interviews conducted by J. Kuiper, 2011: 7.
117 Topa *et al.* (2009) op. cit.; Singer (2008) op. cit.; Cerutti *et al.* (2008a) op. cit.
118 Interviews conducted by J. Kuiper, 2011: 2, 7, 11, 13.
119 Interviews conducted by J. Kuiper, 2011: 4; Cerutti *et al.* (2008a) op. cit.
120 Dkamela (2011) op. cit.
121 Brown and Lassoie (2010) op. cit.
122 Brown, H.C.P., Nkem, J.N., Sonwa, D.J. and Bele, Y. (2009) *Instutional dynamics and climate change in the Congo Basin forest of Cameroon, West Africa*, Copenhagen, Denmark: IARU International Scientific Congress on Climate Change: Global Risks, Challenges and Decisions, 10–12 March 2009.
123 Topa *et al.* (2009) op. cit.
124 Interviews conducted by J. Kuiper, 2011: 7.
125 Brown *et al.* (2011) op. cit.
126 Somorin *et al.* (2012) op. cit.
127 Brown *et al.* (2011) op. cit.
128 Brown *et al.* (2011) op. cit.
129 Cerutti *et al.* (2008a) op. cit.
130 Lang, C. (2011) *Two critiques of REDD in Cameroon, from Forest Peoples Programme and CIFOR*, REDD monitor; Freudenthal *et al.* (2011) op. cit.
131 Freudenthal *et al.* (2011) op. cit.
132 Robiglio *et al.* (2010) op. cit.
133 Brown *et al.* (2010) op. cit.

134 Brown *et al.* (2011) op. cit.
135 Brown *et al.* (2010) op. cit.
136 Sonwa, D., Nkem, J.N., Idinoba, M.E., Bele, M.Y. and Jum, C. (2012) Building regional priroties in forests for development and adaptation to climate change in the Congo Basin, *Mitigation and Adaptation Strategies for Global Change* 17, 441–50.
137 Topa *et al.* (2009) op. cit.
138 Brown *et al.* (2011) op. cit.
139 Lang (2011) op. cit.; Dkamela (2011) op. cit.
140 Sonwa *et al.* (2012) op. cit.

8 Case study
Peru

Felix von Blücher, Nicolien van der Grijp, Joyeeta Gupta and Patricia Santa Maria[1]

8.1 Introduction

Peru has three major natural regions, consisting of the coastal area (*costa*), the Andean Highlands (*sierra*), and the Amazon Basin (*selva*). About 68 million hectares of the Amazonian forests, the world's largest tropical forests, are sited in Peru, covering 53 per cent of its territory.[2] Peru is the fourth largest tropical forest country in the world, and has four million hectares of dry forests located mainly on the Northern coast and in mountain range valleys.[3]

Peru, an emerging economy, is home to approximately 29 million people, of which 73 per cent are urban and 27 per cent rural,[4] including 1,497 indigenous communities.[5] The service industry provides 40 per cent of national income,[6] whereas timber provides 1 per cent.[7] The per capita GDP amounted to USD 8,500 in 2008, its annual growth rate to 9.8 per cent,[8] but 34.8 per cent of the population still lived in poverty in 2009[9] which includes 21.1 per cent of the urban and 60.3 per cent of the rural Peruvians.[10] Peru is a democratic republic with a multi-party system, including 25 regions, 195 provinces and 1,833 municipalities. Since 2002, the government has been decentralizing powers to subnational authorities.

The Peruvian Amazon has large, non-fragmented primary forests with high biodiversity[11] facing increasing deforestation and degradation threats.[12] Deforestation began in the 1960s to liberate local people from poverty,[13] through road construction and agricultural expansion. Terrorism in the late 1980s slowed down economic and agricultural activities.[14] Deforestation rates differ according to sources;[15] between 0.18 and 0.35 per cent of forest cover is lost annually,[16] which is moderate compared to most tropical forest countries. However, these rates may increase as a result of economic growth and investment plans[17] and 91 per cent of these forests may have disappeared by 2041 under a pessimistic scenario.[18] Current deforestation accounts for 47.5 per cent of national GHG emissions.[19]

This layered case study focuses on the national level and the Ucayali region[20] in the Eastern lowlands of the Peruvian Amazon Basin, bordering Brazil in the East (see Figure 8.1), which has the highest deforestation rate in Peru.[21] Other highly deforested regions are San Martin, Amazonas, Loreto and Junin. Ucayali has four provinces (Atalaya, Coronel Portillo, Padre Abad

164 *Felix von Blücher et al.*

Figure 8.1 Map of Peru and the case study area of Ucayali region

and Purús) and 16 per cent of its population are indigenous people.[22] With a 5 per cent annual economic growth rate since 2004 primarily due to the forest industry, poverty has decreased, but immigration from the neighbouring Andean region has increased. Pucallpa, Ucayali's capital, is the centre of the Peruvian timber industry.

In accordance with the analytical framework (see 1.5), this chapter discusses driving forces for deforestation and forest degradation (8.2), the forest policy context (8.3), effectiveness and fairness of policy instruments (8.4), implications for REDD (8.5), and draws inferences (8.6).

8.2 Driving forces of deforestation and forest degradation

As shown in Table 8.1, major proximate drivers of deforestation in Peru include agriculture, infrastructure development, and legal and 'illegal' logging. About 80–90 per cent of deforestation[23] is caused by agricultural expansion by subsistence migrants using slash-and-burn techniques. Such expansion includes small-scale cattle ranching, which requires relatively large feeding grounds, as well as illegal cultivation of coca. In addition, oil palm plantations are being promoted as large-scale agro-industry and could grow in the future.[24] Deforested land yields low-fertility soils, consequently farmers (and ranchers) need to continuously find new land, using at least 15–20 hectares during their lifetime. Sometimes these activities are legal, more often they are not.[25]

Case study: Peru 165

Table 8.1 Drivers of deforestation and forest degradation in Peru and the Ucayali region

Proximate Drivers	Underlying Drivers
Local to national	*Local*
Agriculture: *slash-and-burn techniques; palm oil*, shade-grown coffee, *cocoa*; cattle ranching	Economic: *poverty*
	Local to national
Extraction: *timber and non-timber forest product extraction, unsustainable logging, 'illegal' logging*	Cultural: migrants have different values, for government Amazon is a wilderness to be civilized
Infrastructure: *roads, settlements,* communication and electric infrastructure; hydro power	*National*
	Demographic: increase and *migration from highlands, urbanization*
Industry: oil, gas, and gold mining	Economic: *growth, markets,* commercialization, *low timber and timber product prices*
Biophysical: low soil fertility, forest fires	Technological: *poor agricultural practices, no intensification of ranching, low efficiency of wood processing, mechanization of timber*
	Non-forest policies: *agricultural credits,* spatial policy
	Other institutional factors: *poor policy implementation, regulatory overburden for timber concessions; territorial disputes with indigenous peoples; corruption*
	Global
	Demand/trade/consumption at global level for food, fibre, timber and other forest products; biofuel; climate change

NB. The drivers which apply to the specific context of Ucayali are marked in italics

Infrastructure development is responsible for 5–10 per cent of deforestation.[26] It includes urban settlements, transport and communication infrastructure, and mineral and oil exploitation.[27] Satellite imagery reveals that 75 per cent of total forest damage is found within a 20-km distance from the nearest road.[28] Infrastructure facilitates agricultural expansion and dispersed production by reducing transaction costs. Oil, gas and gold explorations and exploitation could potentially become an important driver of deforestation and cause conflict with indigenous communities.[29]

The timber industry extracts high-yield species sometimes in remote areas, mostly in response to global timber markets.[30] In less remote areas, timber extraction often leads to local extinction of species traded mainly in domestic markets, contributing to forest degradation. There is often no proper forest management and regrowth of these species cannot be assumed. Fuel wood and charcoal demand provides a small contribution to forest degradation, as it is mostly met by waste wood, a by-product of cutting and processing industrial timber.[31] Legal logging through timber concessions should meet forest management criteria, supervised by the Forest Resources Supervisory Agency *(Organismo Supervisor de Recursos Forestales)*. But most logging is 'illegal'.

In Ucayali, the driving factors for deforestation are representative for the processes occurring in the Peruvian Amazon. Infrastructure and roads facilitate access and migration into the rainforest,[32] leading to selective logging, land clearance, agriculture (subsistence, palm oil and coca leaf), fallow lands and then pastures. Energy infrastructure may become important in the future.

Underlying driving factors include rural poverty and the need to exploit forest resources for income generation,[33] and migration of poor people from the Andes into the Amazon region partly in response to climate variability, which exacerbates living conditions in the Andean highlands.[34] In the agricultural sector, low-yield crops, high prices of fertilizers and fragile soils lead to extensive farming and slash-and-burn techniques.[35] The low global price of timber, illegal logging and low levels of local value adding to timber products reduce the incentive for adequate forest management.[36] Technological factors in the logging sector include inefficient cutting and processing (using only about 20 per cent of the trunk[37]) and increased mechanization, as the Free Trade Agreement with the US facilitates the cheap import of large logging machinery.[38]

Further reasons for deforestation and forest degradation include national development and energy policies, former agricultural policies such as subsidized agricultural credits and guaranteed minimum prices, policies for property formalization which create perverse incentives requiring people to deforest and develop agriculture in order to get land titles, and the lack of institutional coordination and planning. Inconsistent national policies are to some extent also caused by the national government's ignorance about the Amazon region.[39] Poor monitoring and enforcement[40] imply that rules about forest management and prohibitions of deforestation in certain areas are not adequately implemented.[41] Poor enforcement can be attributed to low human, financial and technical capacity, poor salaries for officials and widespread corruption.[42] Finally, cultural factors are related to the perception of the Amazon as a region that needs to be civilized and harnessed for economic gain, and the import of Andean mountain agricultural practices into the Amazon region.[43] The views of indigenous and local people have become marginalized in this process.

The underlying driving factors in Ucayali are similar to those at the national level.[44] Timber is still used without value-added processing[45] and subsidized credit schemes are an attempt to attract more investors for palm oil plantations.[46]

8.3 The forest policy context

This section discusses the organizational framework, the evolution of forest policy, and the influence of international policies and bodies.

8.3.1 The organizational framework

Until 2008, the National Institute of Natural Resources (*Instituto Nacional de Recursos Naturales*), linked to the National Ministry of Agriculture *(Ministerio de Agricultura y Ganadería)*, controlled natural resource use and land-use change. Since the concentration of power led to ineffective and inefficient resource management due to lack of interest in self-monitoring,[47] the organizational framework has been restructured.

The Ministry of Agriculture, with competence on agricultural policy, rural development, and forest and wildlife resources, controls the Bureau of Forest and Wildlife *(Dirección General Forestal y de Fauna Silvestre)*. This Bureau formulates and coordinates the planning, management and supervision of forest and wildlife resources and its implementation by regional offices that provide forest resource user rights (e.g. concessions, permits) and supervise remaining forest lands.

The Ministry of Environment, with authority on policies for the strategic development of natural resources, the evaluation, design and establishment of compensation schemes or payments for environmental services including REDD (see Chapter 4), controls the National Agency for the Management of National Protected Areas *(Servicio Nacional de Áreas Naturales Protegidas)*, which manages and supervises protected areas and promotes, regulates, and grants rights over environmental services in the areas concerned. The Forest Resources Supervisory Agency *(Organismo Supervisor de Recursos Forestales)* supervises the compliance by holders of access and user rights with the criteria established for forest areas under concession, private properties and indigenous communities.

Since 2002, decentralization of powers has begun and several competencies of natural resource management were transferred to regional governments and decentralized offices of the Bureau of Forest and Wildlife. While national protected areas still fall under the central government, other forests fall under the Bureau of Forest and Wildlife and the regional governments. Despite the institutional restructuring and decentralization of powers, the Agricultural Ministry remains the most powerful forest agency, often allegedly prioritizing agriculture over forest conservation.[48] Many other national government ministries[49] and agencies[50] are indirectly concerned with forests.

The governmental funding organisation PROFONANPE focuses on conservation projects related to national protected areas *(Sistema Nacional de Áreas Protegidas por el Estado)*, and includes REDD pilot projects within protected areas. Its colleague organisation FONDEBOSQUE finances sustainable forestry management activities. The sources of funds are bilateral and multilateral cooperation (including the Global Environment Facility, the World Bank, KfW and others). The public–private organisation FONAM was created in 1997 and funds plans, programmes, and projects that support environmental protection and sustainable use of natural resources, including forests. It promotes the clean development mechanism (forestation/afforestation) and more recently REDD projects. The funding sources include external debt reconversion, non-refundable contributions of governments, international organizations, and other contributions from public or private entities (national or international).

Other organizations engaged in forest governance include national environmental NGOs *(Derecho, Ambiente y Recursos Naturales, Sociedad Peruana de Derecho Ambiental)*, international nature protection NGOs (World Wildlife Fund, The Nature Conservancy, Conservation International), indigenous peoples' organizations *(Asociación Interétnica de Desarrollo de la Selva Peruana)*, research groups and private companies promoting forest certification, alternative forest practices, and REDD projects (e.g. *Bosques Amazónicos*).

8.3.2 *The evolution of forest policy*

To cope with agricultural policies that encouraged colonization of the Amazon, the Forest Law of 1975 granted timber harvesting rights through permits, authorizations and contracts,[51] and provided private land titles for agricultural and cattle ranching purposes.[52] Forest management was a state task implemented through reforestation financed by a volume-based tax on extracted timber from the contract holders. This contractual practice, a product of Peru's populist political history,[53] resulted in a chaotic forest sector, poor state protection and limited effectiveness of reforestation activities.[54]

The current policy framework is formed by the Constitution, the Organic Law for Sustainable Use of Natural Resources (1997), the Forest and Wildlife Law (2000), and the General Environmental Law (2005).[55] Other relevant laws include the National Protected Areas Law, Land Law, General Law of Hydrocarbons, General Mining Law, and regulations about road construction.

The Constitution labels all natural resources, renewable and non-renewable, as national patrimony. The state is declared responsible for protecting biological diversity in national protected areas and sustainable development of the Amazon Basin. User rights may be granted to third parties, and private or communal ownership is allowed over agricultural land.

The 1997 Organic Law for the Sustainable Use of Natural Resources defines natural resources as those natural components that satisfy human needs, including environmental services, and have a market value. The state controls natural resources, but indigenous and farmer communities are allowed to use local natural resources for subsistence, with preferential rights for indigenous communities within their titled territories. For commercial exploitation, concessions, licences, authorizations, permits, access and exploitation contracts are required. The General Environmental Law establishes principles to ensure the right to a healthy environment, sustainable resource use and sustainable development, and refers to forest and wildlife resources.

The 2000 Forest and Wildlife Law formalizes and professionalizes forest management.[56] It emphasizes sustainable forest management, intensification of the timber industry, 'scientific forestry', and enhancing forest revenues for the state. Forests include natural and planted forests, and exclude those forest lands suitable for potential agricultural activities. All forests are considered national patrimony. This Law replaced contracts based on the 1975 Forest Law by a concession-based system, and volume-based fees by annual area-based taxes to increase the value of production per area logged.[57] Persistent problems in the forest sector, and pressure from the United States to curb 'illegal' logging, led to amendments in 2009. However, indigenous organizations protested about the two main changes proposed (potential privatization of communally held lands and forests and criminalization of informal logging activities).[58] Following violent clashes with the police in Bagua, the amendments were annulled. Since 2009, a new Forest and Wildlife Law has been discussed and should soon enter into force. It will not bring about major substantive changes to the existing law, but introduces institutional reforms such as the creation of the Forest and Wildlife National Service (*Servicio Nacional Forestal y de Fauna Silvestre*), replacing the current Bureau of Forest and Wildlife. This new service will be led by a board of representatives from the public sector and civil society and its implementation will rely on the consent of the indigenous organizations, which demand stronger general rights of consultation and recognition of their remaining territorial claims.

8.3.3 *The influence of international treaties and bodies*

Peru is party to all relevant international agreements: CBD, UNFCCC, KP, UNCCD, ITTA, CITES, RAMSAR, WHC, NLBI, UNPFII; and influenced by UN-REDD (see Chapter 3). Once ratified, the legally binding obligations become national law after implementation through regulations.[59] However, international treaties are considered to have limited influence,[60] except perhaps in terms of reporting responsibilities.[61] Further of relevance is the Forestry Annex of the 2007 US–Peru Free Trade Agreement requiring Peru to comply with minimum requirements and standards for transparency and forest governance.[62] This agreement led to the restructuring of forest governance, the criminalization of illegal logging, the provision of greater funds and capacity to supervise and manage protected areas, and foreign investments.[63] Currently it is influenced by ongoing policies on REDD (see Chapter 4).

8.4 Key forest policy instruments and their analysis

This section presents and analyses a selection of regulatory and market-based instruments[64] (using the numbering system developed in Table 2.3) in terms of their effectiveness and associated fairness issues, given the existing drivers of deforestation and forest degradation, using the [+ + − −?] method developed in 1.5.

Decentralization (ii): The *Organic Law of the Regional Governments* allows regional governments to make policies consistent with national laws.

Furthermore, the regional Bureau of Forest and Wildlife offices, which grant rights over their respective forest and wildlife resources and issue concessions, permits and authorizations, are now directed by the regional governments.[65] Duties are related to respecting property rights of indigenous communities as well as the formulation, development and implementation of environmental programmes in forested areas.[66] These changes imply a transition towards more inclusive management of forest resources involving all key stakeholders. However, the decentralization process is incomplete. Although administrative tasks have been transferred to regional governments, the latter are still normatively, technically and financially dependent on the national ministries.[67] The regional governments' lack of resources impedes their potential to manage the forests sustainably [68] and assume their legal powers, and hence no regional agrarian or forest laws have been formulated so far.[69] Municipalities are not empowered to and generally do not implement or enforce state policy in Peru.[70] However, they facilitate the access to natural resources by clarifying the legal status of the lands, thereby coordinating different land use activities[71] generally on their own authority. They lack natural resource management capacity, and national and provincial bodies are not able to monitor ground-level activities. This leads to a confused situation especially in areas that are managed by indigenous populations who have their own decision-making process.[72]

Decentralization has also led to the creation of forest management committees at the local level to provide information on forest resource use and to coordinate between local stakeholders (farmers, indigenous people, timber industrialists, etc.) and with higher levels of government. However, there is a mismatch between competencies and capacities transferred to lower levels of government. In addition, in Ucayali, regional and local officials do not see deforestation and forest degradation as a problem,[73] and instead promote agro-industrial crops like oil palm. Ways of enlarging financial, technical and administrative capacities need to be sought to allow subnational entities to implement their tasks. [+ −]

Spatial planning (v): According to the Forest and Wildlife Law, land-use change is prohibited without authorization.[74] Spatial planning and zoning is based on categorizing land-use criteria, such as slope, topography, depth of fertile soil, drainage, etc. In this so-called 'major use capacity system' agriculture is prioritized over forestry purposes. Land is declared as suitable for protection when it does not fulfil the minimum requirements for agriculture, cattle ranching, timber, or non-timber forest products. While, in theory, land zoning should be based on *in situ* assessment of land-use criteria (slope, topography, soil fertility, etc.), in practice there is little *in situ* assessment. In addition, zoning plans are often not enforced, which undermines effectiveness and results in agricultural expansion.

The Ministry of Agriculture has divided forests in the Peruvian Amazon Basin into: 1) forests in protected areas and other conservation areas (15.9 million hectares), 2) forests in indigenous and farmer communities and territorial reserves[75] (13.2 million hectares), 3) timber and non-timber concessions (8.6 million hectares), 4) forests for permanent production for concessions (12.3 million hectares), 5) wetlands in the Amazon (3.1 million hectares), and 6) non-classified areas. Currently, 67 protected areas have been designated, covering 27 per cent of the Peruvian Amazon. Protected areas targets the proximate drivers of deforestation and forest degradation, including agricultural expansion, cattle ranching and commercial timber extraction, and also underlying drivers, such as agricultural migration, poor agricultural practices and the perception of the Amazon as an empty space that should be civilized. National protected areas are considered effective in protecting the forests, with only 1 per cent of all deforestation and 2 per cent of all degradation occurring in these areas.[76]

In Ucayali, most areas are classified as forests for permanent production, timber concessions, indigenous territories, or are still non-classified. Only three national parks that border into the region and most biodiversity hotspots are found within community territories.[77] Each protected area should be managed according to an individual management plan, including conservation objectives, management strategies and planning, and coordination and participation arrangements with local people. These protected areas can cover large areas. For example, the relatively small protected area of El Sira that borders Ucayali stretches across 600,000 hectares, and covers 1.6 million hectares including the buffer zone.

Local forests of up to 500 hectares are available for sustainable use by rural and town populations through a system of authorizations and permits. This instrument has marginal significance, as valuable species have usually long been extracted in these easily accessible areas, reducing the benefits local forests may yield to the communities. Ucayali has 3.8 million hectares[78] of forests for permanent production in the provinces of Padre Abad, Coronel Portillo, and Atalaya. However, competing claims for the use of natural resources, by actors involved in mining, conservation and timber extraction, and the lack of inter-sectoral coordinated planning has led to social conflicts.[79]

Although protected areas are quite effective, (a) most protected areas are established in remote areas where there is little threat of deforestation and forest degradation, while many hotspots of biodiversity that face higher risks are left out of the protected area system; (b) human settlements and private properties continue to exist within protected areas, albeit without legal right to use forest resources under the Law of National Protected Areas and with only very little promotion of alternative land-use or income creation options by the state or the park management;[80] (c) protected areas are not always established in consultation with local communities about the disadvantages and advantages it will bring to them and the possibility of opting out. Furthermore, (d) most protected areas have few management officers and park rangers and some parks even have neither a management office nor park rangers, reducing the capacity to supervise and promote alternative land-use options;[81] (e) there may be leakage of activities related to deforestation and forest degradation to other areas, suggesting that low deforestation and degradation rates in protected areas are mainly a result of their inaccessibility or inappropriateness for other uses.[82] However, it may only be a matter of time before protected areas are under increased threat if management capacities are not improved.[83] In addition, management of protected areas is not always equitable, denying settlements in parks and buffer zones the right to access resources. Although the latter are to be compensated, there are no indications of whether this rule is implemented or not. Protected areas are often perceived as coming at the cost of local populations leading to low social acceptance.[84]

The research indicates a need to reconsider the current zoning system in consultation with local people to revise the major use capacity system, to ensure better protection of biodiversity hotspots, and relate better to *in situ* assessment and enforcement. Providing local and indigenous communities with alternative sources of income is essential to ensure that they too have a stake in protection of the area and that protection does not come at their cost. The instrument of local forests was meant to help provide local people with alternatives. However, it has not been effectively implemented. It could be questioned whether this instrument needs further development. Furthermore, enhanced co-management with NGOs may increase the resources for maintenance. Monitoring is essential to ensure that the areas are adequately protected. [+ – –]

Forest logging/concessions (vi): Timber concessions are an instrument of state control of commercial timber extraction, leading to a concentration of concession lands, more efficient timber exploitation, sustainable land management, a long-term perspective, and revenues for the state. Timber concessions target unsustainable logging, illegal/informal logging, and dispersed logging that open up the forest for further exploitation by agricultural migrants. Exclusive rights for sustainable forest exploitation based on a forest management plan are granted for specific public lands. They include forest concessions for different purposes, including timber, non-timber forest products, conservation, reforestation, and ecotourism; 40-year, transferable timber concessions are the most important and popular policy instrument. They are granted in open auction-like competitions based on bids for annual extraction fees per hectare and cover between 10,000 and 40,000 hectares. Annual royalty fees range between USD 0.5–3 per hectare[85] and they must be paid for the whole concession area regardless of whether timber is exploited or not. Logging practices must conform to the forest management plan and annual plan of operation previously approved by the Bureau of Forest and Wildlife. Supervision of logging practices is done by the Forest Resources Supervisory Agency. A continuation of timber extraction rights depends on compliance with the forest management plan and the annual plan of operation made as well as the payment of timber extraction fees.[86] Concessions for other purposes than logging are less popular.

Permits based on the Forestry and Wildlife Law may allow for commercial timber extraction on forest lands of indigenous and farmer communities and privately owned lands.[87] Such permits also require a forest management and an annual operation plan. Authorizations are usually one-time logging permits for state-controlled forest lands, with the purpose of facilitating infrastructure development and urban expansion.[88]

Forest management has improved significantly within active concessions.[89] Deforestation rates within forest concessions have remained low, accounting for approximately 8 per cent of overall deforested areas in the Peruvian Amazon.[90] However, concessions are not an effective instrument under all circumstances. Of the 140 concessions issued in Ucayali between 2002 and 2004, only 12 are still active,[91] while in other regions, the success rate of concessions is only marginally higher.[92] This is because concessions were unable to raise the resources for paying the royalties, or people did not comply with the management and annual operation plans, because they either wanted to reduce costs and increase short-term profits, or they had no expertise in forest management.[93]

Timber extraction according to the state-mandated management conditions requires expensive equipment[94] and seasonal and location-specific profits vary due to weather conditions.[95] Controlling agricultural invasion into forest concessions has been problematic as roads provide easy access to agricultural migrants and there is little state support to keep invaders out. Furthermore, sometimes concession lands overlapped with private or indigenous land, especially in Ucayali, leading to conflict.[96] Finally, the risks involved lead banks to withhold investment loans for concessions.[97]

In addition, logging activities have shifted from concessions to cheaper indigenous lands, private properties and informal logging in unauthorized areas. Only 20 per cent of the timber in the market comes from concessions,[98] and 70–80 per cent is extracted illegally or not in compliance with concession rules because of insufficient supervision of forest areas outside concessions by the poorly financed Bureau of Forest and Wildlife. Illegal 'official' documents are used to make timber from illegal sources appear as legal;[99] these documents are usually not checked *in situ* but on the road, making it impossible to verify the exact origin of the timber. As a result, logging remains dispersed. Furthermore, the control, professionalization and formalization of the timber industry itself is problematic,[100] confirmed by increased deforestation and degradation of 400 per cent just outside the concession areas.[101]

Local forests for small-scale timber extraction are few, and fail to provide a real alternative to small harvesters as valuable timber has long been extracted, pushing most timber extractors into informality.[102]

Research indicates that the design of concessions could be improved by ensuring the quality and economic viability of concession lands; creating low-cost credit schemes; preventing agricultural invasion; protecting the buffer zone; and providing *in situ* supervision of timber extraction occurring in concessions, private properties and indigenous communities to reduce doctoring of official documents. As long as most timber is cut illegally outside concession areas and without payment of the area-based royalty fees, volume-based extraction fees could be levied for all logs transported. This would level the playing field for legally cut wood and reduce the incentive for 'illegal' logging. The increased revenues could be used to finance the other measures. Permits and authorizations need to be accompanied by some clear guidance regarding what a sustainable management plan implies in practice and assistance to help those concerned in implementing these. [+ – –]

Land rights (vii): Land titling is granted as an exception for land suitable for agriculture or livestock on the condition of continuous agricultural activities.[103] However, the condition of continuous profitable use of cleared land is often not enforced, creating an incentive to clear land for the mere purpose of receiving land titles.[104]

Land titling, zoning, and exceptions to the prohibition to change land use in combination provide incentives for deforestation. Once forest land is declared as suitable for agricultural purposes according to the major use capacity system, the prohibition of land use change no longer holds and in order to gain land titles people have to deforest the area. Legally, this process requires prior consultation with the Bureau of Forest and Wildlife and should be guided by scientific criteria and enforced by the regional administrative bodies. In practice, however, migrants deforest areas for agriculture without prior consultation and generally subsequently successfully apply for land titles.[105] The incomplete zoning is often made without *in situ* assessment,[106] and weak enforcement at ground level allows for a posterior legalization of illegal activities.

Providing land titles to indigenous and farmer communities recognizes their existing cultural rights and empowers them, which could indirectly promote forest management.[107] The proportion of land legally assigned to rural communities is limited. Indigenous communities have titles to no more than 15 per cent of the Peruvian Amazon. Technically speaking, indigenous people have 'concessions for specific uses' (and the right to rent land and resources, but not sell),[108] in return for community forest management plans. This tool does not target any specific driver but provides space for indigenous practices that have protected the forests so far,[109] and minimized the risk of forest fires.[110] It recognizes that security of land tenure changes behaviour, and pre-empts (government support for) migration as the lands are occupied and titles have already been claimed.[111] In contrast, in cases where titles were still in dispute, an invasion by agriculturists has occurred.[112]

Overall, only 11 per cent of forest degradation and 9 per cent of all deforestation in 1999–2005 occurred in indigenous territories, despite being populated forest areas.[113] Nevertheless, about 90 per cent of indigenous and farmer communities work with the commercial timber industry, increasing the threat to the forests concerned. The industry often exploits uninformed communities and violates management rules. This can lead to a rise in forest degradation. However, the communities are responsible for the adequate management of these forests and may be vigilant about too much degradation as they depend on them.[114]

Land titling needs to be accompanied by *in situ* assessments which will also be necessary for REDD MRV activities. Land titling for indigenous peoples and local communities has been successful thus far in protecting forests and needs to be completed. Legal recognition of remaining claims (in 2008, 237 indigenous communities were still awaiting recognition and the land claimed by indigenous communities often exceeds their legally recognized territories) could help to prevent forest damage by preventing agricultural invasion and short-term resource exploitation incentives. Additionally the Peruvian government could also consider that similar conservation effects could be achieved by granting rural communities rights over their local forests, as this is an instrument that has not been applied to its full potential. [+ –]

Payment for Ecosystem Services (PES) (xv): The Peruvian government promotes PES as an economic incentive to reduce deforestation and forest degradation. Under the national programme,[115] USD 3.8 per hectare are granted to indigenous and farmer communities (with land titles) that decide to manage their forests complying with a previously elaborated plan contributing to conservation, prevention of illegal logging and sustainable income-generating activities. The areas are chosen by the indigenous communities themselves and the subsidy is guaranteed for five years.

A key bottleneck for PES is the limited resources available, especially in cases where the opportunity costs of land-use change are high. A future extension of the scheme will depend on the availability of international donor funding. These schemes have no MRV procedures in place.[116] [+ –]

CSR/certification (xix): Forest certification targets unsustainable logging practices and low wood prices in the global market by creating demand for sustainable wood, by setting a good example, and by promoting a culture of forest management. It aims at sustainable harvest levels, better regeneration, transparency along the supply chain, and increasing land-use rent from forestry, thus reducing the incentive to change the land use. Forest certification is promoted by the state (by reducing the area-based extraction fees by 25 per cent for concessions whose logging practices are internationally certified[117]) and by international NGOs, consultancies and donors. Forest management standards in certified areas are higher than in the timber concessions, and in Peru FSC decided to only certify forests in concessions and indigenous communities, as they provide long-term security.[118]

In May 2012, 12 forest areas covering 746,275 hectares were FSC-certified in Peru, of which 300,000 hectares are located in Ucayali.[119] In addition, FSC issued 28 chain of custody certificates. However, only 10 per cent of the total area of concessions and 1 per cent of the total Peruvian Amazon Basin are now certified.

Forest certification is potentially effective. After certification, more tree species become lucrative, which reduces the area impacted by timber harvesting.[120] The harvesting of more species and improved forest management and logging techniques lead to higher timber production (6–8 m^3 per ha annually; as against 2 m^3 in non-certified areas)[121] and also better benefits for indigenous communities.[122] Greater supervision of certified areas reduces the risk of invasion by agriculturists or illegal loggers,[123] making certified areas superior to state-run systems.[124] However, certification is expensive, raising the costs of production by more than 25 per cent due to more rigorous infrastructure and planning requirements,[125] is dependent on a market of willing purchasers, and targets only wood for export, which is less than 20 per cent of all timber processed. The recent increase in certified areas in Peru is mainly due to a project[126] led by World Wide Fund for Nature with funds from the US Agency for International Development, which focused on certifying 180,000 hectares in big concession areas.[127] While this project gave financial and technical support, it focused solely on big concessions with a good management record.[128] As certification is expensive, it is out of reach to small-scale loggers and indigenous communities especially in remote areas.[129]

The forest certification potential can be increased by improving access to capital for sustainable logging activities; enhancing credit for logging activities in smaller or remote concessions and indigenous communities; promoting certification in especially vulnerable lands which have high conservation value in cooperation with NGOs and donors; and enhancing access to markets for indigenous communities. [++ −]

NGO-based management (xxv): To open up the possibility of creating public–private partnerships for protected area management, the National Protected Areas Law established the so-called 'administration contracts' to entrust forest activities to non-profit organizations. As of 2011, eight such hybrid management contracts (state-NGOs) were signed and four were under negotiation.[130] NGOs and research institutes are also undertaking capacity building and pilot projects to help indigenous communities make fair deals with timber extraction companies in their territory, adopt modern forest management techniques, and engage in agroforestry projects. However, these efforts have had limited impact thus far.[131]

Management schemes that involve collaboration with third parties enhance the management capacities to protect the forests by increasing the level of expertise and the ability to raise funds and to monitor the area. Furthermore, the participation of NGOs improves the coordination of conservation objectives and the assurance of livelihoods of the local population by introducing alternative land-use programmes and incorporating the local population in active conservation.[132] [++]

8.5 Implications for REDD

In Peru, the implementation of REDD is still in a preparatory stage. During the joint elaboration of REDD-readiness strategies, current policy framework shortcomings have been identified as well as the views of different stakeholders on how to improve this. Furthermore, draft laws with the intention of clarifying property rights over carbon are being elaborated.[133] These are important steps in preparing Peru for REDD, but also for broader reform of the policy framework as different stakeholders are brought together and awareness and capacities are increased. There are 12 REDD pilot projects led by different NGOs and for-profit organizations, mainly in Madre de Dios.[134] These projects are still in an early stage.

REDD can build on lessons learnt from existing forest management (see Figure 8.2). (a) The enabling conditions for REDD implementation need to be enhanced. Peru's national policy framework shows that it is not yet ready for REDD. Protecting forests with MRV instruments requires strong governance structures, which Peru does not yet have. Furthermore, protecting forests at the cost of reducing access to forest resources and employment opportunities for dependent local people will be a critical challenge. REDD requires thus a combination of improved governance at multiple level as well as operating in an equitable manner to enable rural and local development.[135] This calls for engagement with local people,[136] which could be done by paying compensations directly to those managing the land.

(b) REDD needs to be effectively linked with existing mechanisms such as the system of protected areas, concessions, zoning, and land titling, and the associated capacity that has developed to deal with these.[137] Such linking builds on existing capacities in place and allows for multiple goal achievement. For example, (i) protected areas are protected *de jure*, but not always *de facto*. Combining REDD with protected areas could ensure this dual goal achievement. (ii) Combining REDD and concessions could allow for swapping timber for carbon concessions, where REDD payments replace harvest loss,[138] more stringent management criteria are applied, locations in high-deforestation threat areas are chosen, and buffer zones are included. (iii) REDD resources could help support indigenous territories,[139] if these indigenous communities are granted titles that also include carbon rights (which is not the case at present), if indigenous people have *de facto* control over their forests, are well informed, and have access to REDD benefits without huge administrative burdens;[140] especially as implementing REDD activities without indigenous people consent has a high conflict potential.[141] (iv) REDD projects may conflict with national land zoning plans, which focus on whether land is suitable for agriculture or not, rather than zoning land according to threats of deforestation and degradation or conservation value based on biodiversity criteria. Often, forests not suitable for agricultural use face low deforestation risks anyway. REDD project developers must be aware that states might not offer the 'best' areas for REDD projects.[142] (v) REDD resources could be used to strengthen the role of forest management committees at local level. (vi) REDD could support *in situ* assessments which could help both land titling efforts and MRV activities. (vii) REDD could help align different interests (business elites, conservationists, indigenous people), provide incentives for forest conservation and sustainability,[143] and motivate governance reform.[144]

(c) Third, REDD could build upon existing international mechanisms such as the free trade agreement with the US, certification schemes and past

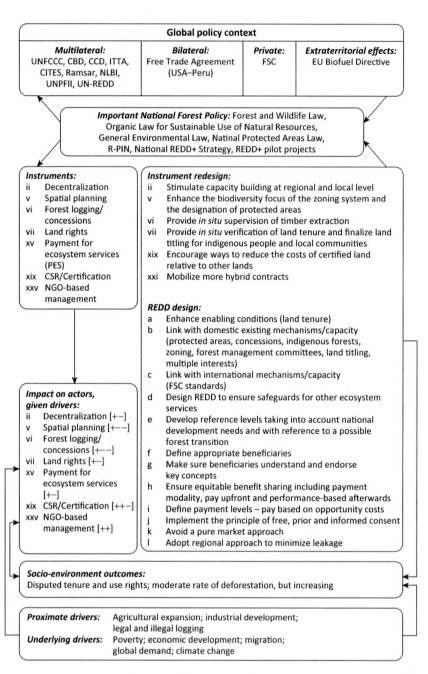

Figure 8.2 Applying the analytical framework to forest and REDD policy in Peru

CDM projects. REDD management should build on and expand FSC guidelines so that both are mutually supportive. (d) REDD also needs to ensure safeguards on other ecosystem services provided by forests and that local equity issues are not compromised. (e) Reference levels need to be based on proposals from NGOs, researchers and other actors taking the forest transition in different parts of the country into account.[145] The elaboration of multiple reference levels combining the subnational and national level can provide incentives for both project level and national level activities. This so-called 'nested approach' is supported by the Peruvian government.[146]

(f) In order to engage in effective benefit sharing, REDD should be able to identify the key direct and indirect beneficiaries. (g) REDD will only be effective if this abstract scheme is explained properly to all stakeholders concerned and endorsed by them. (h) The method of benefit sharing needs to be elaborated upon and could build on upfront payments followed by performance-based activities thereafter. (i) Benefit sharing could also take into account the actual opportunity costs of people in protecting the forests. Performance-based REDD payments imply that REDD activities need to be located in areas under threat of deforestation or degradation. Therefore, payments need to take the opportunity costs into account if they are to effectively reduce deforestation in situations of high risk. This calls for identifying areas and actors with high opportunity costs. REDD may have to promote alternative land uses and distribution mechanisms for compensation payments, which may call for stronger legal rights and responsibilities of the actors involved including forest dwellers. This could alter the incentive structure for land managers to deforest.

(j) The withdrawal of the reformed Forest and Wildlife Law in 2009 due to violent resistance of the indigenous communities has shown that the implementation of REDD activities on land held or claimed by indigenous or rural communities will require close cooperation, consultation and prior informed consent (see Box 4.2) of these communities and other local land managers. (k) Peru is not yet a full market economy and REDD needs to avoid a pure market design. (l) There is need for subnational and/or national approaches to minimize the potential for leakage.[147] Figure 8.2 applies the analytical model to forest and REDD governance in Peru.

8.6 Inferences

This chapter has shown that the effectiveness of Peruvian forest policies thus far has been limited in changing human behaviour. The specific policy instruments aimed at reducing deforestation and forest degradation certainly have an impact, but they are undermined by the incentives set by the general policy framework geared towards agricultural expansion. In addition, there is a lack of strategic planning of long-term development in the Amazon region. Based on the experiences in Peru, it can be argued that the implementation of REDD requires a coherent and consistent national strategy, removal of perverse incentives, and an all-encompassing approach targeting deforestation drivers.

This chapter has identified a number of measures that can be taken to make existing forest protection measures more effective (see 8.4). It has also argued that there are many ways in which potential REDD resources could be used in a tailor-made manner to strengthen the policy process in Peru and other rainforest countries (see 8.5; and Figure 8.2). Rather than redesigning the entire national policy framework, REDD could use existing structures for creating a system that benefits local people instead of facilitating elite capture. It is clear that REDD is far more complex than a simple financial transfer mechanism and will need to be accompanied by incentive structures that alter opportunity cost perception and support long-term thinking.

Notes

1. The authors wish to thank Ms. Sandra Velarde for her useful comments on an earlier version of this chapter.
2. FAO (2010) *Evaluación de Recursos Forestales Mundiales 2010 (Global Forest Resources Assessment 2010)*, Rome, Italy: Informe Nacional Peru.
3. FAO (2010) op. cit.
4. CEPALSTAT. Economic Commission for Latin America and the Caribbean – Database and statistical publications.
5. The figures are from COFOPRI which is responsible for the technical demarcation of communal lands. Information can be found on: www.minem.gob.pe. According to Article 8 of Decree No. 22175, indigenous communities are organizations which originated from tribal groups in the lowland and highland forest, and are constituted by groups of families linked by major elements such as language or dialect, cultural and social characteristics, tenure and common, permanent use of the same territory or in dispersed settlements.
6. MINAM (2010) *El Perú y el Cambio Climático. Segunda Comunicación Nacional del Perú a la Convención Marco de las Naciones Unidas sobre Cambio Climático (Peru and Climate Change. Second National Communication of Peru to the Convention Nations Framework Convention on Climate Change)*, Lima, Peru: Ministerio del Ambiente.
7. FAO (2011a) *State of the world's forests 2011*, Rome: FAO.
8. FAO (2011b) *The state of forests in the Amazon Basin, Congo Basin and Southeast Asia*, Rome: Food and Agriculture Organization of the United Nations.
9. http://www.inei.gob.pe
10. http://www.inei.gob.pe
11. Oliveira, P.J.C., Asner, G.P., Knapp, D.E., Almeyda, A., Galván-Gildemeister, R., Keene, S., Raybin, R.F. and Smith, R.C. (2007) Land-use allocation protects the Peruvian Amazon, *Science* 317, 1233–6; White, D., Velarde, S.J., Alegre, J. and Tomich, T.P. (2005) *Alternatives to Slash-and-Burn (ASB) in Peru, summary report and synthesis of phase II*, Nairobi, Kenya: Alternatives to Slash-and-Burn Programme.
12. Dourojeanni, M., Barandiarán, A. and Dourojeanni, D. (2009) *Amazonía Peruana en 2021. Explotación de recursos naturales e infraestructuras: ¿Qué está pasando? ¿Qué es lo que significan para el futuro?(Peruvian Amazon in 2021. Infrastructure and the Exploitation of Natural Resources: What is happening and what does it mean for the future?)*, Lima, Peru: ProNaturaleza – Fundación Peruana para la Conservación de la Naturaleza.
13. White *et al.* (2005) op. cit.

14 Interviews by F. von Blücher, 2011: 14.
15 Oliveira *et al.* (2007) op. cit.; FAO (2010) op. cit.; FAO (2011b) op. cit.
16 Ramos, N.R. and Domínguez, G. (2008) *Impacto de la producción forestal maderable en la economía de la Región Ucayali, Perú*.
17 MINAM (2011) *Forest Carbon Partnership Fund (FCPF) Peru Readiness Preparation Proposal (R-PP) Form*, Lima, Peru: Ministerio del Ambiente.
18 Dourojeanni *et al.* (2009) op. cit.; cf. less pessimistic figures of Soares-Filho, B.S., Nepstad, D.C., Curran, L.M., Cerqueira, G.C., Garcia, R.A., Ramos, C.A., Voll, E., McDonald, A., Lefebvre, P. and Schlesinger, P. (2006) Modelling conservation in the Amazon basin, *Nature* 440 (7083), 520–23.
19 MINAM (2010) op. cit.
20 Ucayali spans 10.2 million hectares and has about 432,159 inhabitants (2007), which is equivalent to 5 per cent of the national population (http://www.inei.gob.pe).
21 Oliveira *et al.* (2007) op. cit.
22 Ramos and Domínguez (2008) op. cit.
23 Interviews by F. von Blücher, 2011: 4, 14.
24 MINAM (2009) *Causas y medidas de mitigación a la deforestación en áreas críticas de la Amazonía peruana y a la emisión de gases de efecto invernadero (Causes and mitigation measures in critical areas of deforestation in the Peruvian Amazon and the emission of greenhouse gases)*, Lima, Peru: Ministerio del Ambiente.
25 MINAM (2009) op. cit.
26 Interviews by F. von Blücher, 2011: 4, 14.
27 MINAM (2009) op. cit.
28 Oliveira *et al.* (2007) op. cit.
29 Velarde, S.J., Ugarte-Guerra, J., Rügnitz Tito, M., Capella, J.L., Sandoval, M., Hyman, G., Castro, A., Marín, J.A. and Barona, E. (2010) *Reducción de emisiones de todos los Usos del Suelo (Emission reductions from all soil uses)*, REALU Peru Project Report Phase 1. ICRAF Working Paper No. 110, Lima, Peru: ASB, World Agroforestry Centre (ICRAF).
30 Interviews by F. von Blücher, 2011: 25.
31 Interviews by F. von Blücher, 2011: 3.
32 Pokorny, B., Johnson, J., Medina, G. and Hoch, L. (2010) Market-based conservation of the Amazonian forests: Revisiting win–win expectations, *Geoform* 43 (3), 387–401; Ugarte, L.J. (2010) *Migración, carreteras y la dinámica de la deforestación en Ucayali*; White *et al.* (2005) op. cit.
33 MINAM (2009) op. cit.
34 Interviews by F. von Blücher, 2011: 5, 25; see also White, D., Arca, M., Alegre, J., Yanggen, D., Labarta, R., Weber, J.C., Sotelo-Montes, C. and Vidaurre, H. (2005) The Peruvian Amazon. Development imperatives and challenges, *Slash and Burn: The Search for Alternatives,* New York, USA: Columbia University Press.
35 Interviews by F. von Blücher, 2011: 14; see also White *et al.* (2005) op. cit.
36 Pfaff, A., Sills, E.O., Amacher, G.S., Coren, M.J., Lawlor, K. and Streck, C. (2010) *Policy impacts on deforestation lessons learned from past experiences to inform new initiatives*, Duke University: Nicholas Institute for Environmental Policy Solutions.
37 Interviews by F. von Blücher, 2011: 3.
38 Interviews by F. von Blücher, 2011: 11.
39 White *et al.* (2005) op. cit.

40 Contreras-Hermosilla, A. (2011) People, governance and forests – the stumbling blocks in forest governance reform in Latin America, *Forests* 2, 168–99.
41 Interviews by F. von Blücher, 2011: 24.
42 Smith, J., Colan, V., Sabogal, C. and Snook, L. (2006) Why policy reforms fail to improve logging practices: The role of governance and norms in Peru, *Forest Policy and Economics* 8, 458–69.
43 Smith *et al.* (2006) op. cit.
44 Interviews by F. von Blücher, 2011: 25; see also White *et al.* (2005) op. cit.
45 Interviews by F. von Blücher, 2011: 22.
46 Interviews by F. von Blücher, 2011: 26.
47 Interviews by F. von Blücher, 2011: 2, 3, 8, 10, 23.
48 Interviews by F. von Blücher, 2011: 23.
49 These include: the Ministries of Economy and Finance; of Energy and Mines; of Transportation and Communication; of Foreign Trade and Tourism; and of Foreign Affairs.
50 These include: the Agency for the Formalization of Informal Property; the National Centre for Strategic Planning; the National Institute for the Development of Andean, Amazonian and Afro-Peruvian Peoples; and the REDD Technical Group.
51 Small contracts (up to 1,000 hectares) did not require technical or economic feasibility studies (White *et al.* (2005) op. cit.), forest management or administrative control by the state; and large contracts (above 1,000 hectares) required stock surveys, a tree census with topographic maps and some form of forest management (Smith *et al.* (2006) op. cit.), making these contracts expensive. Large companies circumvented the law by acquiring many small contracts (White *et al.* (2005) op. cit.). A volume-based tax on the extracted timber was levied.
52 Smith *et al.* (2006) op. cit.; White *et al.* (2005) op. cit.
53 Granoff, I.M.E. (2008) Peruvian forest law: seeing the people for the trees, *NYU Environmental Law Journal* 16, 533–62.
54 Interviews by F. von Blücher, 2011: 18, 25; see also Smith *et al.* (2006) op. cit.
55 Capella, J. and Sandoval, M. (2010) *REDD en el Peru: Consideraciones Jurídicas para su implementación (REDD in Peru: Legal considerations for implementation)*, Lima, Peru: Sociedad Peruana de Derecho Ambiental (SPDA).
56 Interviews by F. von Blücher, 2011: 22.
57 Smith *et al.* (2006) op. cit.
58 Sears, R.R. and Pinedo-Vasquez, M. (2011) Forest policy reform and the organization of logging in Peruvian Amazonia, *Development and Change* 42 (2), 609–31.
59 Velarde *et al.* (2010) op. cit.
60 Interviews by F. von Blücher, 2011: 6, 12.
61 Interviews by F. von Blücher, 2011: 6.
62 EIA (2010) *Peru's forest sector: ready for the new international landscape?*, Environmental Investigation Agency.
63 Interviews by F. von Blücher, 2011: 25.
64 Other forest policy instruments used in Vietnam, such as community forest management, spatial planning, and protected areas are not included in this analysis.
65 Velarde *et al.* (2010) op. cit.
66 Capella and Sandoval (2010) op. cit.
67 Capella and Sandoval (2010) op. cit.
68 Interviews by F. von Blücher, 2011: 2, 8, 10.
69 Interviews by F. von Blücher, 2011: 2, 25.
70 Interviews by F. von Blücher, 2011: 2.

71 Interviews by F. von Blücher, 2011: 15.
72 White *et al.* (2005) op. cit.
73 Interviews by F. von Blücher, 2011: 26.
74 This was introduced in the Forestry Law of 1975.
75 Areas allocated in favour of isolated indigenous people.
76 Oliveira *et al.* (2007) op. cit.
77 Interviews by F. von Blücher, 2011: 12.
78 Ministerial Resolution N 026-2002-AG.
79 GTCI (2007) *Caracterización del departamento de Ucayali con fines de ordenamiento territorial (Characterization of the purposes of zoning in the department of Ucayali)*, Pucallpa, Peru: Grupo Técnico de Coordinación Interinstitucional – Camisea.
80 One project for example offers initial financing in five NPAs to help develop alternative agricultural practices or crops (e.g. coffee) in order to initiate a new self-sustaining source of income. This model is currently being copied in other NPAs.
81 Interviews by F. von Blücher, 2011: 10, 23, 25.
82 Interviews by F. von Blücher, 2011: 25.
83 Interviews by F. von Blücher, 2011: 10, 11, 25.
84 Interviews by F. von Blücher, 2011: 10, 23.
85 Interviews by F. von Blücher, 2011: 19.
86 Smith *et al.* (2006) op. cit.
87 Velarde *et al.* (2010) op. cit.
88 Sears and Pinedo-Vasquez (2011) op. cit.
89 Interviews by F. von Blücher, 2011: 27.
90 Oliveira *et al.* (2007) op. cit.
91 Interviews by F. von Blücher, 2011: 19.
92 Interviews by F. von Blücher, 2011: 27.
93 Interviews by F. von Blücher, 2011: 27.
94 Interviews by F. von Blücher, 2011: 9, 19.
95 Interviews by F. von Blücher, 2011: 20.
96 Interviews by F. von Blücher, 2011: 27.
97 Interviews by F. von Blücher, 2011: 20.
98 Interviews by F. von Blücher, 2011: 3.
99 Sears and Pinedo-Vasquez (2011) op. cit. and Smith *et al.* (2006) op. cit. explain the structure of the logging industry and how concession documents have become a tradable commodity.
100 Sears and Pinedo-Vasquez (2011) op. cit.; Smith *et al.* (2006) op. cit.
101 Oliveira *et al.* (2007) op. cit.
102 Sears and Pinedo-Vasquez (2011) op. cit.
103 Velarde *et al.* (2010) op. cit.
104 Interviews by F. von Blücher, 2011: 12.
105 Interviews by F. von Blücher, 2011: 12.
106 Interviews by F. von Blücher, 2011: 2, 12.
107 Interviews by F. von Blücher, 2011: 12, 15; see also Velarde *et al.* (2010) op. cit.
108 Interviews by F. von Blücher, 2011: 12.
109 Oliveira *et al.* (2007) op. cit.
110 Nepstad, D., Schwartzman, S., Bamberger, B., Santilli, M., Ray, D., Schlesinger, P., Lefebvre, P., Alencar, A., Prinz, E., Fieske, G. and Rolla, A. (2005) Inhibition of Amazon deforestation and fire by parks and indigenous lands, *Conservation Biology* 20 (1), 65–73.
111 Interviews by F. von Blücher, 2011: 25.

184 *Felix von Blücher et al.*

112 Interviews by F. von Blücher, 2011: 15.
113 Oliveira *et al.* (2007) op. cit.
114 Interviews by F. von Blücher, 2011: 21.
115 The national programme is called *Programa Nacional de Conservación de Bosques para la Mitigación del Cambio Climático (Programa Bosques)*.
116 Interviews by F. von Blücher, 2011: 5.
117 Smith *et al.* (2006) op. cit.
118 Interviews by F. von Blücher, 2011: 16.
119 Interviews by F. von Blücher, 2011: 16; http://www.fsc.org
120 Interviews by F. von Blücher, 2011: 16.
121 Interviews by F. von Blücher, 2011: 16.
122 Interviews by F. von Blücher, 2011: 16.
123 Interviews by F. von Blücher, 2011: 16.
124 Interviews by F. von Blücher, 2011: 25.
125 Interviews by F. von Blücher, 2011: 20.
126 The project is called *CEDEFOR – Certification and Development of the Forest Sector.*
127 Interviews by F. von Blücher, 2011: 25.
128 Interviews by F. von Blücher, 2011: 25.
129 Interviews by F. von Blücher, 2011: 20.
130 Interviews by F. von Blücher, 2011: 10.
131 Interviews by F. von Blücher, 2011: 14, 18.
132 Interviews by F. von Blücher, 2011: 10.
133 The draft bills of the new *Forest and Wildlife Law*, as well as the draft of the *Law of Compensation for Environmental Services* establish the right over carbon and benefits arising from carbon management to the concessionaires, indigenous communities and private property holders.
134 Hajek, F., Ventresca, M.J., Scriven, J. and Castro, A. (2011) Regime-building for REDD+: Evidence from a cluster of local initiatives in south-eastern Peru, *Environmental Science & Policy* 14 (2), 201–15.
135 Velarde *et al.* (2010) op. cit.
136 Interviews by F. von Blücher, 2011: 4, 21, 23.
137 Interviews by F. von Blücher, 2011: 4, 10, 23; see also Pfaff *et al.* (2010) op. cit.
138 Pfaff *et al.* (2010) op. cit.
139 Larson, A.M., Corbera, E., Cronkleton, P., van Dam, C., Bray, B., Estrada, M., May, P., Medina, G., Navarro, G. and Pacheco, P. (2010) *Rights to forests and carbon under REDD+ initiatives in Latin America*, CIFOR.
140 Interviews by F. von Blücher, 2011: 21.
141 Interviews by F. von Blücher, 2011: 4.
142 Interviews by F. von Blücher, 2011: 16.
143 Hajek *et al.* (2011) op. cit.
144 Contreras-Hermosilla (2011) op. cit.
145 One proposal (CATIE) for example specifies a projected, forward-looking baseline for subnational activities, but uses a historical baseline for national-level baselines for national-level activities: GCP (2009) *The little REDD+ book*, Oxford, UK: Global Canopy Programme.
146 Velarde *et al.* (2010) op. cit.
147 White *et al.* (2005) op. cit.

9 Comparative analysis of Vietnam, Indonesia, Cameroon and Peru

Joyeeta Gupta, Nicolien van der Grijp, Léa Bigot, Mairon Bastos Lima, Jonathan Y.B. Kuiper and Felix von Blücher[1]

9.1 Introduction

Building on the case studies of Vietnam, Indonesia, Cameroon and Peru, this chapter provides a comparative analysis revealing insights in terms of the driving forces of deforestation and forest degradation (9.2), the effectiveness of national forest policy instruments dealing with these drivers (9.3), their equity aspects (9.4) and the implications for REDD (9.5).

9.2 Driving forces of deforestation and forest degradation

9.2.1 Direct drivers of deforestation and forest degradation

The data on forests in all four countries is contested (see 9.3) and there are differing definitions of deforestation. For example, 71 per cent of Indonesian land is officially forested and 29 per cent is non-forest land; but only 68 per cent of the forest land has forest cover and 15 per cent of non-forest land has forests.[2] However, what appears clear is that, in absolute terms, deforestation has been the highest in the past two decades in Indonesia, followed by Cameroon and Peru;[3] Vietnam has afforested on balance.[4]

Based on the framework on drivers (see 2.3), and the country-specific case studies (see Chapters 5–8), the proximate driving factors of deforestation and forest degradation can be compared (see Table 9.1).

The key driver of deforestation, in all four countries is cash crop production (Vietnam: rubber, coffee, pepper, cashew, shrimp farming; Indonesia: palm oil, coffee, cocoa, rubber; Cameroon: cocoa, coffee, bananas, palm oil, sugar cane, rubber; Peru: palm oil, coffee, cocoa, cattle ranching). In Peru, a substantial percentage (80–90) of deforestation[5] is caused by subsistence migrants using slash-and-burn techniques for agricultural expansion, small-scale cattle ranching (which requires relatively large feeding grounds) and illegal cultivation of coca. Commercial logging is very important in Indonesia (for acacia and eucalyptus). Vietnam logs less because of stricter rules and its furniture export business is dependent on imported wood, most of which is not certified. Commercial logging should be differentiated from over-harvesting and

Table 9.1 Proximate drivers of deforestation and degradation in the case study countries

	Vietnam	Indonesia	Cameroon	Peru
Agriculture				
Commercial	rubber, coffee, pepper, cashew, shrimp farming	oil palm, rubber, coffee, cocoa	cocoa, coffee, bananas, oil palm, sugar cane, rubber	palm oil, shade grown coffee, cocoa; cattle ranching
Non-commercial	shifting cultivation	shifting cultivation	shifting cultivation	slash and burn techniques
Extraction	commercial logging, overharvesting, fuelwood, 'illegal' logging	commercial logging (for acacia, eucalyptus), violation of SFM rules in concessions, small-scale encroachment	industrial wood, fuelwood, chainsaw milling, illegal logging	timber and non-timber forest product extraction, unsustainable logging, 'illegal' logging
Infrastructure	hydropower, roads, settlements	hydropower, roads, settlements	hydropower, roads, settlements	hydropower, roads, settlements
Industry	–	coal, gold mining	mining	oil, gas, gold mining
Biophysical	forest fires	forest fires	poor soil	forest fires, poor soil

'illegal' logging which degrades forests; logging provides 75 per cent of the Cameroonian domestic timber and energy needs, in Vietnam about 20–25 per cent of fuel wood needs. Agriculture and commercial logging cause significant land-use change. Infrastructure such as dam construction is a driver in Vietnam and Cameroon and in the future also in Peru; while forest roads allow for greater encroachment into forest areas. Mining is a driver in Indonesia and Peru and is projected to become important in Cameroon. Expansion of settlements is relevant in Vietnam and Peru while urbanization in Cameroon leads to increased demands for timber and non-timber forest products. Forest fires, accidental and deliberate, are common to all four countries.

Degradation is more difficult to measure, but implies a reduction in the quality of the forests, through changes in the type of forest and trees and density. It thus affects the interactions between the components of a forest ecosystem and can lead to a loss of many ecosystem services including biodiversity. Primary forests have been cut down and replaced by plantations encouraged through concessions (Indonesia, Cameroon, Peru) or by reforestation schemes as in Vietnam. Furthermore, degradation results from the search for, and often local extinction of, commercial species especially in less remote areas as in Peru. In Cameroon, forest degradation is mostly the result of unauthorized small-scale chainsaw milling and artisanal logging.[6] Forest fires, and over-use and access to non-timber forest products (e.g. mining) also lead to degradation.

9.2.2 Underlying drivers of deforestation and forest degradation

The direct causes of deforestation and degradation can often be related to underlying causes. The poverty and dependence of local people living in and around forests on forest products and services for their lives and livelihoods is a key underlying factor. This, in combination with an ongoing competition between the value of forest protection and that of exploiting forests for development, between the desire for isolation of some indigenous communities and the desire to be connected with the outside world, amongst different social actors – forest dwellers versus the state, agricultural immigrants and mining companies – has put traditional conservation values under stress.

In all four countries a greater part of deforestation and forest degradation can be attributed to the growing local through to global demand for food, fibre, timber, other forest products and biofuel, caused by population density and migration (Vietnam, Peru), urbanization (Cameroon) and evolving consumption patterns in consumer countries. Furthermore, national development goals to enhance agricultural production, industrialization, and mining through issue-specific targets (e.g. agricultural targets in Vietnam), concessions and support (e.g. agricultural credit) are critical factors leading to land-use change. Technology is a double-edged sword. Low agricultural productivity and poor technological skills lead to greater deforestation to generate revenue; but access to heavy agricultural and forest machinery can also make it easier to deforest and invest in commercial extensive agriculture. In general, issues such as poor interagency cooperation, high corruption (Peru ranks 80th, Indonesia 100th, Vietnam, 112th and Cameroon 134th on the Transparency International list)[7] and weak policy implementation are indirect drivers of deforestation.

The different drivers reinforce each other. For example, forest roads increase access for local villagers and tourists. This leads to increasing encroachment and exploitation of forest peripheries along the roads. Agriculture in buffer zones is short-lived as forest soils are exhausted, fast leading to increasing incentives for slash and burn. In Peru, the Environment Ministry sees this as the cycle of deforestation.[8]

9.3 Forest policy instruments assessed

Over the years all four countries have developed complex forest governance structures. Although Indonesia has a dedicated Forest Ministry, forest conservation may not be the overarching goal of the ministry. In Vietnam there is conflict of interest between the Ministries of Agriculture and Forests on spatial planning and forest-related tasks, which is exacerbated by the fact that the Ministry of Planning and Investment coordinates Official Development Assistance projects on forests and the Ministry of Industry and Trade regulates trade in forest products including biofuels. In Cameroon, the Ministry of Flora and Fauna regulates the sustainable management of, and law enforcement in, all forest estates[9] and the Ministry of Environment and Nature Protection regulates environmental impacts including deforestation and forest land degradation[10] and coordinates climate policy including REDD, while the Ministry of Economy and Finance controls the chain of custody of timber products and the collection of the annual area fee from concessions and community and council forests. In Peru, the Agricultural Ministry has competence over agricultural, rural and forest policy, while the Environment Ministry has authority over the strategic development of natural resources and compensation schemes for environmental services including REDD. The governance framework at national level in these countries resembles the diffusion of roles and responsibilities at international level; it is fragmented, leading to plural and competing policies, and excludes many ecosystem services from the purview of the instruments that have been created.

Forest policies in all four countries are affected by trade restrictions. The EU-FLEGT agreements apply to Vietnam and Indonesia while the EU biofuel standards influence Indonesia, Cameroon and Peru. The US Lacey Act influences for example Vietnamese exports and the US–Peru Free Trade Agreement influences the governance structure and forest policy in Peru. These trade restrictions address the drivers of global demand for, and trade in, unsustainable wood (see Table 2.3). Where such restrictions are generated in cooperation with exporting countries it has a more positive influence on governance (e.g. the US–Peru FTA and the EU–Vietnam FLEGT/VPA), but where this results from unilateral measures, there is resentment and opposition (e.g. the perceived impact of the US Lacey Act on Vietnam, and of the EU Biofuel Directive on Indonesia). The design implications are that bi- and multilateral agreements are politically more acceptable than unilateral ones, and trade restrictions combined with capacity building are likely to be more effective.

Internally, all four countries are decentralizing their overall and forest governance. In Vietnam this process started in 1992, in Cameroon in 1996, in Indonesia in 1999 and in Peru in 2002. This is potentially an improvement on centralized, exclusive top-down policy processes and can address local drivers. However, while decentralization in Vietnam has led to the transfer of powers and responsibilities to the local level, empowering local people to a limited extent, it has enlarged state control over local areas.[11] In Indonesia, decentralization relies on local capacity, which is often missing, to implement policies in line with national law. However, local governments see this devolution as equivalent to local autonomy which may translate into forest land use change without permission augmented by their enhanced mandate to raise their own revenues.[12] In Cameroon, the decentralization process has been in progress since 1996, but is still incomplete. Decentralized entities took effect in 2004; yet by 2011 no regional council had been elected by a democratic vote. If policy is dictated from above, this may lead to disenfranchising local people while allowing elite capture.[13] In Peru, the subnational authorities remain dependent on the central state[14] and lack the resources to implement their mandate. Municipalities do not have the mandate to implement or enforce state policy. While, *de facto*, they manage activities at the local level without the necessary resources, the national state has the capacity and knowledge but no *in situ* monitoring takes place. Municipal Forest Management Committees are legally recognized but have little support, while most local officials do not see deforestation and forest degradation as a problem.[15] The design implication is that decentralization per se is not a panacea for all problems as the process is incompletely executed; local officials may have vastly different visions of development; and decentralization may disempower local people either through greater control or through elite capture of the benefits.

The principle of sustainable forest management (SFM) has been explicitly adopted in the national policies in the four countries and is also part of many international certification processes. This principle aims to address unsustainable (and 'illegal') logging drivers. It provides a unifying idea although interpretation and implementation vary. This is partly because of varying definitions at global level (see 3.3.2). There is need for greater harmonization and context-relevant interpretations of this principle between the international certification companies, national policy arenas, and specific arrangements such as FLEGT, the US Lacey Act and the EU Biofuel Directive.

Spatial planning is important in all four countries. In Vietnam there are two overlapping land-use classifications and terminologies used by the Ministries of Environment and Agriculture.[16] Bare land is either classified as unused land or as forest land without forests leading to variations in the forest cover statistics. In Indonesia, there is forest land (which is partly unforested) and non-forest land (which has some forests); the former is divided into Conservation Forests (biodiversity conservation; national parks, reserves, etc.), Protection Forests (protection of hydrology, life-support systems, watershed protection, flood prevention, soil fertility, etc.) and Production Forests. Peat areas that are deeper than 3m or located in Protection or Conservation Forests cannot be converted for other uses. However, land zoning has been done without *in situ* verification, leading to mismatches between maps and actual forest cover. The lack of recognition of the specific lands under indigenous control often leads to plural rights of access to land, where land previously used by indigenous people and recognized by authorities is leased to other actors for other purposes. In Cameroon, the forests include the permanent forest estate (80 per cent) and non-permanent forests under state control; the latter consist of remaining forested lands and may be converted for community crops, plantations, private forests and community forests, all of which do not require sustainable land management.[17] In Peru, land is zoned based on its major use capacity – which means that forest land suitable for agriculture, cattle ranching, timber or non-timber forest product extraction can be converted, although it formally requires prior authorization from the Ministry of Agriculture (de facto, prior authorization is seldom requested). The public forests include production forests, forests for future use, protected areas, forests for protection, indigenous, farmers and rural community forest land and local forests (see Table 9.2). This zoning system has prioritized forest land conversion for agriculture, and the protected areas are not always the regions rich in biodiversity or at highest risk of biodiversity loss.

Although spatial planning efforts are under way, there are challenges with the choice of protected areas; different ministries have overlapping mandates on the same parcel of land, multiple actors receive access rights to the same parcel of land, maps are not verified through *in situ* analysis, spatial planning is scarcely ever undertaken in consultation with local actors, and buffer zones suffer from encroachments because of poor land allocation, monitoring and enforcement. Spatial planning has brought to surface ideological conflicts between environmental protection and neo-liberalism, free trade and the growth paradigm. It has exposed the jurisdictional conflicts between the different ministries and levels, and has made explicit the conflict of interests between states that wish to control the resources as part of national patrimony and the time-honoured customary rights of access of indigenous peoples and local communities. Solving this is not simply a matter of 'getting the maps right'; it is rather a matter of national debate regarding how stakeholders define sustainable development and their growth path, and how they negotiate with the powerful and less powerful stakeholders in society as to how these maps are drawn. This means that systems of access to land (whether through tenure rules or concessions and permits) are likely to remain contested for some time (see 9.4).

Forest concessions potentially address the driver of unsustainable wood extraction and are used in Indonesia, Cameroon and Peru. In Vietnam, state forestry enterprises manage half of the forests. Management boards manage protection forests whereas forestry corporations manage production forests self-financing their activities. The latter have not been very successful because of limited annual harvest quotas and high taxes. Indonesia has industrial logging concessions (on both natural forests and timber plantations), ecosystem restoration concessions and social forestry concessions for community-based management, all expected to lead to sustainable forest management. However, weak enforcement has rendered these concessions largely unsustainable. The ecosystem restoration concessions can address deforestation. However, they have to deal with 'illegal' encroachment, dubious benefits for local people (some included, some fenced off) and, if done without consideration for equity issues, may become a form of 'green land grabbing'.[18] Community-based timber plantations face low timber prices and skewed political and market power in a vertically integrated sector. Village forest concessions have the potential of social, ecological and economic benefits but face institutional challenges as local authorities earn relatively little from

Table 9.2 Some of the instruments used in different countries

	Instrument	Vietnam	Indonesia	Cameroon	Peru
Regulatory	Trade restrictions (i) EU FLEGT Other	US Lacey Act	EU Biofuel Policy	EU Biofuel policy	US–Peru FTA
	Spatial planning (v): forest belongs to	State; however, partial tenure rights to local people have been allocated since the land law (1993)	State	State; since 1976, land certificates granted	State; since 2000 land zoning and concession-based system
	Spatial planning; type of forests	Special-use forests (protected areas and reserves), protection forests and production forests (natural and planted)	Conservation, production and protection forests	Permanent forests including protected areas and forest reserves (also for production), and non-permanent forests	Production forest, forests for future use, National Protected Areas, forests for protection, indigenous, farmers and rural community forest land and local forests
	Land rights (vi)	Planted forest land managers (use, transfer, rent, inherit, give away and mortgage rights)Natural forest land managers (use rights)	Customary rights recognized ('45/ '60); however, rights are subordinated to state interests, frequent absence of land titles	Customary use rights of indigenous people recognized ('94), no rights for commercial purposes	Since '97, forest dwellers have subsistence use rights; land titling for agriculture/ livestock possible; limited legal title for indigenous people
	Forest logging/ concessions		Concessions for industrial logging, ecosystem restoration, and community-based management	Concessions for production forests and council forests	Concessions on timber production, other forest products, ecotourism, conservation, afforestation and reforestation

them in comparison to the opportunity costs of not expanding plantations. In Cameroon, production forests are managed through concessions that require sustainable forest management plans. Foreign companies can comply with these requirements more easily[19] than domestic operators, forcing the latter often into the informal sector.[20] Although concessions must be monitored, poor enforcement makes room for unsustainable practices and encroachment in the peripheries.[21] Additionally, the 15-year operating licences are not synchronized with the 30-year harvesting cycles, providing little incentive for operators to ensure regeneration of key commercial species.[22] Peru has a concession-based system for timber, other forest products, ecotourism, conservation, afforestation and reforestation. Timber concessions are most important, each covering up to 40,000 hectares for up to 40 years. Concession holders must adhere to forest management plans, pay area-based taxes and report on activities. However, regulatory overburden, limited monitoring and rule enforcement undermines the system. In recent years, concessions may have opened the door for 'land grabbing'.

Forest land ownership potentially addresses policy and institutional drivers and local drivers. In Vietnam, the 1992 Constitution vested all land in the state. However, eight groups of stakeholders were allocated long-term management rights. Households and individuals can receive a land-use certificate for up to 30 hectares of planted forests for up to 50 years and have the right to use, transfer, rent, inherit, give away and mortgage the land. Natural forest landowners can use, but not transfer, give away, rent or mortgage natural forest land. This policy aims at local empowerment. However, disagreements between the Agricultural and Environmental Ministries and lack of funding and implementation imply that land types and land rights are uncertain and contested. In Indonesia, forest land belongs to the state; customary rights are recognized but subordinated to 'national interests'. The implementation has led to plural ownership/access rights causing confusion, local insecurity and land conflicts.[23] The allocation of land-use concessions arguably favours large actors: in 2008 the government granted more than 4.3 million hectares of forest land to large timber companies growing probably to 9 million hectares by 2016,[24] 26 million hectares to private forest-logging concessions[25] but only 1 million hectares to local communities.[26] Cameroon nationalized unclaimed forest lands in 1974, allowing people to obtain land titles since 1976 (but not on ancestral or customary grounds). In 1994, a limited right of indigenous peoples to use their land for subsistence purposes was recognized. Granting land tenure has still to be completed, creating uncertainty for local communities. In Peru, all forests are national patrimony. Forest lands suitable and used for agriculture can be owned by individuals. Forest dwellers have subsistence use rights, and some indigenous people have been granted land titles over their lands. This system has probably exacerbated deforestation because if forest land is formally seen as suitable for agricultural purposes, according to the major use capacity system, people can deforest the land and then apply for land titles.

All four countries have rules on reporting, monitoring and enforcement, but have low capacity, funding and implementation potential. Local actors and authorities may not necessarily agree that deforestation and degradation are a problem.

They all participate in trade liberalization following from their membership of the WTO in 1995 (for Indonesia, Cameroon and Peru) and 2007 (Vietnam). The trade liberalization process and open-market system has given all four countries an incentive to benefit from the export of timber, non-timber forest and mining products. They receive resources from aid agencies and international bodies for investing in forests. For example, Cameroon receives help from German, French, Canadian and Dutch development cooperation agencies and the Global Environment Facility.[27]

Taxes are a common feature. For example, Cameroon has an annual area fee levied on concession areas and a sawmill entry tax, raising resources while promoting production efficiency. The tax is redistributed to the Ministry of Flora and Fauna (50 per cent), local councils (40 per cent) and communities (10 per cent) and empowers them in the process. Enforcement is however a challenge.[28] Vietnam provided a USD 940 million subsidy to promote reforestation from 1998 to 2010 (Programme 661) to stakeholders participating in national reforestation schemes which had 80 per cent success;[29] but paid little attention to the quality of afforestation, provided low subsidies (USD 5–10 per hectare),[30] was subject to elite capture of revenues, and displaced subsistence farming to other areas.[31] However, Programme 661 is seen as a model for payment for forest ecosystem services in Vietnam. With foreign support, two pilot schemes were implemented between 2008 and 2010 where four hydropower and water supply plants have raised USD 3 million to pay 52,000 local communities[32] for their forest protection efforts. Starter problems included inadequate funds, land allocation challenges, low and sometimes delayed payments to households relative to forest management boards, and lack of farmer understanding of the basic concept. Payments to groups of households were more successful than payment to individual households – a lesson for the future.[33] In Peru, a payment for ecosystem service scheme has been launched paying USD 2.7 per hectare per year to forest communities. In Vietnam a small A/R CDM project rehabilitated degraded land (365 hectares) which will lead to potential benefits for local farmers from carbon credits and wood harvesting revenues after 16 years. However, complex methodologies and high transaction costs[34] limit its impact. Two debt-for-nature swaps were carried out between the US and Indonesia, with limited impact.

Financial instruments that subsidize reforestation through paying local people may enhance local income; if reforestation is well designed it may even enhance biodiversity protection. The question is, how does one finance this? The experiments in Vietnam with collecting funds from water supply companies and hydropower plants in order to compensate local forest protection efforts are one way to generate local resources. The potential for replication needs to be sought, but may be limited. Carbon funds could finance the carbon sequestration aspects as in A/R CDM and in REDD. But, with no targets in the post-2012 period, offset funds may be limited. The ultimate goal is to make these schemes self-sustaining and this may be a challenge.

Certification potentially addresses the driver of unsustainable and 'illegal' logging (see 9.4). Vietnam, Indonesia, Cameroon and Peru have 41,409, 904,089, 820,630 and 761,253[35] hectares of forest land that have been certified respectively. Vietnam aims at having 30 per cent of its production forests certified by 2020,[36] and Peru reduces the area-based extraction fees by 25 per cent for certified concessions.[37] Certification bypasses state shortcomings to directly target the export market. However, certified forests represent a small percentage of the total forested area; the larger companies benefit more from this relatively expensive system which sometimes opens up forest areas for greater commercial exploitation and exploitation of the peripheries; and global demand for certified wood is limited (possibly solvable through public procurement policies and education in the consumer countries).

Our interviews revealed that all four countries are influenced by international policy processes and instruments which lead to converging state forest policies as shown above, even though implementation remains challenging. Furthermore, these instruments are not able, by themselves, to change country calculations of the opportunity costs of protecting their forests, except to a limited extent. Few of these countries use suasive mechanisms of their own to create greater domestic public awareness on forests.

Having nationalized the forests, the governments concerned are now slowly beginning to encourage, engage with and empower local communities (see Box 2.6) to participate in forest management. This is being done through land tenure schemes, reforestation and payment for ecosystem services in Vietnam; limited tenure rights to indigenous people, village forest management and community-based timber plantations in Indonesia; through limited tenure rules, council and community forests in Cameroon; and through tenure rules, 'local forests' and administration contracts in Peru. In general there is no clear picture with respect to community-based management of forests. The literature indicates that community-based management is associated with less deforestation (more carbon storage) and livelihood benefits.[38] At the same time, in some places communities welcome the end of isolation, the possibility to develop further and earn an income. Thus, if these communities wish to develop with the meagre resources they have, they must be provided with alternative employment if they are to stay from deforesting their lands. Land grabbing has already become a big challenge (see Box 9.1).

9.4 Equity issues: impact on access and allocation

The case studies have discussed the equity implications of specific forest instruments. This section integrates this information using the access (of the poorest and most vulnerable to resources) and allocation framework (distribution of

> *Box 9.1* **Land grabbing worldwide**
>
> Land grabbing is a global phenomenon which refers to the large-scale purchase of land in developing countries by foreign investors.[39] This has been driven by growing world demand for biofuels, food, fodder and carbon sequestration. This leads to the conversion of forest and agricultural land for mainly domestic uses to the production of goods and services for export markets. These purchases and/or leases are often of land with contested ownership – where the state claims ownership, but local people may have had customary rights for many generations. Sometimes, formal local owners do not have the means to resist land purchases by foreign owners. This practice has come under considerable critique – is seen as neo-colonialist, as using neo-liberal principles to get control over the land in other countries and is seen to violate the UN proposed code of conduct on principles for responsible agricultural investment. REDD could potentially be a contributing factor to land grabbing ostensibly for carbon sequestration.

forest resources, risks from poor forest management, and responsibilities for forest management)[40] (see 1.5).

Access and allocation issues have gone through five phases. First, in early societies, local communities had customary access rights to forests. Second, during colonization and post-colonization periods, these rights were often taken over by the state in law if not *de facto* (see case studies). Following the onset of social and indigenous peoples' movements (see Boxes 3.3 and 4.2),[41] and the rise of decentralization, the third phase of partially rhetorically recognizing the rights of local communities has emerged. In the fourth phase, during the implementation of these rights, 'bundles' of rights on use, control and alienation (see Chapter 2) are being distinguished allowing for plural control. When these rights are shared between different actors, they may disempower some while empowering others. One can anticipate a fifth phase. If the commodification of forests and its services continues as a logical follow-up to the globalization agenda and REDD implementation, this may lead to more stress over claims about who has access rights over forests and what this means.

In these discussions, some local actors have become relatively more advantaged while others possibly face some disadvantages. The international discussions on indigenous peoples' rights have brought their plight to the global agenda. However, there is less focus on other local communities including farmers. The stress between agricultural migrants and local communities has also intensified. This may lead to situations in which parts of the local community find some of their rights protected at the cost of other parts of the community.

The above discussion is gender neutral; but gender-related issues are vital to understanding how best to design effective policy instruments.[42] The lack of gendered approaches in forest policy reflects poor gender balance in policy processes; this, in turn, leads to policies that may not adequately take the diversity of interests into account.[43] Gender always transects with other forms of inequality such as class, race, ethnicity, age and disability. This transectionality determines to what extent people have differentiated access and control of resources and their management. Recent literature brings up some issues about differential use, control and alienation rights in different tropical forest countries[44] or where some rights are subordinated to other rights.[45] For example, men and women differ in terms of how they access forests, what they use forest products for, and their status under inheritance law. Further, gender approaches and roles are different in relation to direct forest management and the use of forest products[46] and in participation in forest policy processes.[47] Finally, decentralization is not in itself a guarantee of gender equity.[48] Gender interests, roles and patterns are not homogeneous and contextual validation is needed.[49]

Access and allocation examine how policy instruments (see 2.4) influence the distribution of rights, responsibilities and risks between various actors. Based on the case studies, regulatory instruments have four major impacts on equity. First, the allocation of forest resources among different social actors is affected by the adoption of legal principles – nationalization in the case of Vietnam, Indonesia and Cameroon and state patrimony in Peru which reallocates the forests to the state. In Indonesia, the rights of indigenous peoples are now recognized but they are not given formal ownership papers. The UN Committee on the Elimination of Racial Discrimination has issued a warning to Indonesia to ensure that it takes indigenous peoples' rights into account.[50] Second, the above principles in combination with how forests are defined and spatial planning tools like protected areas can be used to empower some at the cost of others. In Vietnam, the differing types of forest land give local 'owners' differing rights. Planted forest land 'owners' have five land-use rights including transfer, rent, inheritance, gift and mortgage rights[51] while natural forest 'owners' cannot transfer, give away, rent or mortgage natural forest land[52] or get loans from banks. If a local person happens to live in a planted forest land area, he or she will have more rights than those who live in natural forest lands. Forest parks in Cameroon have marginalized access rules for local communities,[53] sometimes leading to migration with or without compensation. Third, rules regarding concessions also have an impact on allocation – mostly empowering richer groups further at the cost of poorer communities. In Cameroon, for example, concessions are granted to private companies with technical and financial capabilities often in collusion with state officials. These companies benefit from the concessions and the lax monitoring thereafter. These concessions displace or negatively impact local populations or are given with local populations still living in the concession area. Fourth, many customary access activities have over time been labelled as 'illegal' in all four countries. For example, local people living in concession areas in Cameroon lose their rights to hunt and farm.[54] This conversion of customary rights into illegal activities affects local access and gives it a sinister twist. This is not just limited to the 'labelling' of activities; reporting, monitoring and law enforcement instruments may be used to penalize these 'criminals'.

The use of economic and financial instruments may also affect equity. First, some instruments such as micro-credit can empower the poorest although their interest rates are high relative to bank interest rates. However, they are often lower than the interest rates charged by local moneylenders and can sometimes help those engaged in small-scale enterprises. The contextual application of this scheme may determine whether it is successful or not. In theory, the instrument of payment for ecosystem services can benefit those who look after ecosystems, but there are practical challenges here, not least connected with ownership rights to ecosystems.[55] Second, the distribution of funds, subsidies, tax returns, resources from Debt-for-Nature swaps, offset schemes and REDD funds can empower some actors in relation to others. However, the payments are relatively small in all four countries – mostly less than USD 5 per hectare. Third, certification squeezes out small domestic companies who cannot compete with larger companies and treats access by local communities living in certified areas as illegal. Fourth, even if some of these instruments are designed to deal with equity issues, the risk of elite capture (see Chapters 2, 4–8) is present in all four countries. For example, the case study on Vietnam discusses elite capture of reforestation subsidies, and in Cameroon, the council forests which were meant to empower the local communities ended up benefitting the local officials and mayors who wield power at local level.[56]

The bottom line is that some land-use policy and instruments have benefitted large commercial timber, oil palm and other such companies and some indigenous groups but not others. The gap between official maps and *in situ* situations leads to plural rights to the land and exacerbates local conditions through elite capture and corruption. At the same time, if the spatial planning system is corrected to take account of *in situ* verifications, this may not necessarily address elite capture and corruption, but may further exacerbate access and equity issues by criminalizing customary access. As forests become a global and national agenda item and are commodified, the struggle for control over forests through definitions, spatial planning, land ownership rules and market mechanisms becomes more intensive.

Policies and instruments are inspired by the three dominant discourses in forest governance[57] (the neo-liberal ecological modernization theory,[58] green governmentality[59] and civic modernization).[60] These discourses provide alternative storylines about how to allocate forests and among whom. While ecological modernization would empower market actors, green governmentality would give government strong control, and civic modernization would empower civil society and non-state actors. Proponents of these visions are marketing their views in the case study countries. REDD with its strong focus on commodifying forests is being institutionalized in all four countries, while the views of civic modernization theory and green governmentality are being incorporated into the 'co-benefits' and 'safeguards' discussion at the international level. Ecological modernization at the international level is being translated into green governmentality leading to stronger state control at the national level. Thus access and allocation is affected by competing ongoing underlying discourses, actors and processes.

9.5 Implications for REDD

It is within this existing context that REDD schemes are likely to be implemented. Chapter 4 discussed the history and implementation challenges of REDD. The four governments of Vietnam, Indonesia, Cameroon and Peru are in various stages of completing the documentation and administrative process, required for participation in REDD projects. REDD, however, is continuously evolving in the international arena – the definitions (RED to REDD++), criteria, reference levels, assessment of safeguards and co-benefits are evolving. This section does not make recommendations about how the general REDD discussions should evolve, but instead uses the case studies to discuss context-specific lessons on how REDD implementation can be enhanced in these countries.

First, there is need to enhance the enabling conditions and build capacity in these countries as the existing REDD readiness programmes are doing. Second, countries may wish to integrate REDD into existing national policies and programmes and use existing capacity. It may be necessary to fine-tune these existing programmes to deal with current design and implementation challenges and make them more equitable and effective. In Vietnam, this implies linking REDD with the institutions and capacity developed to implement Programme 661 and protected areas; in Indonesia, to sustainable community forest management schemes. In Cameroon, this means supporting council forests through sustainable forest management capacity building to increase revenues while ensuring revenue distribution to riparian populations, in addition to the permanent forest domain strategy and the use of forest management plans. In Peru, this implies linking with community forests, the concessions system, protected areas and ecosystem restoration. A key challenge here is whether REDD will be then seen as additional in terms of reducing GHG emissions. The research shows that although countries have instruments in place, they have major difficulties in actually implementing these. To that extent, if REDD can support existing processes and build on existing capacity, it may have a higher chance of success, although the extent of additionality may remain contested. Furthermore, it is also vital to consider forest buffer zones as potential REDD areas. Buffer zones face higher levels of threat and, once their resources are depleted, pressure on the protected areas increases as well.

Third, similarly there is a need to integrate REDD into existing international policies and programmes. This implies linking it effectively with existing trade and governance programmes (e.g. for Vietnam and Indonesia EU-FLEGT, for Peru US-FTA) and with certification schemes to optimize mutual learning and leverage resources. For Vietnam, the elaborate methodologies and processes created under AR-CDM and the PFES experiences could come in useful. For Indonesia, drawing parallels with its ongoing bilateral agreement with Norway is critical.

Fourth, REDD needs to be designed to ensure that access to food, fibre, timber and water for local people is not compromised. There is ongoing discussion on ensuring that REDD has some ecosystem and equity safeguards (see 4.4.6) in place to ensure that the local proximate drivers of deforestation are dealt with and equity considerations taken into account. However, section 9.4 has shown how complicated the equity challenges are; and many governments are resisting the inclusion of these safeguards in practice. In all four countries, when local people lose access rights and are marginalized, they 'encroach' on peripheral lands and buffer zones, engage in slash-and-burn and shifting cultivation in order to survive. Either these people need to have access to local forests (as in Indonesia), council forests (as in Cameroon) or reforestation funds (as in Vietnam) so that they can earn from protecting the forests, and/or they have to be provided alternative employment or income for protecting the forest. But, whatever the policy outcome, this needs to be based on participatory approaches with the local people in order to ensure local relevance, legitimacy and effectiveness. Designing REDD in a way that the safeguards are in place will therefore be a major challenge.

Fifth, REDD reference levels will need to draw on an understanding of the current state of development in a country, the specific development challenges and opportunities it faces and the way societies debate and discuss the role of forests in their development process. The forest transition curve (see 2.3) reflects past developed country experiences. It does not predict the future of developing countries, but shows how a combination of drivers (see Figure 2.2) and instruments (see Figure 11.4) can influence the path a country can take in terms of managing its forests. However, financing these instruments to compete with the drivers in an effective and equitable manner will be very challenging.

Sixth, for REDD to be successful, it is critical that the beneficiaries are clearly defined and identified based on criteria; if that is not the case, the system may become subject to elite capture.

Seventh, it is important to discuss the scheme with beneficiaries. Experiences from Vietnam with AR-CDM suggest that if beneficiaries do not understand the concept of ecosystem services, and the role of carbon, they may not be able to effectively participate in REDD schemes.[61] Explaining such an abstract and politically charged concept as REDD to local communities is not easy.[62] However, this does not imply that such participatory discussions should not be encouraged.

Eighth, benefit sharing including payment levels should be defined. Assuming that REDD funds are generated, it is not clear how these will be shared, by whom and with whom and in what way. There are major ongoing discussions on this issue. Without entering into these discussions, a key issue is that large investors who participate in deforestation argue that their opportunity costs should be covered. This leads to a curious paradox. Those who do not deforest normally do not perceive high opportunity costs. Those who do deforest face high opportunity costs of not doing so. If the resources paid to or the benefits shared with these people are to serve as an incentive for changing behaviour, they need to be in line with local calculations. However, ironically this will relatively speaking disadvantage the poor and reward the rich. Defining payment levels will be a critical challenge in the future.

Ninth, an appropriate modality of benefit sharing including payment which addresses issues of elite capture needs to be explored. Another lesson learnt from all four countries is that if there is no easy modality of payment there are likely to be serious problems. It is important to reduce the number of intermediaries and administrative steps involved. The payments to household groups in Vietnam may be one system that could be replicated elsewhere. The Annual Area Fee in Cameroon may be a possible route through which payments could be channelled if the challenge of elite capture is dealt with.

Tenth, all four countries have indigenous populations. Many of them were dispossessed through the nationalization processes in these countries and are now in various stages of gaining access to their lands. Indigenous groups have lobbied heavily to have free prior and informed consent (see Box 4.2). Vietnam is considered as a pilot country regarding the development of a Free Prior and Informed Consent (FPIC) mechanism; implementing such a process is complicated in a country with a limited civil society and freedom of speech, so implementing this fully is likely to be challenging.[63]

Eleventh, there have been discussions about designing REDD either as a market instrument, a regulatory instrument or as a hybrid instrument. The official markets in the four case study countries are far from perfect, nor does their regulatory climate provide an atmosphere conducive to supporting market instruments. This suggests that REDD should not be designed purely as a market instrument but more as a hybrid instrument. At the same time, keeping government engaged in this process is vital for the long-term sustainability of government procedures and for promoting the rule of law. Ensuring stakeholder participation and a gendered approach (see 9.4) is vital to help in dealing with equity issues.

Twelfth, adopt a subnational, national, regional or even a global approach to deal with the challenge of leakage and scale (see 4.4.1 and 4.4.5). It is important to move beyond a project level to a regional or national approach to minimize leakage within countries. But if the major driver is global demand, there will be global leakage, unless a global approach is adopted. The UN-REDD programme in Vietnam is developing a regional REDD+ approach in order to avoid the risk of leakage in the Lower Mekong Basin.

Three caveats to the above are necessary. First, although some argue that REDD should only be introduced after there is coherence between national policies and all underlying drivers are dealt with, this ignores the long-term opportunity costs a country may face and delays implementation indefinitely. A pragmatic approach may be needed. Second, payments in REDD are only made following monitoring, reporting and verification (MRV; see 4.4.4) of deforestation and degradation. Earlier analysis has shown that forest data is contested, deforestation and degradation data is contested and there is little *in situ* verification of information that can complement remote sensing data (see Indonesia case; in Cameroon, remote sensing data is difficult to gather because of cloud cover). Organizing accurate and incontestable MRV data is expected to be expensive (as the experience with AR-CDM in Vietnam shows) and may lead to inequities in how REDD resources are being spent with more going to professional MRV companies than those who actually take care of the forests. Third, if REDD needs to take care of all ecosystem services and equity issues through safeguards and create an ideal system for benefit sharing, the mechanism may become very complex in design. Perhaps REDD should not be seen as the overall solution based on comprehensive political reform. Broadening the scope of REDD too much may overburden and paralyze this international mechanism.

9.6 Conclusion

The four case study countries are more or less typical large developing countries with huge forest resources; but with different political systems and policy approaches. Much can be learnt from them. The proximate driving factors in these countries are commercial (and subsistence) agriculture, commercial extraction of wood and non-wood products in an unsustainable (and 'illegal') manner, mining, dams and roads. The underlying factors include the conflict between the conservation values of some and the desire of others to exploit the forests for subsistence and commercial gain to support national economic growth. Such national growth strategies have often resulted in concessions, credit and other forms of support for agriculture, mining, and timber activities at the cost of the forest sector.

Hence, for national governments forest policy competes with other policy areas (that are often underlying drivers of deforestation and degradation) in terms of its contribution to the national economy. It is not self-evident from the research that forest conservation will survive as a notion in the short- to medium-term if there are no financial resources forthcoming from outside the country. This means that other ministries have more power in resisting forest policy and there is heavy competition between the different sectors. This is most clear in the case of Peru where agriculture clearly takes precedence over forests in land-use strategy; in Cameroon where mining is becoming more lucrative; in Vietnam and Indonesia where the focus on commercial crops brings easy profit. However, the transformation of the Vietnamese saying *Rừngvàng, biênbạc* [The sea is silver, the forest is gold] into *Rừngkhôngvàng, biênkhôngbạc* [The sea is not silver, the forest is not gold] shows the growing awareness about forests.

Forest policy in all four countries is being dealt with in the context of primarily regulatory instruments, some economic instruments, and the gradual sharing of management with other social actors. Suasive instruments to create public awareness and commitment are of a much smaller order. These countries are influenced by the international policy processes and national discourses and policies are shaped by international ones. However, the adoption of the sustainable forest management principle and the use of instruments show considerable variation in interpretation. Most have been incompletely implemented and reflect national struggles. Decentralization challenges represent the reluctance of the state to hand down mandates, resources and skills, while lower governments may often see this as a way to enhance autonomy of decisions and increase revenue from local forest exploitation. Spatial planning policy has exposed the struggles between different ministries over who has control over the land and its resources including minerals. The granting of land rights has exposed the struggle between the state which wants to use spatial planning to gain control over the land and local people who find their customary land rights first nationalized, then contested, then only partially, if at all, returned.

The choice of which forests are to be protected is also a critical issue. While forests with biodiversity hotspots are perhaps the most important to protect, zoning efforts in these countries (e.g. Peru) have not always taken this issue into account. Definitional issues (see Vietnam and Indonesia), prioritization (see Peru), literal turf battles (see all four countries) and the lack of consistency between maps and *in situ* assessments continue to plague the policy process. A common challenge in all four countries is the land tenure rules. All four countries face the problem of overlapping land rights resulting from unclear *in situ* information and/or a short-term appeasement strategy – pleasing indigenous people and the mining sector; as well as the international community and local stakeholders. While uncertain land tenure leads to local insecurity and reduces the incentive to manage the lands, land titling enhances access to credit and security, but is no guarantee that the land will be managed as anticipated. The increasing commodification and gender-neutral policy approaches tend to empower some actors in relation to others, legalize some activities over others, and create new untenable situations. The research concludes that the bulk of the instruments deal with some proximate drivers, but few with underlying drivers.

Notwithstanding the evolving global discussions on REDD design, this chapter presents 12 possible recommendations about improving the design of REDD for country application (see 9.5). These include enhancing enabling conditions, making better use of and building upon national institutions and capacities, building on existing international mechanisms operating in the country, maximizing the potential for safeguards, developing reference levels in relation to growth aspirations, defining beneficiaries, including free, prior informed consent as a principle for including indigenous peoples and local communities, avoiding a pure market approach, and developing a national approach to minimize subnational leakage.

Chapter 3 postulated that most international agreements protect some ecosystem services more than others. The analysis in the case study chapters suggests that ecosystem services that can be priced (food, fuel, timber, fodder, fibre, and now carbon) are more likely to be protected than the non-priced services such as biodiversity protection, soil conservation and habitats for indigenous communities. The commodification of forests risks prioritizing high market value services over low market value services. This will remain an enduring challenge for forest management in coming decades.

Notes

1. We would like to thank Simone Lovera and Andrea Brock for their comments, suggestions and research support on this chapter.
2. Ministry of Forestry (2009a) *Forestry Statistics of Indonesia: 2008*, Jakarta: Ministry of Forestry.
3. FAO (2011a) *The state of forests in the Amazon Basin, Congo Basin and Southeast Asia*, Rome: Food and Agriculture Organization of the United Nations.
4. FAO (2011b) *State of the world's forests 2011*, Rome: FAO.
5. Interviews by F. von Blücher, 2011: 4, 14.
6. Cerutti, P.O., Lescuyer, G., Assembe Mvondo, S. and Tacconi, L. (2010) *Les défis de la redistribution des bénéfices monétaires tirés de la foret pour les administrations locales (The challenges of the redistribution of monetary benefits from the forest for local governments)*, Bogar Barat: Center for International Forestry Research; Dkamela, G.P. (2011) *The context of REDD+ in Cameroon: drivers, agents, institutions*, Bogor Barat: Center for International Forestry Research.
7. Transparency International (2011) *Corruption perceptions index 2011*.
8. MINAM (2009) *Causas y medidas de mitigación a la deforestación en áreas críticas de la Amazonía peruana y a la emisión de gases de efecto invernadero (Causes and mitigation measures in critical areas of deforestation in the Peruvian Amazon and the emission of greenhouse gases)*, Lima, Peru: Ministerio del Ambiente.
9. REM (2010) *IM-FLEG Cameroon: Progress in tackling illegal logging in Cameroon*, Cambridge, UK and Yaoundé: Cameroon: Resource Extraction Monitoring.
10. Decree No. 2005/0577 /PM of 23 February 2005, Arête No. 0070/MINEP of 22 April 2005, Arête No. 0001/ MINEP of 3 February 2007, all on EIA.
11. Fritzen, S.A. (2006) Probing system limits: Decentralisation and local political accountability in Vietnam, *The Asia-Pacific Journal of Public Administration* 28 (1), 1–23.
12. Interview by M. Bastos Lima, 2011: 6, 13, 14, 15, 16, 19, 22, 48.
13. Cerutti, P.O., Ingram, V. and Sonwa, D. (2008a) *The Forests of Cameroon in 2008*, Food and Agricultural Organization.; Dkamela (2011) op. cit.
14. Capella, J. and Sandoval, M. (2010) *REDD en el Peru: Consideraciones Jurídicas para su implementación (REDD in Peru: Legal considerations for implementation)*, Lima, Peru: Sociedad Peruana de Derecho Ambiental (SPDA).
15. Interview by Felix von Blücher, 2011: 4, 5, 6, 7.
16. Hoang, M.H., Do, T.H., van Noordwijk, M., Pham, T.T., Palm, M., To, X.P., Doan, D., Nguyen, T.X. and Hoang, T.V.A. (2010) *An assessment of opportunities for reducing emissions from all land uses. Vietnam preparing for REDD, Final national REALU report*, Nairobi, Kenya: Partnership for the Tropical Forest Margins.
17. Dkamela (2011) op. cit.
18. Interview by M. Bastos Lima 2011: 13; 14, 18, 21.

19 Topa, G., Karsenty, A., Megevand, C. and Laurent, D. (2009) *The rainforests of Cameroon: experience and evidence from a decade of reform*, Washington DC: The World Bank; REM (2010) op. cit.
20 Interview by J.Y.B. Kuiper, 2011: 18.
21 Dkamela (2011) op. cit.; Topa *et al.* (2009) op. cit.
22 Topa *et al.* (2009) op. cit.
23 Interview by M. Bastos Lima 2011: 5, 6, 7, 8, 12, 13, 14, 15, 16, 51; Colchester, M., Jiwan, N., Andiko, Sirait, M., Firdaus, A.Y., Surambo, A. and Pane, H. (2006) *Promised Land: Palm oil and land acquisition in Indonesia – Implications for local communities and indigenous peoples*, Moreton-in-Marsh, England and Bogor, Indonesia: Forest Peoples Programme, Perkumpulan Sawit Watch, HuMA, and the World Agroforestry Centre.
24 Obidzinski, K. and Dermawan, A. (2010) Smallholder timber plantation development in Indonesia: what is preventing progress? *International Forestry Review* 12 (4), 339–48.
25 Ministry of Forestry (2009a) op. cit.
26 Obidzinski and Dermawan (2010) op. cit.; interview by M. Bastos Lima 7, 52.
27 GEF projects with forestry in the title include for Vietnam: four projects (total USD 14,499,545); Indonesia: six projects (total USD 15,905,053); Cameroon: two projects (total USD 13,843,850); and Peru: one project (1,820,000 USD). See GEF (2012) *GEF projects for Vietnam/Indonesia/Cameroon/Peru*.
28 REM (2010) op. cit.; Dkamela (2011) op. cit.; Cerutti *et al.* (2008a) op. cit.
29 GSO (2011) *The official national site for statistics*, General Statistics Office of Vietnam; Meyfroidt, P. and Lambin, E.F. (2009) Forest transition in Vietnam and displacement of deforestation abroad, *PNAS* 106 (38), 16139–44.
30 Interview N1 conducted by L. Bigot on 19 April 2011.
31 McElwee, P. (2009) Reforesting 'Bare Hills', in Vietnam: Social and environmental consequences of the 5 Million Hectare Reforestation Program, *Ambio: a Journal of the Human Environment* 38 (6), 325–33.
32 Hess, J. and To, T.H. (2010) *GTZ accompanies Vietnam in development and implementation of policy on payment for environment services (PFES)*.
33 Interview by L. Bigot, 2011: N1, N5.
34 Phuong, V.T. and Hai, V.D. (2011) *Report on Forest Carbon Trade activities in Vietnam*, Hanoi, Vietnam: Forest Science Institute of Vietnam.
35 FSC (2012) *Global FSC certificates: type and distribution*, Bonn, Germany: FSC, AC.
36 Government of Vietnam (2007) *Vietnam Forestry Development Strategy 2006–2020. Promulgated and enclosed with the Decision No. 18/2007/QD-TTG of the Prime Minister*.
37 Smith, J., Colan, V., Sabogal, C. and Snook, L. (2006) Why policy reforms fail to improve logging practices: The role of governance and norms in Peru, *Forest Policy and Economics* 8, 458–69.
38 Chhatre, A. and Agrawal, A. (2009) Trade-offs and synergies between carbon storage and livelihood benefits from forest commons, *PNAS* 106 (42), 17667–70.
39 See, for more information: de Schutter, O. (2011) How not to think of land-grabbing: three critiques of large-scale investments in farmland. *Journal of Peasant Studies* 38 (2), 249–79; Robertson, B. and Pinstrup-Andersen, P. (2010) Global land acquisition: neo-colonialism or development opportunity? *Food Security* 2 (3), 271–83; Cotula, L., Vermeulen, S., Leonard, R. and Keeley, J. (2009) *Land grab or development opportunity? Agricultural investments and international land deals in Africa*, London/Rome: IIED/FAO/IFAD.

40 Gupta, J. and Lebel, L. (2010) Access and allocation in global earth system governance: water and climate change compared, *INEA* 10 (4), 377–95.
41 Savaresi, A. (2012) The human rights dimension of REDD, *RECIEL* 21 (2), 102–13.
42 Agarwal, B. (2010) *Gender and green governance: The political economy of women's presence within and beyond community forestry*, Oxford, UK: Oxford University Press.
43 Bandiaky-Badji, S. (2011) Gender equity in Senegal's forest governance history: why policy and representation matter, *International Forestry Review* 13 (2), 177–94.
44 Sun, Y., Mwangi, E. and Meinzen-Dick, R. (2011) Is gender an important factor influencing user groups' property rights and forestry governance? Empirical analysis from East Africa and Latin America, *International Forestry Review* 13 (2), 205–19.
45 Bose (2011) cited in Mwangi, E. and Mai, Y.H. (2011) Introduction to the special issue on forests and gender, *International Forestry Review* 13 (2), 119–22.
46 Sun *et al.* (2011) op. cit.; Mai, Y.H., Mwangi, E., and Wan, M. (2011) Gender analysis in forestry research: looking back and thinking ahead, *International Forestry Review* 13 (2), 245–58; Purnomo, H., Irawati, R.H., Fauzan, A.U. and Melati, M. (2011) Scenario-based actions to upgrade small-scale furniture producers and their impacts on women in Central Java, Indonesia, *International Forestry Review* 13 (2), 152–62; Westermann, O.J., Ashby, J. and Pretty, J. (2005) Gender and social capital: The importance of gender differences for the maturity and effectiveness of natural resource management groups, *World Development* 33 (11), 1783–99; Agarwal (2010) op. cit.; Molinas, J. (1998) The impact of gender composition on team performance and decision-making: Evidence from the field, *World Development* 26 (3), 413–31; Agarwal, B. (2007) Gender inequality, cooperation, and environmental sustainability, in Baland, J.M., Bardhan, P.K. and Bowles, S., *Inequality, cooperation, and environmental sustainability* New York: Russell Sage Foundation; Princeton: Princeton University Press; Agarwal, B. (2009) Gender and forest conservation: The impact of women's participation in community forest governance, *Ecological Economics* 68 (11), 2785–99.; Agrawal, A., Yadama, G., Andrade, R. and Bhattacharya, A. (2004) *Decentralisation, community, and environmental conservation: Joint forest management and effects of gender equity in participation*, CAPRI Working Paper No. 6, Washington DC: International Food Policy Research Institute.
47 Sun *et al.* (2011) op. cit.; Agarwal (2010) op. cit.
48 Bandiaky-Badji (2011) op. cit.
49 Mwangi and Mai (2011) op. cit.
50 Communication of the Committee adopted pursuant to the early warning and urgent action procedures, 13 March 2009, available at:<http://www2.ohchr.org/english/bodies/cerd/docs/early_warning/Indonesia130309.pdf
51 Article 113, Land Law 2003.
52 Article 63, section 4, Land Law 2003.
53 Singer, B. (2008) *Cameroonian forest-related policies: A multisectoral overview of public policies in Cameroon's forests since 1960*, Institut d'Etudes Politiques.
54 Singer (2008) op. cit.
55 Farrell, K. (2012) *Freedom to Serve: a critical exploration of the injustices of international payments for ecosystem services*, Paper presented at Lund Conference on Earth System Governance – Towards a Just and Legitmate Earth System Governance: Addressing Inequities, Lund, 18–20 April 2012.

206 *Joyeeta Gupta et al.*

56 Topa *et al.* (2009) op. cit.; Singer (2008) op. cit.
57 Bäckstrand, K. and Lövbrand, E. (2006) Planting trees to mitigate climate change: contested discourses of ecological modernization, green governmentality and civic environmentalism, *Global Environmental Politics* 6 (1), 50–75; Nielson, T.D. (2012) *Governing forests as cost-effective mitigation, carbon sinks or providing livelihood activities: a discursive overview*, Paper presented at Lund Conference on Earth System Governance – Towards a Just and Legitmate Earth System Governance: Addressing Inequities, Lund, 18–20 April 2012; Arts, B., Appelstrand, M., Kleinschmit, D., Pülzl, H., Visseren-Hamakers, I., Eba'a Atyi, R., Enters, T., McGinley, K. and Yasmi, Y. (2010) Discourses, actors and instruments in international forest governance, in Rayner, J., Buck, A. and Katila, P., *Embracing complexity: meeting the challenges of international forest governance, a global assessment report prepared by the global forest expert panel on the international forest regime*, International Union of Forest Research Organizations.
58 Ecological modernization theory is neo-liberal and focuses on growth, efficiency, win–win options, and promotes the use of voluntary and flexible mechanisms and economic and market instruments that encourage the private sector to take a major role.
59 Green governmentality focuses on effectiveness, administrative rationality and the role of institutions in addressing complex problems.
60 Civic modernization focuses on equity issues, the north–south divide, local participation, and ecological safeguards.
61 Doets, C., Son, N.V. and Tam, L.V. (2006) *The Golden Forest: practical guidelines for AR-CDM project activities in Vietnam*, Hanoi: SNV Vietnam.
62 The authors are indebted to Simone Lovera for this point.
63 Interview by L. Bigot and N.L. Bui, 2011: N15.

10 REDD policies, global food, fibre and timber markets, and 'leakage'

Onno Kuik[1]

10.1 Introduction

This chapter complements the political, policy and legal approach of the previous chapters by taking an economic and global look at REDD policies. The analytical framework (see Chapter 1) stressed the links between drivers and incentives in understanding complex issues like deforestation and forest degradation. This chapter narrows the analysis while broadening it at the same time. The analysis is *narrow* because it is confined to the economic dimensions of the problem. It is *broader* because it tries to capture economic relations that, in principle, span the whole world. It specifically considers the growing demands for food, fibre, fuels, timber and other products of agriculture and forestry of a growing and more prosperous world population as drivers of land-use change and deforestation, and, vice versa, the potential impacts of a policy to halt deforestation on these drivers. It tries to say something on the two-way relationship between drivers and incentives.

This may generate useful insights to policymakers and the general public. Although the economic model employed abstracts from many of the day-to-day realities of forest dwellers and government officials at ground level, it aims to capture the no less real macro-economic linkages at national and international levels. The recent global economic turmoil from the financial crunch shows how vulnerable countries are to the cold winds of 'anonymous' economic market forces, and this is also increasingly true for seemingly isolated forest dwellers – and certainly true for the bigger commercial interests in forests.

From an economic perspective, deforestation is the result of the competition for the scarce resource 'land'. A parcel of land that is presently forested has alternative potential uses. It can be clear-cut and converted to agricultural land, it can be used to extract timber and non-timber forest products, and it can be left alone to produce local, national and global ecological services, such as water catchment, biodiversity maintenance, and carbon sequestration and storage. A rational landowner will choose the alternative use that maximizes the land's rent. Rent is the annual reward to a factor of production that is (quasi) fixed in supply. A classical model of the spatial allocation of land over alternative uses is the Von Thünen model.[2] Two basic assumptions of the model are that 1) profit-maximizing landlords allocate the land to the use that generates the highest rent, and 2) rents decline with distance to the market because of transportation costs. In the simplest model,[3] areas of alternative land uses lie around the central market (central city) in concentric rings. Figure 10.1 shows a quarter of these rings and the associated rent curves.

The central market is in the origin of the diagram. At close distance land is allocated to intensive agriculture, with products yielding the highest prices in the central city. Rent decreases with distance because of transportation costs that are relatively high for these expensive (e.g., fresh) products. Further away (at the intensive margin), it is more profitable to produce goods that have lower per-unit transportation costs (e.g. cereals, meat). Still further away (at the extensive margin), forest products become attractive. Finally, there are inaccessible forests with zero land rents (the old-growth forests in Figure 10.1). At current technology and market conditions, it is not a profitable option to develop these forests in any way.

Using the Von Thünen model, Angelsen shows how changes in the demand for agricultural and forestry products and for labor, and changes in production and transportation technologies, affect the level and slope of the rent curves, thereby affecting land use and land cover.[4] Increases in agricultural land rents push the agricultural frontier outwards and tend to increase deforestation. Causes of increases in agricultural land rents might include: (i) higher output prices, caused by increased demand due to higher income, changes in preferences, opening up markets for international trade, devaluation of currency, lower taxes, etc.; (ii) better roads and transport infrastructure will reduce transport costs and hence increase land rents; (iii) lower off-farm employment

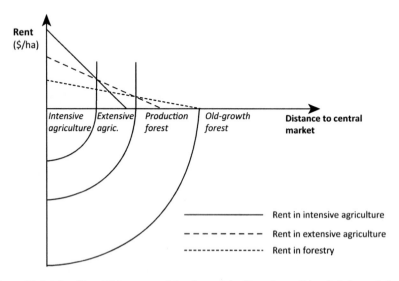

Figure 10.1 The Von Thünen model of spatial allocation of land (adapted from Angelsen 2007)[4]

opportunities and wages reduce the opportunity costs of farm labor and therefore tend to increase land rents and agricultural expansion; and (iv) the effects of changes in technology, input prices, access to credit, and tenure security are ambiguous: they can lead to agricultural expansion but they need not.

Increases in forest land rents push the agricultural frontier inwards and tend to reduce deforestation. Causes include: (i) higher prices of forest products because of higher demand; (ii) technical progress in forestry industries (e.g. logging) except when market prices for forest products decrease significantly as a result; (iii) payments for environmental services, carbon sequestration credits; and (iv) community-based forest management or other ways to manage open access.

Angelsen argues that the effects of these causes (e.g. technology, changes in input and output prices) *and their interactions* on deforestation need to be carefully examined in a general equilibrium framework that takes account of market structure, endogenous prices and quantity changes, and constraints at the relevant levels (local, national, international).[5]

This section discusses some elements of a simple economic model of deforestation based on Von Thünen's model of land allocation. Central to this model is land rent and the factors affecting the relative rents of agriculture and forestry, both over space and over time. These factors include changes in markets, technology and institutions, which, as in Young's Institutional Dimensions of Global Environmental Change (IDGEC) framework (see 1.5), interact in complex causality.

The next section implements the model using an existing computable general equilibrium (CGE) model of the world economy adjusted for forest analysis. A brief description of the adjusted model and the data on which it operates are presented in the next section (see 10.2). Section 10.3 develops a baseline scenario of deforestation over the period 2010–30. It pays special attention to the effect on deforestation of the growing demand for biofuels. Section 10.4 introduces a simple REDD mechanism to mitigate deforestation. It assesses the effects of reduced deforestation on global crops and timber markets and on economic activity and growth. Section 10.5 looks at the unilateral reduction of deforestation in the case study countries and assesses the rate of 'leakage' or 'displacement' in terms of additional deforestation in countries that have no anti-deforestation policies in place. Section 10.6 concludes this chapter.

10.2 Methods and data

The simulations in this chapter were carried out with the Dynamic GTAP model,[6] a global, multi-region, multi-sector *Computable General Equilibrium* model that was slightly adjusted for the analysis. The most important adjustment concerns the land market. This is explained in some detail below, but first this section introduces the geographic regions and economic sectors of the model. The world economy is disaggregated into eight regions that either represent a single country or a composite consisting of several countries. Criteria for regional aggregation were, first, being a case study country in this book and, second, the importance of the region in the supply or demand of tropical timber.

Hence the model includes the case study countries Indonesia, Peru and Vietnam; other tropical timber producing regions such as sub-Saharan (tropical) Africa (including Cameroon) and the 'producing countries' of the International Tropical Timber Organization (ITTO);[7] and major timber-consuming countries and regions such as China and the industrialized countries (IC). The remainder of countries and regions, including Argentina, Chile, Uruguay, Russia, North Africa and the Middle East, is labelled Rest of the World (ROW).

Each region's economy is divided into eight sectors or commodity groups with an emphasis on agriculture and food, and forestry and forestry-related products. According to its comparative advantages each region produces its own variety of goods using primary factors of production such as skilled and unskilled labor, capital, land, and other natural resources, and intermediate inputs from domestic or foreign sources. Final demands for goods produced in each region include government consumption, private consumption, investments, and exports. The model keeps track of all bilateral trade flows of goods between the distinguished countries and regions.

In this model, land is a critical factor of production and four types of land are distinguished:

1. Cropland
2. Grazing land
3. Timberland ('production forest' in Figure 10.1)
4. Unmanaged forest land ('old-growth forest' in Figure 10.1)

Each type of land can be transformed into another type, but this transformation is not costless.[8] At the start of the simulation, 'unmanaged forest land' is not used for the production of goods. If, for some reason, land rents of the cultivated types of land rise it may become attractive to transform some area of unmanaged forest land into productive land, either cropland, grazing or timberland. If unmanaged forest land is transformed into crop or grazing land it is called deforestation. If unmanaged forest land is transformed into timberland it is referred to as (potential) degradation. The degradation caused by forestry activity can range from total deforestation (clear-cutting without regeneration) to negligible (sustainable forestry). Unfortunately the model cannot predict the type of management regime that is likely to be in place on new timberlands. Therefore it is referred to as (potential) degradation with the understanding that this can range from zero to major destruction of the natural vegetation.

10.3 A scenario of future deforestation

10.3.1 Introduction

This section develops a simple scenario of future deforestation. It starts with projections of the growth of population and incomes in the regions of the model. It then focuses on the associated increases in the demands for crops, timber and fuels. It discusses the effects of increasing demand for land-intensive goods on land markets and deforestation. It presents a baseline scenario of tropical deforestation for the period 2010–30.

10.3.2 Growth of population and income

According to demographic projections, the world population grows from almost 6.8 billion people in 2010 to 8.2 billion in 2030.[9] The population will also become richer: in the baseline scenario per capita income increases by almost 50 per cent.[10] The population growth is concentrated in developing countries; almost 60 per cent of the global population growth is in the rainforest countries. The highest population growth is projected for Tropical Africa, with an increase in population of 50 per cent. The world's income is unequally divided: in 2010, per capita income in rainforest countries was only 5.4 per cent of that in developed countries. This gap is not expected to close in the next two decades: in 2030 the ratio has only increased to 5.6 per cent. There are, however, large differences between the rainforest countries: while income per capita is projected to double in Indonesia (119 per cent) and Vietnam (100 per cent), the growth in ITTO countries (62 per cent) and Peru (50 per cent) is moderate, while the per capita income growth in Tropical Africa is projected to be negative (–2 per cent), caused by sluggish economic growth combined with high population growth.

10.3.3 Demand for food and timber

This larger and richer population will demand more food, especially livestock products such as meat and dairy products.[11] Currently, meat consumption per capita is more than four times as large in developed countries as in developing countries. It is generally expected that as incomes in the developing world increase, the gap in meat and dairy consumption will narrow. For example, while population doubled in China between 1961 and 2006, meat consumption grew by a factor of 33.[12]

The demand for timber and other forest products is also expected to increase. From all the land-using sectors, the strongest growth in demand is projected for forest products, due to rising demands for furniture, construction and paper products.[13] China has become a major importer of timber, and is presently the world's leading importer of industrial roundwood.[14] The trend of booming domestic consumption and growing exports of processed wood products, combined with restrictions on domestic timber harvests, will likely increase China's share in global demand in the foreseeable future.[15]

10.3.4 Demand for biofuels

An additional pressure on agricultural land is the growing demand for biofuels from the US and the EU. The US Renewable Fuel Standard mandates a minimum use of 36 billion gallons of biofuel by 2022 (up from 11 billion gallons at present). The EU, in its climate and energy package (the '20-20-20 targets'), require that at least 10 per cent of transportation fuels should be biofuels by 2020. Currently, biofuels such as ethanol and biodiesel are produced from crops like corn, rapeseed, sugar cane and cassava. The inputs to the so-called 'first-generation' biofuels are land-intensive and are diverted from use as food or feed. 'Second-generation' biofuels, produced from waste or grown on degraded and abandoned agricultural lands planted with perennials, are far more efficient and less land-consuming, but they are not yet competitive.[16] Even in the most optimistic scenario, their introduction is not expected before 2015 and after that date their deployment will only be gradual. In a more pessimistic scenario, second-generation biofuels will not enter the market before 2030.[17] Nevertheless, the US mandate requires an increasing share of second-generation biofuels in the overall mandated use.

Especially since the global food price spike of 2008, scholars have become concerned about the implications of biofuel demand on food prices. Early publications on this issue,[18] especially those that were prominently featured by the magazine *Science*, aroused a lot of attention and media coverage, as they predicted that the biofuel policies of the US and the EU could increase food prices by 30 to 60 per cent and, through their impact on land use and deforestation, would, on balance, *not* reduce the global emissions of CO_2 for a very long time.

Some scholars have questioned the magnitude of these early estimates of food price increases. At a high-level FAO meeting on the future of world food and agriculture, Fischer presented estimates of food price increases of 5–26 per cent in 2030, depending on the stringency of the biofuel mandates and the speed of market penetration of second-generation biofuels.[19] The earlier studies have not fully taken account of expected dietary changes towards animal products and they often neglected the option of expanding the cultivated area of farmland.[20] In terms of the Von Thünen model this would mean that the extensive frontier is pushed outwards (to the right in Figure 10.1). Taking account of projected dietary changes and an *endogenous* land endowment, Chakravorty et al. estimate the full effect of the biofuel mandates of the US and the EU on world food prices at 5–17 per cent in 2025, again depending on the speed of market penetration of second-generation biofuels.[21]

10.3.5 *Future demand and supply of land*

In the 2010 Presidential Address to the US Agricultural and Applied Economics Association, Thomas Hertel critically reviewed the evidence on the future supply and demand for agricultural land up to the middle of this century.[22] A number of the factors affecting the future *demand* for agricultural land have been discussed above: population growth, income growth and the attendant changes in food demand that accompany changes in per capita purchasing power, and the increasing demand for biofuels and perhaps other forms of bioenergy. The 'mediator' between the demand for food and timber and the demand for agricultural land is the productivity of the land. What are the future prospects for land productivity, i.e. what is the scope for increasing yields?

Hertel cites a paper to the aforementioned high-level FAO meeting, where Bruinsma calculated that over the 1961–2005 period more than three-quarters of the growth in world crop production was due to increasing yields and another one-tenth due to increased cropping intensity.[23] Only 14 per cent of the historical growth was due to land expansion. Will this trend of ever-increasing yields continue in the future? There is evidence of a declining growth rate for agricultural yields over the past two decades. There can be different reasons for that: it can be (economically) linked to a declining rate of growth in demand; it may be because of less expenditure in agricultural research, and a diversion from yield-enhancing projects; or it may be because yields reach some natural limit, i.e. in many regions of the world actual yields reach 'maximum attainable yields'. Hertel concludes that the technical potential may not yet be in sight for many crops in many regions, but that for a variety of reasons, for the individual farmer the economic optimum is often far below the technical optimum.

The future *supply* of land is determined by another set of factors, including land lost to urbanization (especially in densely populated regions of China and India, but also a factor in the case study countries), the removal of land from commercial production for recreation and the preservation of biodiversity, the negative effects of climate change on the productivity of land, and increasing water scarcity. In addition, the supply of agricultural land could be further reduced by increasing forest cover for carbon sequestration, the subject of this chapter.

If land will be in short supply globally, the extensive margin may shift into the old-growth forest. This does not necessarily imply, however, that total forest cover would decrease in every country. The 'forest transition' literature has pointed out that despite alarmingly high overall rates of tropical deforestation, some tropical countries have recently been through a forest transition – a shift from net deforestation to net reforestation (see 2.2).[24] The causes of such a transition include overall economic development that pulls labor away from marginal agriculture and increasing scarcity of forest products. A forest transition is not the same as forest conservation, however. A forest transition may be concomitant with a continuing degradation and clearing of natural forests. In a summary of various estimates and scenarios of global land use for the 2000–2030 period, Lambin and Meyfroidt report *both* global expansion of industrial forestry (56–109 million hectares) and clearing of natural forests (152–303 million hectares).[25]

10.3.6 *Baseline scenario of land-use change and deforestation*

Based on the above considerations and estimates, the baseline scenario of land-use change and deforestation for the 2001–30 period has been constructed. It uses land-use change data from FAO for the period 2001–10 to calibrate the

214 *Onno Kuik*

model.[26] The period 2010–30 is the projection period. Over this period, the model predicts that world market prices of crops and timber increase by 1.5 and 1.0 per cent per year, respectively.[27] This is partly due to an increased demand for biofuels by the US and the EU. Land rents increase because of the rising crop and timber prices, and idle lands, including unmanaged forest areas, are converted into crop, grazing and timberland. In the baseline scenario, total deforestation over the period 2001–30 is 235 million hectares, which is in the middle of the range suggested by Lambin and Meyfroidt. Deforestation over the period 2001–10 has already happened; over the period 2010–30 additional deforestation at 138 million hectares is expected, of which 111 million hectares are in tropical forest countries. Figure 10.2 shows deforestation as the permanent conversion of unmanaged forest land into crop, grazing and timberland. In Peru and Africa there is also conversion of timberland into crop and grazing land (indicated by the bars below the x-axis).

Figure 10.2 suggests that agricultural expansion will be the primary cause of deforestation in Peru, Africa and ITTO countries, while the expansion of forestry activities will be a relatively more important cause in Indonesia. This is in line with the findings of a recent report of FAO and ITTO on the three rainforest regions Amazon Basin, Congo Basin and South East Asia.[28] The report identifies 'excessive extraction and poor harvesting techniques' as major causes of forest degradation in South East Asia, while the 'expansion of the agricultural frontier' and 'slash-and-burn agricultural expansion and illegal or unsustainable firewood extraction and charcoal production' are primary causes for the Amazon Basin and the Congo Basin, respectively.[29]

10.4 Economic effects of REDD-induced forest conservation

10.4.1 Introduction

This section addresses the question, what would happen if agricultural expansion and deforestation would be significantly slowed down by forest conservation measures in the tropical forest countries, possibly induced by an international REDD mechanism?

First, the model considers the effect of forest conservation measures on the *prices* of crops and timber and on the total economy of the regions concerned, for two reasons: first, to analyse whether there is a possible trade-off between forest conservation and food security and, if there is, how serious it is and who is likely to be most affected; second, in order to say something on the two-way relationship between drivers of deforestation and incentives to halt deforestation. That is, the growing demands for crops, fuels and timber are major drivers of deforestation. These growing demands express themselves through increasing rents for productive uses of land and thereby increasing the *opportunity costs* of forest conservation. The opportunity costs of forest conservation are the forgone returns of the next best alternative use (e.g. agriculture). If the

REDD policies, global food, fibre and timber markets, and 'leakage' 215

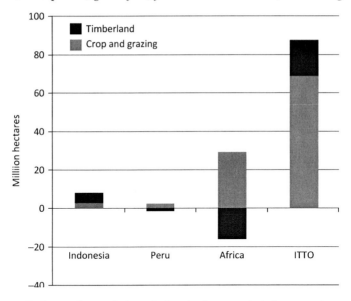

Figure 10.2 Deforestation and degradation in four tropical forest regions over the period 2010–30

conservation of one parcel of forest would *increase* the prices of crops and timber, thereby increasing the land rent, the opportunity cost of conserving the next parcel of forest would increase. It is this dynamic interaction between *the amount* of conservation and *the cost* of conservation that can be assessed.

This section first describes a forest conservation policy scenario (section 10.4.2), then describes the effect of this policy scenario on the world market prices of crops and timber (section 10.4.3), and finally describes the overall effects of the policy scenario on the economies of the regions concerned (section 10.4.4). This section is closed by some concluding remarks.

10.4.2 A forest conservation policy scenario

In the forest conservation policy scenario, we assume that forest conservation policies are successfully implemented from 2015 onwards. In the period 2015–20, these policies reduce baseline deforestation rates in tropical forest countries by one quarter, and between 2020 and 2030 they reduce these rates by one half. In this scenario, by 2030 the conservation measures would have saved 28 million hectares of forest.

10.4.3 Global food and timber markets

The forest conservation policies would restrict the expansion of crop, grazing and timberland into tropical forests. This would limit the supply of crops, livestock products and timber relative to the supply in the baseline. This would put a pressure on their prices, creating incentives for intensification of production of existing lands and possibly some substitution in consumption. The latter will be limited as the price elasticity of food is generally low, meaning that the overall demand for food is not much affected by a change in its price.

A simulation of the forest conservation policy with the economic model suggests relatively small world market price increases for crops, livestock and timber. By 2030, the average price of crops is about 0.8 per cent higher than in the baseline, that of livestock 0.3 per cent, and that of timber 1.4 per cent. To get an idea of the magnitude of these price increases it is good to compare them to the price increases of these goods in the baseline. Crops and livestock products are projected to increase by 34 and 18 per cent over the course of the projection period (2010–30). Timber prices are projected to increase by 21.5 per cent. Hence, the *additional* global price increase of food due to the forest conservation policy is rather limited.

There are regional differences in the price increases, though. Due to the forest conservation policies, crop prices in Indonesia increase by 1.8 per cent by 2030 and those in ITTO increase by 2.3 per cent above already impressive baseline growth rates of 60 and 52 per cent. Price increases for crops in Peru and Africa are less pronounced at the global average (0.8 per cent). Market prices of timber increase in all regions, from 3.8 per cent in ITTO to 1.1 per cent in Africa and Peru.

10.4.4 Economy-wide effects and environmental benefits

What are the effects of the forest conservation policy on the overall economy? This is measured by the change in Gross Domestic Product (GDP) which gives an idea of the opportunity costs of the forest conservation policy. That is, it measures the value of the economic activity that is forgone because of the policy, taking into account optimal economic adjustments by economic agents (such as intensification of production and adjustment of consumption patterns because of increased prices of food and timber, as discussed above). The opportunity cost encompasses the cost element that Sohngen labels as 'system-wide adjustment cost'.[30]

Next to the opportunity costs there are other costs of forest conservation, especially in the REDD context. Sohngen distinguishes between implementation and management costs, measurement, monitoring and verification costs, and other transaction costs (e.g. search costs, learning costs, the costs of drafting contracts, legal procedures, etc.).[31] There is not much empirical evidence on the magnitude of these costs. On the basis of the available literature, including estimates for projects in developing countries involving many smallholders, Sohngen takes transaction costs (including measurement, monitoring and verification costs) of 20 per cent of the carbon price as a 'simplifying assumption'.[32]

Figure 10.3 shows that the effect on GDP is highest in Indonesia, suggesting the opportunity costs of forest conservation are highest in that country;

higher than in the ITTO region, and higher than in Africa and Peru. By 2030, the opportunity cost of the forest conservation policy in Indonesia is almost 0.7 per cent of GDP, which is not insignificant.

It is interesting to note that the projected GDP change of the industrialized countries (INC) is positive. This is due to the fact that the industrialized countries have a surplus on their agricultural trade balance: the value of their overall exports of agricultural commodities is larger than the value of their imports. An increase in the prices of agricultural commodities increases the values of exports and imports in equal proportion, but because of the trade surplus, the total benefits of the increase in export value exceed the total costs of the increase in import value. The total benefit for industrialized countries is so large in the simulation that they could pay the tropical forest countries an annual rental value for avoided carbon emissions equivalent to up to USD 23/tCO_2, without being worse off in comparison to a situation without a forest conservation programme.

The benefits of the forest conservation policy may be significant for other reasons than carbon storage and sequestration as well. Forest conservation has many environmental and social benefits, including supporting, provisioning, regulating and cultural services (see Table 1.1). How much CO_2 emissions would be avoided is subject to scientific uncertainty. The available evidence suggests that this would depend on many factors, of which especially the fate of carbon stored in the soil is still poorly understood. Ongoing studies are addressing this issue through modelling and field experiments.[33] FAO/ITTO presents estimates of above-ground and below-ground carbon in tropical forests in the three major tropical forest basins. The carbon stocks range from 580 tons of CO_2-equivalent in Africa to 800 tons of CO_2-equivalent in South America.[34] Not all carbon may be lost due to conversion, however. Kindermann et al. present three global forestry and land-use models that predict rates of CO_2 avoided between 300 and 400 metric tons per hectare of tropical forest.[35] Based on these estimates, Sohngen[36] uses a global-average estimate of 350 tons, very close to the estimate of Searchinger.[37] The estimate is used here as a conservative estimate of CO_2 emissions avoided by deforestation.

Under this assumption, REDD payments are calculated as an eternal annuity on the value of the CO_2e-emissions avoided. The net benefits are the REDD payments minus the GDP opportunity costs minus fixed transaction costs per ton of CO_2. The different regions have different 'break-even' carbon prices, above which the net benefits of REDD would become positive. For Africa, the break-even price would be USD 12/tCO_2e, for Peru USD 28/tCO_2e, for ITTO USD 43/tCO_2e, and for Indonesia USD 59/tCO_2e.[38]

Table 10.1 below presents estimates of the costs of avoiding emissions through deforestation. The estimates of opportunity costs (excluding transaction costs) are compared with those of Grieg-Gran and Kindermann et al.[39] All estimates refer to the year 2030. There is general agreement among the studies on the costs for Africa, there is some agreement for Peru, but there is wide disagreement for Indonesia. For Indonesia, the cost estimate is 40 per cent

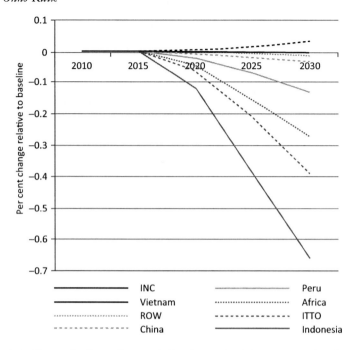

Figure 10.3 Change in Gross Domestic Product (%) with respect to baseline due to forest conservation policy

higher than the high estimate of Kindermann and more than four times higher than the high estimate of Grieg-Gran. This suggests that the uncertainty on the cost of avoiding deforestation in Indonesia is particularly high.

For most regions (except for Africa) the cost estimates are higher than those found in earlier literature. One reason for this difference could be the assumption of rising real food and timber prices, whereas the models described in Kindermann do not explicitly model agriculture and the estimates of Grieg-Gran are static, based on 2007 outputs and prices. Another reason may be that the two alternative estimates do not take system-wide adjustment costs (knock-on effects on the rest of the economy) into account. For example, a lower supply of timber will also negatively affect the wood-carving industry, and so on. It is, of course, also the case that the model is rather coarse, especially regarding the spatial level. The model implicitly assumes that all land in a region has the same quality and it also does not take account of the 'remoteness' of land, as was crucial in the Von Thünen model, recall Figure 10.1. It may be that the model overestimated the productivity of deforested land, both because of quality and remoteness, and that a more spatially explicit model is needed to derive more accurate results.

Table 10.1 Estimates of the costs of avoiding emissions through deforestation in 2030 (USD/tCO$_2$)

Country	Author	Cost per ton of CO$_2$ avoided (USD)
Indonesia	Grieg-Gran	5 – 12
	Kindermann (Southeast Asia)	8 – 38
	This study	54
Peru	Grieg-Gran (Bolivia)	7 – 9
	Kindermann (Central and South America)	10 – 24
	This study	24
Africa	Grieg-Gran (Cameroon, DRC, Ghana)	5 – 10
	Kindermann	5 – 12
	This study	8

10.4.5 Inferences

A REDD-type of forest conservation policy in tropical countries could reduce CO$_2$ emissions. It would also raise crop and timber prices, however, thereby amplifying the major driving forces of deforestation. Opportunity costs of the conservation policy would be lowest in Africa and highest in Indonesia. Industrialized countries would benefit economically from the forest conservation policies through an advantageous terms-of-trade effect. This would help them to compensate the tropical forest countries for their opportunity and policy costs. In the simulation, the compensation needed would be relatively high for some regions, especially Indonesia. While we think that the general patterns that we found are valid, the exact number may have to be validated by a more spatially explicit model that would take account of within-country differences in land productivities.

10.5 Leakage

10.5.1 Introduction

Leakage occurs when efforts to control emissions in one place cause emissions to shift to another place that is not subject to the policy and where the emissions are not accounted for. This is a problem as it makes the *net* contribution of a local emissions reduction project or policy to global emissions reduction uncertain. Leakage as a result of forest conservation policies is difficult to measure. To a limited extent, it is possible to monitor land-use changes in the immediate vicinity of projects. But there is no possibility of monitoring and measuring leakage directly if leakage is transferred through international markets. It has been argued that forest and agricultural projects that aim to reduce the emissions of carbon are *susceptible* to leakage because of the fixed land base and the broad scope of markets for forest and agricultural commodities.[40]

Thus, a change in land use in one region can easily have consequences on land use in faraway regions. This is not directly observable, but can only be inferred from more or less complex economic land-use and carbon emissions models.

Leakage can occur within the borders of a country or between countries. Most current REDD proposals focus on national-level approaches, partially to avoid leakage problems that have been found in project-based approaches like CDM. Nevertheless, REDD may be susceptible to international leakage. The risk of leakage is greater, the greater the effects of national REDD policies on world commodity markets, for example on the world markets of timber and agricultural commodities. Often multiple markets will be affected simultaneously by REDD policies. Assessing the rate of leakage in such circumstances 'requires a [...] modeling framework that simultaneously solves for multiple markets and integrates carbon accounting with the simulated changes in market outcome'.[41]

> In the context of REDD, van Noordwijk and Minang identified four types of leakage:[42]
> 1. Land-based leakage: if land-use restrictions in one area shift those land uses to other areas where associated emissions are not controlled and accounted for.
> 2. People-based leakage: if land-use restrictions reduce livelihoods of people in the area, people may seek alternative employment outside the area that generates emissions.
> 3. Commodity-based leakage: if the supply of carbon-intensive commodities (such as charcoal) is diminished, these commodities may be produced elsewhere.
> 4. Cross-sectoral leakage: if leakage from the above types is to be avoided by increasing the productivity of alternative employment in the area (e.g. agricultural intensification) the associated emissions should be accounted for.

Although there is clearly some overlap between these types of leakage (e.g. between the land-based and commodity-based types), the typology clearly illustrates that leakage is the end result of many economic decisions of producers and consumers. This emphasizes the need to apply broad, multi-sectoral models to assess leakage.

Leakage is generally considered to be a major problem in carbon sequestration policies, including REDD policies. There is a small body of research that has quantitatively assessed displacement and leakage effects of forest conservation and carbon sequestration policies, mostly in the USA and other developed countries. There are as yet no comprehensive assessments of international emissions leakage due to forestry-based carbon sequestration policies or REDD policies.

A number of studies have addressed the issue of displacement effects from forest conservation policies. The term displacement effect is used for the geographical shifts in the supply of commodities from the policy region to other regions. Wear and Murray carried out an *ex-post* econometric study on the displacement effects of timber harvesting restrictions on US federal lands in the North-West Pacific starting from the late 1980s.[43] These timber harvesting restrictions were in part driven by efforts to protect the habitat of the northern spotted owl as mandated by the Endangered Species Act of 1973. Wear and Murray estimated that over 40 per cent of the forgone harvests were shifted to private lands within the region, to other regions in the USA and to Canada.

Lang and Chan suggest that the 1998 logging ban in China resulted in a substantial displacement of forest exploitation to South East Asian countries with weak regulatory regimes, such as Indonesia, Myanmar, Cambodia and Papua New Guinea.[44] Meyfroidt and Lambin suggest that Vietnam's forest transition (increasing forest cover since 1992) has been partially made possible by the displacement of forest exploitation to and legal and illegal imports from Cambodia and Laos in the early 1990s and later Malaysia, Myanmar and Indonesia.[45] They estimate on the basis of various sources and detailed material flow analysis that 39.1 per cent of the regrowth of Vietnam's forests over the period 1987–2006 was made possible by the international shift in forest exploitation and imports. The authors also estimate that approximately half of the wood imports to Vietnam in this period were illegal.[46]

Gan and McCarl used a Computable General Equilibrium (CGE) model to simulate displacement effects of unilateral timber supply restrictions internationally, distinguishing ten countries/regions.[47] They found displacement rates between 42 per cent (Canada) to 95 per cent (Russia). Displacement rates for regions with tropical forests ranged from 70 per cent (East Asia) to 87 per cent (Sub-Saharan Africa). Studies on within-country leakage due to regional carbon sequestration forestry projects have been carried out for the USA[48] and for Bolivia.[49] The estimated rates of displacement of various projects in these countries varied between 0 and 100 per cent.

Although interesting, these 'displacement' studies might not provide much insight into the question of carbon leakage. First, they do not quantify the areas actually deforested and their associated emissions, and second, they are concerned with forestry only and do not take the important role of agriculture into account. In order to try to begin to fill this gap, some 'leakage' simulations with the economic model are carried out. Section 10.5.2 discusses the leakage simulations. Section 10.5.3 discusses what can be learned from the simulations.

10.5.2 Leakage simulations

The leakage simulations are carried out to shed some light on the complexities of leakage in the REDD and forestry conservation context. The exact estimates of the rates of leakage that are generated by the simulations should not be taken too seriously, as will be explained in more detail below.

In the leakage simulations it is assumed that the tropical forest countries and regions implement forest conservation policies unilaterally. The conservation rates for the countries and regions are the same as in the previous REDD policy scenario. Hence, in the period 2015–20, these policies reduce baseline deforestation rates by one quarter, and in 2020–30 they reduce these rates by one half. The difference with the previous scenario is that all countries and regions act unilaterally, so we have four scenarios that we just name after the country or region that implements the conservation policy: Indonesia, Peru, Africa and ITTO.

Three alternative 'displacement' metrics are distinguished. The first one is displacement of forestry output. This metric corresponds to the metrics used by the studies that were discussed above. It measures the ratio of the increase in forestry output in all other regions and the decrease in the output in the 'policy' region. Hence, a displacement percentage of 20 means that of the decrease of the forestry output in the 'policy' region, 20 per cent is offset by increases in forestry output in other regions. The second metric is displacement of forest land. It measures the ratio of the increase in timberland in all other regions and the decrease in timberland in the 'policy' region. The third metric is the displacement of unmanaged forest land. It measures the ratio of the decrease in unmanaged forest land in all other regions and the increase in unmanaged forest land in the 'policy' region.

Table 10.2 presents the results of the leakage simulations in terms of the displacement metrics that were just discussed. Displacement ratios for forestry output range from 13.4 per cent (Africa) to 28.5 per cent (ITTO). Displacement ratios for timberland are smaller, from 0.9 per cent in Peru to 14.3 per cent in Indonesia. Displacement ratios for unmanaged forest are the smallest ratios (except for Peru) and they range between 0.6 per cent (Africa) and 3.8 per cent (Indonesia).

The results of Table 10.2 suggest that the percentage increase in forestry output in non-conservation countries as a result of conservation efforts in the 'policy' country is larger than the percentage increase in forest area (timberland plus unmanaged forest). There are two major reasons for this difference.

First, different regions have different options to increase forestry output. While in some regions output is mainly increased by increasing all factors of production (including land) in equal proportions, other regions increase forestry output by increasing output per hectare, i.e. by intensifying production. For example, in the 'Peru' experiment, more than 90 per cent of forestry output is displaced to industrialized countries.[50] The displacement of forestry land to industrialized countries is only 25 per cent, however. Looking closely at the change in forestry production in industrialized countries reveals that the increase is indeed made possible by an increase of labor and capital employed per hectare of timberland. The reason for that is that the forest conservation policies did not only raise the world market price of forestry output, but also that of agricultural outputs. The competition for land in the other regions therefore increases and its price will rise. All land-using activities, including forestry, will try to economize on its use. Thus, land is to some extent substituted for labor and capital.

Table 10.2 Alternative displacement metrics for unilateral forest conservation policies

	Indonesia	Peru	Africa	ITTO
Displacement, forestry output	19.7	16.5	13.4	28.5
Displacement, timberland	14.3	0.9	4.1	10.1
Displacement, unmanaged forest	3.8	1.2	0.6	1.4

Second, the simulations reveal a complex interaction between timberland and agricultural land. In the Indonesia, Africa and ITTO simulations, the increase in timberland in the other regions is larger than the increase in new land through the conversion of unmanaged forests. This is possible because some agricultural land, specifically grazing land, is converted to forestry. In the Peru simulation, the reverse seems to be true: some timberland is converted to agricultural land. These complex interactions also make it difficult to translate displacement into carbon emissions. The conversion of grazing land into timberland surely has a different emissions profile than the conversion of unmanaged forest into cropland, and the sign of the change in emissions may even be different.

Which displacement metric is the best approximation to carbon leakage? Without knowledge of the type of management of forestry (e.g. predatory, sustainable) in the countries to which forestry is 'displaced', it is hard to tell. If all forestry on timberlands were sustainable, then the only concern would be the encroachment of forestry operations into unmanaged forest land. If all forestry were based on predatory logging, carbon leakage could be much higher. Without additional knowledge of management, carbon leakage may be anywhere between the unmanaged forest displacement metric and the forestry output displacement metric. As an example, leakage because of unilateral Indonesian conservation measures would be between 3.8 and 19.7 per cent.

If more than one country or region were to implement conservation measures, the leakage *rate* would not increase but the volume of avoided emissions *would*. Hence, the leakage effects (due to additional deforestation and degradation) in the rest of the world would increase.

Finally, we briefly address the issue of 'people-based leakage' that was identified by van Noordwijk and Minang.[51] People-based leakage can be inferred from changes in sectoral employment and associated emissions. Table 10.3 shows model results of employment changes in the forest conservation scenario of section 10.4 for the tropical forest regions. For three regions, the scenario shows small percentage employment changes away from the land-based sectors (crops, livestock and forestry) into mining and manufacturing. This will in general signify smaller land-based emissions and higher energy-based emissions. The exception is Africa, where land-based employment increases despite the land-use restrictions that follow from the forest conservation policy. In Africa, the higher crop prices because of the forest conservation policies outweigh the higher land costs and the smaller land area is farmed more intensively. Of course, as van Noordwijk and Minang note, this *intensification* of agricultural production may increase its associated emissions.[52]

224 *Onno Kuik*

It is not possible to evaluate the employment changes in terms of emissions. A model would be needed that would account for land-based as well as energy-based emissions. This is a challenge for further research. For now it seems that the resulting employment changes are rather small (less than 1 per cent of sectoral employment for all sectors) and associated emissions will therefore also be relatively small.

10.5.3 *Inferences*

Leakage is a somewhat elusive concept, especially in the contexts of REDD and forest conservation policies in general. The causal chain from conservation in one place, demand and supply reactions on multiple commodity markets at national and international levels, land-use changes in other places, and finally the consequences of these land-use changes for emissions is long and complex. The above simulations were meant to give some insight into the complexities rather than to provide accurate predictions of the rate of leakage. Major complexities and uncertainties involve the interactions between agricultural and forestry markets, the elasticity of the supply of land, forestry management, and the net emissions of carbon due to location-specific land-use changes.

10.6 Inferences

Economic modelling of the consequences of REDD is a complex issue. To a large extent this is due to difficulties of correctly modelling the complex dynamics of forest management that involve very long planning horizons up to 50 years or longer. However, in recent years, significant improvements have been made in the economic modelling of forestry activities within an economy-wide (general equilibrium) context.[53] Global databases of forest activities and forest carbon have been developed that can be used in CGE models[54] and novel approaches for dealing with the dynamics of forestry management, land use and land supply (encroachment of hitherto inaccessible lands) have been developed.[55] Although there seems still a long way to go, these developments make the modelling of the economy-wide effects of forestry-related greenhouse gas mitigation policies, such as REDD, an exciting enterprise.

Table 10.3 'People-based' leakage: employment changes of unskilled labor in different sectors (%)

	Indonesia	*Peru*	*Africa*	*ITTO*
Land-based	−0.66	−0.03	0.04	−0.64
Mining	0.29	0.02	0.03	0.26
Manufacturing	0.21	0.12	−0.08	0.31
Services	0.10	−0.03	−0.01	0.01

This chapter has reviewed some of the literature and presented the results of relatively simple simulations with a dynamic CGE model to illustrate some of the complexities and interactions. It started out with a brief exposition of Von Thünen's classic model of land allocation to emphasize the central role of land rent in the analysis of the economic effects of REDD. It then introduced the dynamic CGE model with which REDD policy scenarios were simulated. It paid attention to the modelling of a baseline scenario for future deforestation. In the scenario, deforestation is driven by the economic forces that determine the future supply of and demand for land. Factors that influence this future demand are the growth of population and income and the derived demand for food, fuel and fibre, with a special role for the increasing demand for biofuels. To what extent can this growing demand for land-based commodities be met by increasing yields and to what extent will it be necessary to convert hitherto unproductive lands, including old-growth forests, into productive use? The baseline scenario for the period 2010–30 projected that globally 111 million hectares of tropical forest would be lost.

A REDD-type of forest conservation policy was simulated, conserving 28 million hectares of tropical forest that would otherwise be lost. This forest protection policy would increase crop and timber prices by a few per cent, potentially to the disadvantage of the poorest consumers. In terms of overall economic impact, the loss of income would be largest in Indonesia where an income loss of almost 0.7 per cent of GDP is projected for 2030. Perhaps somewhat surprisingly, income in industrialized countries increases because of the forest conservation measures in tropical countries. This was the result of an improvement of the term-of-trade of the industrialized countries. This benefit would make it easier for industrialized countries to pay the tropical countries for the opportunity costs and measurement, monitoring and verification costs of REDD conservation policies. Using simple assumptions on carbon avoided through forest conservation, it was calculated that the costs of the conservation policies would vary from USD 12/t CO_2e in Africa to USD 59/t CO_2e in Indonesia. A comparison with earlier assessments showed that these cost estimates for most regions are on the high side, especially for Indonesia.

In the literature on REDD much attention is paid to the possibility of leakage. This issue was addressed by carrying out some additional simulations. These simulations gave some insight into the complexities of estimating the rate of leakage. Major complexities involved the interactions between agricultural and forestry markets, the rate at which forests are cleared as a result of rising land rents, forestry management, and the calculation of net emissions of different land-use changes. It was emphasized that displacement of forestry activity that is sometimes used in the literature as a proxy for carbon leakage is not a very good proxy, and that the analysis should be focused as much as possible on land use. The chapter also suggested some further work on the analysis of the interactions between land-based and energy-based emissions in the analysis of leakage.

Notes

1 The author thanks Marjan Hofkes, George Dyer and Patrick Meyfroidt for their constructive comments on earlier drafts of this chapter.
2 Thünen, J. H. von (1842) *Der isolierte Staat in Beziehung auf Landwirtschaft und Nationalökonomie*, Rostock.
3 In the simplest model, there is one isolated market that has no interactions (trade) with other markets. The land is flat and of homogeneous quality. There are no transport infrastructures such as roads or rivers and farmers transport their produce by horse and cart. Transportation costs depend on the type of commodity and distance.
4 Angelsen, A. (2007) Forest cover change in space and time: combining the Von Thünen and forest transition theories, *World Bank Policy Research Working Paper 4117*, Washington DC: The World Bank.
5 Angelsen (2007) op. cit.
6 Ianchovichina, E. and McDougall, R.A. (2001) Theoretical Structure of Dynamic GTAP, West Lafayette, Indiana, USA: Purdue University.
7 Producer countries of ITTO comprise 33 countries from Latin America, Africa and Asia & Pacific. The ITTO region includes these countries minus Peru, Indonesia and all African countries except the Democratic Republic of Congo.
8 The ease of transformation is determined by the elasticity of transformation. The elasticity of transformation measures the relative change in land use because of a relative change in rents. The smaller this elasticity, the larger the difference in land rent should be to shift one hectare of land from a lower-rent use to a higher-rent use.
9 United Nations (2009) World population prospects: The 2008 revision.
10 Walmsley, T. (2006) A baseline scenario for the dynamic GTAP model, West Lafayette, IN, USA: GTAP, Purdue University.
11 Chakravorty, U., Hubert, M.-H., Moreaux, M. and Nostbakken, L. (2011) Will biofuel mandates raise food prices? Edmonton AB, Canada: University of Alberta.
12 Roberts, M.J. and Schlenker, W. (2010) Identifying supply and demand elasticities of agricultural commodities: implications for the US ethanol mandate, NBER Working Paper #15921.
13 Golub, A., Hertel, Th., Sohngen, B. (2009) Land use modeling in recursively-dynamic GTAP framework, in: Hertel, Th. W., Rose, S. K. and Tol, R. S. J. (eds) *Economic analysis of land use in global climate change policy*, 235–78, Oxon, UK: Routledge.
14 Zhang, J. and Gan, J. (2007) Who will meet China's import demand for forest products? *World Development* 35 (12), 2150–60.
15 Zhang and Gan (2007) op. cit.
16 Fargione, J., Hill, J., Tilman, D., Polasky, S. and Hawthorne, P. (2008) Land clearing and the biofuel carbon debt, *Science* 319 (5867), 1235–8.
17 Fischer, G. (2009) World food and agriculture to 2030/50, Paper prepared for the FAO expert meeting on 'How to feed the world in 2050', FAO, Rome, 24–26 June 2009.
18 Searchinger, T., Heimlich, R., Houghton, R.A., Dong, F., Elobeid, A., Fabiosa, J., Tokgoz, S., Hayes, D. and Yu, T.-H. (2008) Use of US croplands for biofuels increases greenhouse gases through emissions from land-use change, *Science* 319, 1238–40; Rosegrant, M. W., Zhu, T., Msangi, S. and Sulser, T. (2008) Global

scenarios for biofuels: Impacts and implications, *Review of Agricultural Economics* 30 (3), 495–505; Fargione *et al.* (2008) op. cit.
19 Fischer (2009) op. cit.
20 Chakravorty *et al.* (2011) op. cit.
21 Chakravorty *et al.* (2011) op. cit.
22 Hertel, Th.W. (2010) *The global supply and demand for agricultural land in 2050: A perfect storm in the making?* GTAP Working Paper No. 63, West Lafayette: Purdue University.
23 Bruinsma, J. (2009) The resource outlook to 2050. By how much do land, water use and crop yields need to increase by 2050? Paper prepared for the FAO expert meeting on *How to feed the world in 2050*, FAO, Rome, 24–26 June 2009.
24 Meyfroidt, P. and Lambin, E.F. (2011) Global forest transition: Prospects for an end to deforestation, *Annual Review of Environment and Resources* 36, 343–71.
25 Lambin, E.F. and Meyfroidt, P. (2011) Global land use change, economic globalization, and the looming land scarcity, *Proceedings of the National Academy of Sciences* 108 (9), 3465–72.
26 FAO (2010) *Global forest resources assessment 2010*, FAO Forestry Paper 163, Rome: Food and Agriculture Organization of the United Nations.
27 These prices are price indices of global exports. Crop prices are endogenously determined by the model. They are in the range projected by Fischer (2009) op. cit. when the biofuel-induced price increases of Chakravorty *et al.* (2011) op. cit. are taken into account. Timber prices are exogenously shocked based on Golub *et al.* (2009) op. cit.
28 FAO/ITTO (2011) *The state of forests in the Amazon Basin, Congo Basin, and Southeast Asia*. A report prepared for the Summit of the Three Rainforest Basins, Brazzaville, Republic of Congo, 31 May–3 June 2011, Rome: FAO.
29 FAO/ITTO (2011) op. cit., pp. 47, 39 and 42, respectively. Infrastructure development, expansion of industrial agriculture and population growth are also important drivers in South East Asia.
30 Sohngen, B. (2010) Forestry carbon sequestration, in Lomborg, Bjorn (ed.) *Smart solutions to climate change*, 114–32, Cambridge: Cambridge University Press, p.116.
31 Sohngen (2010) op. cit.
32 Sohngen (2010) op. cit., p.126.
33 Powers, J.S., Corre, M.D., Twine, T.E. and Veldkamp, E. (2011) Geographical bias of field observations of soil carbon stock with tropical land-use changes precludes spatial extrapolation, *Proceedings of the National Academy of Sciences* 108 (15), 6318–22; Farmer, J., Matthews, R., Smith, J.U. and Singh, B.K. (2011) Assessing existing peatland models for their applicability for modelling greenhouse gas emissions from tropical peat soils, *Current Options in Environmental Sustainability* 3, 1–11.
34 FAO/ITTO (2011) op. cit.
35 Kindermann, G., Obersteiner, M., Sohngen, B., Sathaye, J., Androsko, K., Rametsteiner, E., Schlamadinger, B., Wunder, S. and Beach, R. (2011) Global cost estimates of reducing carbon emissions through avoided deforestation, *Proceedings of the National Academy of Sciences* 105 (30), 10302–7.
36 Sohngen (2010) op. cit.
37 Searchinger *et al.* (2008) op. cit.
38 Without transaction costs, the break-even prices would be USD $4/tCO_2e$ lower.

39 Grieg-Gran, M. (2008) The cost of avoiding deforestation. Update of the report prepared for the Stern Review of the Economics of Climate Change, London: International Institute for Environment and Development; Kindermann *et al.* (2011) op. cit.
40 Murray, B.C. (2009a) Leakage with forestry and agriculture offsets: What do we really know? *Biological Sequestration through Greenhouse Gas Offsets Conference*, Washington DC: Nicholas Institute for Environmental Policy Solutions, Duke University.
41 Murray, B.C. (2009b) Leakage from avoided deforestation compensation policy, in Palmer, C. and Engel, S. (eds), *Avoided Deforestation*, 151–72, London and New York: Routledge, p.160.
42 Van Noordwijk, M. and Minang, P.A. (2009) If we cannot define it, we cannot save it, in Van Bodegom, Arend Jan, Savenije, Herman and Wit, Marieke (eds), *Forests and climate change: adaptation and mitigation*, 5–10, Wageningen: Tropenbos International.
43 Wear, D.N. and Murray, B.C. (2004) Federal timber restrictions, interregional spillovers, and the impact on US softwood markets, *Journal of Environmental Economics and Management* 47 (2), 307–30.
44 Lang, G. and Chan, C.H.W. (2006) China's impact on forests in Southeast Asia, *Journal of Contemporary Asia* 36 (2), 167–94.
45 Meyfroidt, P. and Lambin, E.F. (2009) Forest transition in Vietnam and displacement of deforestation abroad, *Proceedings of the National Academy of Sciences* 106 (38), 16139–44.
46 Meyfroidt and Lambin (2009) op. cit.
47 Gan, J. and McCarl, B.A. (2007) Measuring transnational leakage of forest conservation, *Ecological Economics* 64, 423–32.
48 Murray, B.C., McCarl, B.A. and Lee, H.C. (2004) Estimating leakage from forest carbon sequestration programs, *Land Economics* 80 (1), 109–24.
49 Sohngen, B. and Brown, S. (2004) Measuring leakage from carbon projects in open economies: a stop timber harvesting project in Bolivia as a case study, *Canadian Journal of Forest Research* 34 (4), 829–39.
50 These detailed results are not shown in this chapter. Information is available from the author.
51 Van Noordwijk and Minang (2009) op. cit.
52 Van Noordwijk and Minang (2009) op. cit.
53 Hertel, T.W., Rose, S.K. and Tol, R.S.J. (2009) Land use in computable general equilibrium models: an overview, in Hertel, T.W., Rose, S.K. and Tol, R.S.J., *Economic analysis of land use in global climate change policy* (pp.3–30), London and New York: Routledge.
54 Sohngen, B., Tennity, C., Hnytka, M. and Meeusen, K. (2008) *Global forestry data for the economic modeling of land use*, GTAP Working Paper No. 41, West Lafayette, IN, USA: GTAP, Purdue University.
55 Gouel, C. and Hertel, T.W. (2006) *Introducing forest access cost functions into a General Equilibrium Model*, GTAP Research Memorandum No. 8, West Lafayette, IN, USA: GTAP, Purdue University.

11 The future of forests

*Joyeeta Gupta, Robin Matthews,
Patrick Meyfroidt, Constanze Haug,
Onno Kuik and Nicolien van der Grijp*

11.1 Global forest governance: a twenty-first-century myth of Sisyphus?

Forests, 'the quintessential peripheral spaces',[1] are complex entities. They are interlocked in multiple levels of governance, engaging multiple actors with different contexts, understandings, approaches and resources, at local through to global level.

Forests have been on the global agenda for decades (see Chapter 3). There have been many direct and indirect efforts at global forest governance. These efforts have been quite incoherent and often not effective at ground level. Hence, experts refer to global forest governance as a failure; and that such efforts have 'become notorious in diplomatic circles for their apparent futility'.[2] It is even suggested that:

> Finally, the forest episode of world politics suggests that norms, institutions and governance are not coterminous. [...] Global forestry institutions provide no mechanisms for governance—not because they fail in implementation but because they are 'decoys' deliberately designed to pre-empt governance.[3]

Despite repeated failures, efforts to protect the global forests continue in differing forums and by different actor coalitions.[4] The latest effort has been initiated in the climate change regime. After early discussions on the link between climate change and forests preceding the UN Climate Convention, tropical forests in the Kyoto Protocol were only addressed through the inclusion of afforestation and reforestation projects into the Clean Development Mechanism (CDM). This changed with the emergence of REDD (reducing emissions from deforestation and forest degradation), whose aim is to incentivize developing countries to protect their forests. REDD has evolved to include not only deforestation and forest degradation, but also the enhancement of carbon stocks (see Chapter 3). REDD constitutes a way to 'enlarge the negotiation pie'; it is an integrative bargaining strategy to enhance the potential of win–win opportunities within the climate negotiations.[5]

Seven years into the international negotiations on REDD, the question is: is REDD the latest 'decoy', is it a 'REDD herring'? Will REDD be a short-lived hype or will it succeed in producing lasting change, effectively preserving forest carbon stocks in the tropics and contributing to development and poverty alleviation? These questions are part of the broader question raised in Chapter 1 about whether the political situation has changed so much over the past 20 years that the global community is ready to deal with forests; and whether such a focus can help buy time in mitigating climate change. At a more direct level, the question is whether an assessment of existing multi-level institutional forest arrangements from global to local level can help in designing more effective and equitable forest governance.

These questions are addressed by discussing the lessons from global forest governance (11.2), national forest governance (11.3) and REDD experiences (11.4). The potential for mainstreaming forest governance into national and international policies is discussed (11.4.4), and final conclusions are drawn (11.5).

11.2 'Glocal' forest governance

11.2.1 Evolutionary phases in forest governance

Forest governance is constantly evolving and can arguably be seen to have passed through nine partly overlapping phases in history (see Figure 11.1). Phase 1 (till the onset of settled agriculture) focused on the customary use of forest and non-forest products by forest dwellers and local communities. Phase 2 (settled agriculture) led to forest land use change in favour of agricultural production. Phase 3 (the onset of industrialization and ship building) led to colonial and post-colonial forest exploitation for timber extraction, mining, and medium-scale extraction of non-forest timber products. Phase 4 (post World-War II) is characterized by trade liberalization, starting with the adoption of the General Agreement on Tariffs and Trade, and countries increasingly specialized in exporting products in which they had a competitive advantage. At the same time, the ex-colonies continued the forest policies of the colonizers and used forests as a source of economic revenue. Phase 5 (the onset of indirect forest governance with the adoption of multilateral environmental agreements in the 1970s) focused on specific forest ecosystem functions – such as providing a habitat for species in wetlands and elsewhere, providing cultural heritage, a home for indigenous people, and subsequently on carbon sequestration. Phase 6 (the onset of direct forest governance) saw the establishment of the International Tropical Timber Organization and is moving gradually towards looking at all ecosystem services. Phase 7 concurrently emphasized forests in the context of national development policy, democracy and decentralization politics. Phase 8 focuses on developing global instruments within increasingly hybrid and neo-liberal governance patterns.[6] A potential phase 9 could be a shift towards 'glocal', comprehensive, multi-level

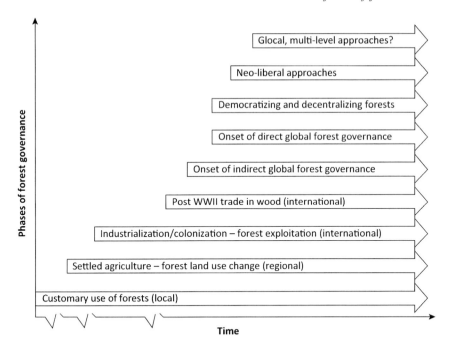

Figure 11.1 The timeline of the phases of forest governance through human history

(NB The interruptions in the x axis denote major discontinuities in time. Customary use goes back hundreds of years. Between settled agriculture and industrialization, there is also a huge time gap.)

interactive forest governance (see 11.2.2). At present governance is pluralistic as all these different phases continue to co-exist in many different parts of the globe.

11.2.2 *The politics of scale: Should there be 'glocal' forest governance?*

The historical evolution of forest governance suggests a progressive evolution of forest governance from local origins to the global level. Physically, forests are located within specific locations, mostly falling under the jurisdiction of nation states. This would, prima facie, imply that these resources, like others such as land, minerals, water and air space that fall within territorial limits, are subject exclusively to state control. The question then is: what is the rationale of managing forest resources at global level?

The 'politics of scale' refers to the way countries frame the forest issue. Its scaling typology presents arguments why a country may wish to scale an issue up (to global level) or down (to national or local level).[7] Applying this typology to forests, the research reveals reasons why actors attempt to scale up forest issues to the global level (see Table 11.1). First, there is the need to enhance problem understanding. There are four aspects to this:

(a) From a biophysical perspective, forests are part of the global earth system. Forests affect global environmental change (in terms of biodiversity, the global hydrological cycle and climate change) and are at the same time affected by it. In particular, the relationships between forests and climate are very complex. Forests influence climate through biogeochemical and biogeophysical processes.[8] Biogeochemical effects refer mainly to the role of forests as carbon sinks. This carbon is released when plant life decomposes, forest vegetation is burnt and when forest wood products disintegrate. As of 2010, forests stored 289 Gt of carbon, and deforestation and forest degradation released about 5 Gt of carbon annually over the last decade.[9] There are also carbon losses from forest soils through the burning of forest vegetation, erosion, and in the form of methane through mineralization of organic soils – e.g. after peatland drainage. Forest changes also affect nitrogen fluxes. Biogeophysical effects of forest changes include changes in surface energy fluxes through changes in albedo, latent and sensible heat fluxes, among others, and changes in water cycling. Deforestation usually leads to an increase in the surface albedo, but can also decrease evapotranspiration, both having opposed effects on energy fluxes. In tropical regions, the evapotranspiration effect generally dominates, so deforestation often results in further warming of the local climate, whereas in boreal regions the opposite may occur. Hydrological changes also influence the development and presence of clouds and, hence, the precipitation regime.

(b) Deforestation and forest degradation, even though occurring locally, is increasingly subject to global drivers.[10] These include global trade and investment regimes, consumption patterns and the rise in global welfare leading to increased demands for food, land, non-timber forest products, and minerals (see Tables 2.1, 5.1, 6.1, 7.1, 8.1 and 9.1). The need to repay foreign debt in the post-colonial era has often led to export-driven policies to enhance income, for instance through exports of cash crops, timber and non-timber forest products. The latest in the series of global drivers is the impact of climate change on forests, a point closely related to point (a) above. Changing local weather patterns including precipitation, rainfall intensity and distribution, temperature distribution and the potential for forest fires, can affect forest resilience by affecting forest growth and density and the species distribution within forests. There are positive feedback loops where deforestation affects climate change which in turn may further affect forests. Furthermore, to the extent that the local drivers of deforestation can benefit from international support (science, monitoring systems, technology, financial resources, the recognition of local people's rights), global-level actions and governance can contribute to deal with these drivers.

(c) There is a need to determine global impacts and thresholds of a problem based on a collective understanding of the issue. The challenges facing forests may be local in nature but the cumulative and collective impacts lead to significant global trends.[11] Globally there are four billion hectares of forests. According to the Global Forest Resource Assessment,[12] which relies on reporting by individual countries, deforestation has decreased from 16 million hectares per year in 1990–2000 to 13 million hectares in 2000–2010. However, this data is not confirmed by the FAO's Remote Sensing Survey[13] which shows an increase in the loss of forest cover for the period 2000–2005 in relation to 1990–2000. The baseline scenario of Chapter 10 projects that globally 111 million hectares of tropical forest would be lost between 2010 and 2030 if more stringent action is not taken. Many of the goods and services provided by forests benefit local or national populations, or are traded as commodities, and do not directly affect the global Earth system. Yet, in a world with increasingly scarce resources in terms of land,[14] soil, water, and even timber,[15] the ways by which local resources are managed and goods and services are produced bear increasingly global relevance.

(d) Legal scholars, social scientists and movements often urge an understanding of how dominant global ideologies (e.g. ecological modernization, green governmentality and civic modernization)[16] and framing of instruments (e.g. decentralization, spatial policy; see 2.4) drive decision-making.

Second, global governance arguably improves the legitimacy, equity and effectiveness of decision-making on issues that are impossible to govern within national boundaries. Only at the global level is it possible to adopt global goals and targets (see Chapter 3), based on scientific research. At the global level, it is easier to put collective pressure on countries to change their behaviour with respect to forests in general or the rights of indigenous peoples in particular (see Boxes 3.3 and 4.2). It is also possible to level the playing field in terms of global prices for forest products and to reduce the potential for leakage (see Chapter 10, and 11.2.3). Furthermore, countries often wish to increase the negotiating space in international negotiations on one issue by making links to others (see 1.4).

Third, countries often wish to promote domestic interests, avoid domestic action and focus the attention elsewhere by globalizing issues. Some (mostly forest-rich) countries see global action on forests as a means to delay domestic action as global action takes a long time. Similarly, some (mostly developed) countries use global processes to distract attention from industrial emission reductions which are a major bottleneck in the climate change negotiations and focus instead on the role of forests. Some countries see the need for cost-effective measures and argue in favour of global offset mechanisms (e.g. CDM, REDD). Some countries wish to minimize free-riding; states that have already addressed their own forest-related challenges and have fewer costs associated with these challenges may be more willing to scale up the issue to focus attention on other countries' challenges.[17]

Finally, there is the wish to promote extraterritorial interests through global governance. Countries and actors often want to gain access to resources (land, water, forest products, minerals) in other countries. By promoting policies at global level through trade and investment regimes, or through promoting market instruments, such access can often be gained. This has led in the 1990s and 2000s to multinational companies and other actors gaining access to water resources and to the problem of 'land grabbing' (see Box 9.1). Some countries wish to control the management of resources in other countries. The extraterritorial impacts of the US Lacey Act in Vietnam (see Chapter 5) and of the EU biofuel policy in Indonesia (see Chapter 6) are first steps in this direction. Globalizing these policy measures is the next step. Alternatively, forest-rich countries may choose to support framing forests as a global issue if this can lead to protecting domestic interests while being able to have a say in the policies of others.[18] Countries and other international actors may sometimes wish to bypass fellow governments in order to gain direct access to actors at lower levels of governance. Some actors promote transnational instruments to control action – e.g. debt-for-nature swaps (see Box 2.4), certification (see Box 2.5) and payment for ecosystem services (see Box 2.3) (see Table 11.1).

Similarly, governments and actors have reasons to scale down forest issues to national/ local level (see Table 11.2).

First, downscaling enhances problem understanding for the following reasons. Enhanced knowledge of contextual ecosystem links improves the grain and resolution of problem understanding, allowing for better focus on national drivers e.g. population migration, economic goals (see Tables 2.3, 5.1, 6.1, 7.1, 8.1). National to local impacts and priorities can better be accounted for at these levels – for example, the need for watershed management may lead to a choice for forest protection, or the need for agricultural products or expanding urbanization may justify deforestation. Similarly, downscaling allows for better opportunities to account for local diversity in culture, policy context, and domestic discourses, allowing for a better appreciation of gendered roles in forest use, control, management, and participation in policy processes.

Second, downscaling enhances policy design and effectiveness. A key argument for downscaling is that forests fall under national jurisdiction and thus national sovereignty. The concept of absolute territorial sovereignty, first articulated by US Judge Harmon in his Harmon Doctrine, provided states absolute control over the resources within their own territory. This doctrine was common in the area of water resources although it has been losing its credibility.[19] The doctrine of permanent sovereignty over natural resources was the response of developing countries in the post-colonial era to try and regain control over resources within their own territory[20] in a time before the impact that local land use change could have on global processes was fully appreciated. A competing doctrine is that of limited territorial sovereignty where sovereignty is limited to ensure that substantial harm is not caused to others.[21]

Table 11.1 Reasons for scaling up forests

Type: to	Motivation: to	Application to forests
Enhance problem understanding	Understand global ecosystems links	E.g. water, climate
	Account for global drivers	E.g. trade, consumption
	Determine global impacts and thresholds	E.g. on climate, biodiversity
	Understand global ideologies/discourses	E.g. ecological modernization
Improve governance	Protect the common good	E.g. a global afforestation target
	Include governments, enhance legitimacy	E.g. producers and users of wood
	Put collective pressure on countries	E.g. isolate deforesters
	Level the playing field, common instruments	E.g. SFM, Protected Areas
	Increase the negotiating space	E.g. UNFCCC deadlock
Promote domestic interests	Postpone decisions, avoid taking measures	E.g. delay domestic forest action
	Distract attention from domestic challenges	E.g. industrial CO_2 emissions
	Make domestic decisions cost-effective, create level playing field	E.g. through carbon offsetting (CDM, REDD)
	Reduce free-riding, scale up domestic policy	E.g. countries with good forest policy
To promote extra-territorial interests	Access resources	E.g. land, water, forests
	Control resources	E.g. forest management
	Bypass an agency	E.g. problematic national govts
	Via transnational instruments	E.g. Debt-for Nature swap, certification

Source: adapted from Gupta (2008) op. cit.

Many developing countries have rejected the notion of making forests a global issue based on the idea of sovereignty and the right to determine their own future.[22] A second argument is that there are already national institutions in place to deal with forests (see 5.3, 6.3, 7.3, 8.3) and new policies will only be effective if they are linked to existing frameworks. Forests are either owned by the state or by different actors within the domestic context. Hence, this justifies treating forests as a national issue. Moreover, including national/local stakeholders in policymaking enhances the legitimacy and equity content of such decision making. It also increases the chances of mobilizing these people to help solve their own forest-related problems and challenges (see Box 2.6).

Table 11.2 Reasons for scaling down forests

Types: to	Motivations: to	Application to forests
Enhance understanding	Understand contextual ecosystems links	E.g. local biophysical conditions
	Account for national drivers	E.g. migration, economic goals
	Account for national impacts and priorities	E.g. on watershed management
	Account for contextual discourses	E.g. local cultures and contexts
Improve policy effectiveness	Protect the national good and territory	E.g. through sovereignty arguments
	Use local institutions to enhance efficiency	E.g. through existing forest frameworks
	Include local stakeholders to enhance legitimacy and equity	E.g. forest product and services users, producers, traders and consumers
	Mobilize stakeholders to enhance effectiveness	E.g. through community management
Strategize	Avoid liability for externalized effects	E.g. for the impacts on carbon emissions
	Divide and control or include and exclude	E.g. allows selective choice of partner countries
	Manage and protect national and local interests	E.g. allows promotion of the right to develop and national choices
	Bypass an agency which is perceived as a hindrance	E.g. allows bypassing specific international agencies

Source: adapted from Gupta (2008) op. cit.

Third, downscaling is a good strategic tool. Tackling forest issues at a national level helps to avoid liability for externalized effects. It also allows for strategic use of divide and control of policies in specific countries and the possibility to include or exclude others in the policy process. Local governance allows exclusive options for managing a 'territorial' resource; to argue in favour of the national right to develop and promote economic growth. Making forests a matter of uncontrolled global management would open the door for foreign intervention in national policy, foreign control of what is possible within the domestic context, and foreign condemnation if such policy is not implemented. For many developing countries who did not like the initial exclusive focus on tropical forests and the implication that scaling up would give other countries control over how states managed their own resources, this is a problem.[23] Downscaling also allows opportunities to bypass agencies perceived as problematic.

Given that the bulk of the world's forests are located in ten countries,[24] the need to globalize forest governance is also contested.

Arguments for scaling up and down forest governance have been used opportunistically by countries over time.[25] For example, although Canada and the US favoured global forest governance in 1992, the US subsequently reversed this view as a result of pressure from the American Forest and Paper Association[26] who opposed this; at the same time Canada felt that harmonizing global standards would lead to a level playing field and help in exports.[27] While India and Brazil continued to oppose global forest governance until REDD came on to the agenda, Malaysia changed its position after 1992 (see Figure 11.2). The advent of REDD and the possibility of accessing large sums of money to protect forests led many developing countries to accept forests as a global issue. At the same time, the promise of forests as a cost-effective climate mitigation option has increased developed country support for this framing. This is demonstrated by the discussions on REDD within and outside the UNFCCC regime (see Chapter 4). Some NGOs have remained sceptical. However, if carbon sequestration in the tropics proves more complex and expensive than assumed by its proponents, and if this implies decreasing funds for forest conservation, this may once more lead many countries to see forests as a local to national issue.

Thus, the scale of forest governance has been contested. Recent efforts at framing forest governance as a global issue include labelling forests as global commons, or part of global stewardship, as an issue of common concern, common heritage, or global public goods. In legal terms, states have avoided adopting the concept of common heritage at the global level. Climate change and biodiversity are seen only as issues of common concern.[28] In the case of forests, international forest-related consensus documents have emphasized the sovereignty principle.[29] To the extent that forests are

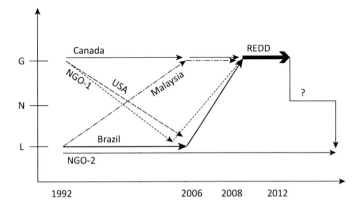

Figure 11.2 Scaling up and down forests: a stylized representation
(Source: adapted and updated from Gupta 2008; G = Global; N = National; L = Local)

included in the Biodiversity and Climate Conventions, it is seen as a common concern.

There are clearly reasons to both scale-up and scale-down decision-making on forests. Only at the global level can a systemic understanding lead to better analysis of how to manage forests. Only at global level can global drivers be addressed. But it is only at national to local level that contextual issues can be adequately understood. The term 'glocal' is used here to describe the interactive multi-level policymaking process that is needed to ensure that policies are science based, legal, legitimate, accountable, transparent, equitable, take a gendered approach, and are context relevant. A general principle in such a process must be to devolve decision-making to the lowest level relevant to the issue in question – decisions affecting the global environment need to be made at the appropriate global forums, whereas those affecting individual livelihoods must of necessity be at that level. The challenge is in developing and implementing conflict resolution mechanisms for when these decisions do not align. Only such a process can mobilize people at all levels of governance.

11.2.3 Current global forest governance

The characteristics

Current global forest governance can be characterized as incoherent, incremental, fragmented and a Möbius web system, pluralistic, limited in coverage, and moving in the direction of administrative law.

Forest governance is incoherent. For example, the World Bank was promoting logging projects while there were global discussions on protecting the forests; the World Trade Organization promotes free trade in wood products, which might conceivably clash with certification schemes for sustainably logged timber. And while forest conservation, sustainable forest management, and the establishment of protected areas are supported by international agreements on Wetlands (Ramsar) and Biological Diversity, the International Tropical Timber Association continues to be dominated by the economic interests of timber producing countries.[30]

Forest governance demonstrates 'creeping *ad hoc* incrementalism'[31] in that the process moves slowly and decisions are mostly incremental in nature (see Chapter 3).

Forest governance is fragmented and diffuse in that direct and indirect forest governance is spread among many intergovernmental, non-governmental and private agencies and takes different forms such as global, regional, bilateral and extraterritorial control. This fragmentation with no overall coordination leads to duplication and loss of efficiency.[32] Forest governance currently resembles a 'mobius' web type governance;[33] including both formal and informal, multi-directional linkages and hierarchical as well as networked interactions.[34] This fragmented character makes forest governance pluralistic in that many policy rules and instruments of different agencies may have relevance to specific jurisdictions. Thus, Forest Stewardship Council certification has different criteria than ITTO and these differ from REDD safeguards currently under discussion.

This has resulted in a relatively piecemeal and uneven approach to the security and management of forest ecosystem services. For example, cultural services are the least addressed, but there is much more activity on provisioning, regulating and supporting services, mainly because the latter are more easily quantified in economic terms. Many initiatives focus on trade in forest products, human livelihoods and biodiversity protection. Particularly since the advent of REDD, the carbon sink function of forests has dominated global level activity above all others (see Chapter 4).

Finally, global forest governance is moving from legally binding to non-legally binding instruments (see Chapter 4). Current governance includes rules and targets but none that are legally binding. There is a gradual convergence in many different forums (from the ITTO to the World Bank) to the concept of sustainable forest management – at the same time the ecosystem approach is being explored and may or may not come to dominate the way forests are managed in the future. However, the interpretation of both concepts varies across contexts. In terms of policy instruments, while early arrangements included also hard law regulatory instruments, later arrangements are primarily soft law and suasive (see Table 3.1) or economic instruments such as forest certification and forest carbon offsetting. This shift towards abstract ideas, non-binding policies[35] and economic instruments can be explained thus: first, there is a general trend at international level to move towards administrative law, a system where law emerges more from the executive than the legislative or judiciary.[36] This has long existed at national level and is now developing at international level (e.g. through Conference of the Party decisions in international treaties). Second, as the global community increasingly takes a systemic, multi-level governance perspective to managing forests, the law is not able to define simplistic targets and timetables; it can only provide a normative framework whose suasive intention is to persuade states to act accordingly. This is consistent with a democratic, inclusive approach that allows states some freedom in interpreting the collective wisdom of states. Third, this takes into account the politics of scaling – that countries may be reluctant to participate in global policymaking for the reasons discussed earlier in this Chapter (see 11.2.2).

Focusing on the politics of forest definitions

A symptom of the contested nature of forest politics (see 11.2.2) and the complicated and convoluted forest governance process (see 11.2.3) is the issue of forest definitions. Figure 1.1 shows that as forest definitions change, so do the statistics of annual forest deforestation rates. FAO defines a forest as having a 10 per cent forest cover of trees more than 5 metres in height and no other major land use in that area. It excludes agroforestry, but includes 'temporarily unstocked' forests.[37] The CBD defines a forest as being more than 0.5 hectares in extent, a non-agricultural tree canopy of more than 10 per cent, of trees that can reach 5 metres in height.[38] This is not enough to protect biodiversity which requires at least a 40 per cent tree cover if not substantially more,[39] does not protect natural forests, as this does not specify the nature of trees and could encourage natural forest-to-plantation conversion, includes agroforestry which may not be considered a forest in the domestic context (and yet still contains significant quantities of carbon), and includes 'temporarily unstocked' forests.[40] The Climate Convention defines a forest as having a tree crown cover of 10–30 per cent of trees which can reach 2–5 metres in height with a minimum area of 0.05–1.0 hectares.[41] Box 1.1 illustrates that the definition of forests in the context of the Clean Development Mechanism is counter-intuitive as conversion of forests to oil palm plantations is not considered deforestation (see also Chapter 6), the conversion of natural forests into plantations is not seen as problematic, and temporarily unstocked lands can be considered as forests. However, all vegetation and soils have a carbon storage function. But the treaty excludes other tree-based systems – smaller agroforestry, urban forestry, and soil-related emissions. The choice of definition from RED, through REDD, REDD+, and others can also imply great variations in the carbon emissions that become subject to discussion.[42] This is a political choice governed by the practical feasibility of being able to measure carbon stocks and emissions. As at global level, national forest definitions are contested (see case study chapters; 9.2).

Mopping up with the tap open: The issue of leakage

Another symptom of fragmented and pluralist forest governance is that local forest governance has to compete with growing global demand for forest products. The economic analysis (see Chapter 10) shows, in a hypothetical example, that if REDD were able to conserve 28 million hectares of tropical forest it could potentially reduce CO_2 emissions by 9.8 Gt. However, this could inadvertently raise food and timber prices, thus further marginalizing the poorest consumers, and amplifying the major underlying deforestation drivers. In terms of overall economic impact, the loss of income would be largest in Indonesia where an income loss of almost 0.7 per cent of GDP is projected for 2030. For some countries the opportunity costs of not deforesting would be very high (e.g. Indonesia), for others lower (e.g. Cameroon); where the opportunity costs are high, effectively addressing deforestation would be quite difficult. Using simple assumptions on carbon avoided through forest conservation, it was calculated that the costs of the conservation policies would vary from USD 12/tCO_2e in Africa to USD 59/tCO_2e in Indonesia. A comparison with earlier assessments showed that these cost estimates for most regions are on the high side, especially for Indonesia. At the same time, addressing deforestation in tropical countries could help industrialized countries benefit through an advantageous terms-of-trade effect. This would again feed into the developing country fear that their right to develop (see Chapters 3–4) is compromised in order to help the developed countries develop. Perhaps it would not be unreasonable to ask the industrialized countries to use a part of this 'windfall profit' to compensate tropical countries for their opportunity costs and for the measurement, monitoring and verification costs of REDD conservation policies.

If countries implement forest conservation policies unilaterally, there can be leakage of 'avoided emissions' to countries without forest conservation policies. Chapter 10 discussed some of the complexities of measuring leakage, even in economic models. The simulations with the economic model show country-specific rates of leakage: for Indonesia between 3.8 and 19.7 per cent, for Peru between 1.2 and 16.5 per cent, and for Africa (including Cameroon) between 0.6 and 13.4 per cent. These are not accurate predictions, however. At this stage, the most that we can say is that leakage might be a problem for REDD and that maximum participation of tropical forest countries in REDD would be the best way to avoid or mitigate it.

11.3 National forest governance

11.3.1 The forest transition

Countries pass through many transitions – demographic, economic, environmental and forest transitions (see Chapter 2). Each transition is influenced by issue-specific drivers and policies, and other transitions that occur in a country. The forest transition curve[43] represents how some developed countries have dealt with their forests over time, but it is important to realize that there are many pathways to stabilizing forest levels and the forest transition curve is not necessarily deterministic of the way in which forest protection will proceed in the future.[44] Chapters 5–9 have collected information regarding the drivers of deforestation and degradation and the policy instruments used in the four case study countries and the selected provinces and local areas per country. Applying this knowledge, a rough estimation of where these countries might be in terms of a possible relative forest transition curve can be derived (see Figure 11.3).

The case studies show that Cameroon has started its deforestation process, Peru is further along the line, while Indonesia is furthest still. Vietnam has, at the national level at least, started reforesting. This does not imply that these countries will definitely move along this pathway. The confluence of drivers and policy instruments will determine the direction in which they are likely to move.

Within individual countries too there are different transitions taking place. For example, in Indonesia, Papua still has virgin forests, Kalimantan and Sumatra have heavy deforestation, while Java is approaching stabilization.[45] In Vietnam, although at the national level net reforestation is taking place, heavy deforestation is ongoing especially in the Central Highlands region,[46] which serves as a 'facilitating region' to support the reforestation elsewhere in

242 *Joyeeta Gupta et al.*

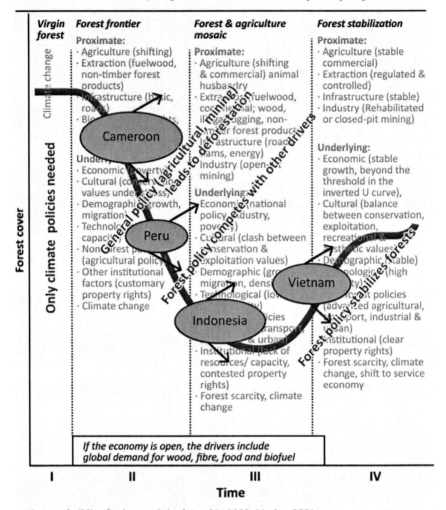

Source: building further on Geist & Lambin 2002, Mather 2001

Figure 11.3 Case study countries in different stages of the forest transition curve: a stylistic representation

the country. But, given recent national trends – a decreasing share of natural regeneration in the total reforestation and recent acceleration in deforestation in the Central Highlands, the country might well shift back to a trend of net losses of natural forest over a few years. Policies thus need to take into account not only subnational differences in forest trajectories and drivers of deforestation, but also the interactions among regions in a country.

11.3.2 Forest transitions, drivers and policies

Figure 2.3 postulated that the nature of drivers changes as and when societies pass through specific phases in the forest transition. Thus, in the forest frontier phase, a country normally has shifting agriculture; in the forest and agriculture mosaic phase, it has shifting and commercial agriculture; while in the forest stabilization phase; there is stable agriculture. Table 2.3 identified 26 forest policy instruments from the literature. Based on the analysis in this book, it is also possible to inductively postulate that as countries develop and as they progress along the forest transition they (a) tend to use different types of instruments; and (b) the effectiveness of these instruments in dealing with local to global drivers changes. In the first phase of virgin forests, no policies are needed; possibly some form of community-based management systems exist and community rights of access and control are recognized. In the forest frontier phase, there is implicit or explicit state policy that sees forest land as necessary for conversion to settled agriculture, fuel wood extraction, and basic infrastructure. There is scarcely any forest policy at this stage, although some states may decide to nationalize forests for multiple reasons. Cultural conservation values may come under stress in this period (see Figure 11.4).

In the forest and agriculture mosaic phase, government policy may explicitly support commercial agriculture, animal husbandry, fuel and commercial wood extraction, open-pit mining and large-scale agricultural development.

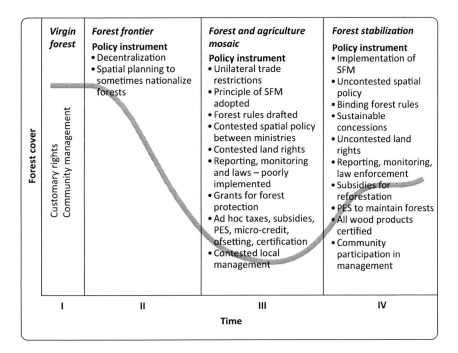

Figure 11.4 Policies compatible with different stages of the forest transition curve

Global drivers of demand for forest products may become more important especially if this phase coincides with the opening up of the national economy. In this phase, forest policy instruments are developed. There is competition between the different ministries and forests may even come under the (partial) jurisdiction of agricultural ministries, as in the case of Vietnam and Peru. The policy instruments are mostly regulatory in nature. Spatial policy and labelling areas as forests (Protected Areas) and providing concessions are popular instruments at this stage. As forest cover declines and local forest ecosystem services are recognized and valued, policy instruments to protect them may become more important. At this stage, regulatory instruments such as decentralization begin to become important; and market instruments such as subsidies for reforestation, and payment for ecosystem services, may become more significant. At this stage the need to reconcile plural, confused and contested ownership patterns of land may receive greater attention. *Ad hoc* application of instruments to deal with local proximate drivers becomes increasingly important. *Ad hoc* instruments may be launched internationally to deal with some global and local drivers such as international funds, 'unilateral trade restrictions' by states and 'certification' by private actors. Making reporting, monitoring and law enforcement effective becomes more important in this stage. Policy instruments to protect forests can scarcely compete with the other dominant policies and are often designed in a weak manner or poorly implemented.

In the fourth stage, forest policy is mainstreamed into national development strategy and leads to stable or growing forests. The local drivers of deforestation are kept under control and policy development and implementation is effective. The global drivers of deforestation are managed through strong domestic forest policy.

11.3.3 National forest policies

Chapter 9 has comparatively assessed the policy instruments being used in the four countries. Most instruments identified in Chapter 2 are being implemented in these countries; however, the emphasis is on regulatory instruments (spatial planning, land tenure, concessions) and economic instruments (taxes and subsidies) and not on suasive instruments that might be able to create broad-based awareness in these countries. Although the four case study countries are large and have massive forest resources, their governance systems range from a socialist-oriented market economy under a communist regime (Vietnam) through transitioning democracies (Indonesia and Peru), to poorer, more unstable economies (Cameroon). A key common feature is the drive to develop rapidly and to optimize national resource use (land, forests, minerals and water) profitably. At local level, the driving factors of deforestation and forest degradation are subsistence and economic profit. These two driving factors are more dominant than forest policy and its instruments which address primarily local driving forces (often at the cost of social concerns, see 9.4 on access and allocation). Chapter 9 argues that it is not self-evident from the research that forest conservation will survive as a notion in the short to medium term if there are no financial resources forthcoming from outside the country. This means that other ministries have more power in shaping government policy and there is heavy competition between the different sectors. This is illustrated differently in the four countries: in Peru where agriculture clearly takes precedence over forests in terms of land-use strategy; in Cameroon where mining is becoming more lucrative; in Vietnam and Indonesia where the focus on commercial crops brings easy profit.

At the local level, communities and indigenous peoples run the risk of losing their lands, cattle and access rights depending on the way in which customary land rights are recognized and whether they happen to live on lands that are part of protected areas. The possibility of receiving funding for carbon sequestration in these forests has intensified the competition between actors for control over the forests. At the same time, land-use change to other more lucrative uses – such as agroforestry, industry and mining – make the opportunity costs of maintaining forests very high. In many cases, the compensation for carbon sequestration might not match these high opportunity costs.

From a biodiversity perspective, critical hotspots need protection. From a carbon sequestration perspective, the only focus is on carbon. Carbon protection need not thus go hand-in-hand with biodiversity protection. The case study on Peru illustrates this risk.

Although there is considerable discussion about integrating forest protection within NAMAs, the question is whether forest protection is likely to take precedence over other uses of forests in the absence of additional sources of income for protecting them.

11.4 REDD revisited

11.4.1 Practical options for implementing REDD in countries

The concept of REDD, introduced in 2005, aims to promote emission reduction through reduced deforestation and forest degradation either through a fund-based mechanism, a market mechanism or a hybrid approach. Payments are to be made only after monitoring, reporting and verification of the emission reductions. Discussions are currently under way, not only on how best to operationalize REDD itself, but also on how to maximize the co-benefits and ensure safeguards (see 1.4). REDD discussions within the UNFCCC process are ongoing but slow; REDD discussions outside the UNFCCC process are moving ahead. The key challenges are the issue of scale (at what level would accounting be most effective?), reference level (against what level should performance be measured?), financing (how can adequate financing be generated?), MRV (how can monitoring, reporting and verification processes be developed, maintained and made cost-effective?), permanence (the risk that subsequent deforestation can release emissions), additionality (that REDD payments are made for emission reductions over and above what would have been made anyway), leakage (that forests are maintained in one place at the cost of deforestation elsewhere), and safeguards (protecting other ecosystem services and indigenous people's rights through governance, social and environmental safeguards) (see 4.4).

All four case study countries – Vietnam, Indonesia, Cameroon and Peru – are seeing the potential of participating in REDD, REDD Readiness programmes and REDD pilot projects. REDD does not operate in a governance vacuum; the forest sector in each country is influenced by drivers of deforestation and instruments to deal with forests (see Chapters 4–8). The role that REDD could play in each of these countries is thus different. In Vietnam, the role of REDD would be to help stabilize forest levels and to improve the quality of the forests being stabilized through ecosystem safeguards. In Cameroon, Peru and Indonesia, the role of REDD is to compete actively with other drivers of deforestation in order to make forest stabilization attractive.

However, lessons learnt from past experiences with forest-related instruments lead to the following conclusions (see 9.5 for details). First, REDD is only likely to work if there are enabling conditions in a country. This is partly the purpose of REDD Readiness activities. Second, REDD needs to be integrated into or linked to existing relatively successful domestic institutions, rather than distract resources and manpower away from other goals. In Vietnam, this would imply linking to the institutions of Programme 661 and the national Protected Areas; for Indonesia to existing concepts of community forests, ecosystem restoration, the moratorium and peat land policy, in addition to the existing zoning of conservation, production and protection forests; for Cameroon, the Council Forests, and the existing tax distribution opportunities to build upon for REDD payments; in Peru, the community forests, concession system, the Protected Areas, and ecosystem restoration.

Third, REDD should build upon other international policies and programmes and not replace them. This implies it could build upon international certification standards through mutual learning. It could try and link the MRV to the existing FLEGT programmes. Lessons learnt from the implementation of CDM projects may be helpful. Fourth, there has been much discussion about the need to ensure co-benefits and local safeguards at the international level. The case studies re-emphasized the need to ensure that access to food and water by local and indigenous communities in the forest zones is not compromised. Some instruments tend to marginalize or even criminalize the existing customary rights of local communities and force them to 'encroach' on peripheral lands and buffer zones, or to move to marginal lands where they may have to engage in slash-and-burn activities which represent a form of leakage. If people lose their access to customary rights, their survival may be affected unless alternative incomes and resources are made available to these people. Fifth, reference levels need to balance the goal of reducing and eventually eliminating tropical deforestation with countries' needs for development. Sixth, if REDD aims at financing those who protect the carbon content of forests, it is important to identify what the possible categories of beneficiaries are and who in particular is eligible to be a beneficiary. Seventh, the beneficiaries need to understand what they are being compensated for. The Vietnam case study shows that local A/R-CDM participants did not quite understand the CDM and this affected their ability to effectively participate. Eighth, the benefit-sharing system, including payment levels, needs to be clear. It is highly likely that if these payment levels for protecting forests are lower than the opportunity costs of using the land for other purposes, they will not be effective. At the same time, if differentiated payments are made, this may ironically provide more resources to those who are already relatively resource-rich and wish to exploit the land for commercial purposes, and less to those who are poorer and have subsistence needs – thereby exacerbating local inequities. International NGOs suggest a Fair and Efficient REDD Value Chain Allocation to ensure equity in the process of benefit sharing.[47] Ninth, it is important to clarify how the payments are made and how the benefit-sharing system with the local people works. This needs to be undertaken carefully to avoid elite capture, a key challenge in all four countries. Experiences in individual countries – for example, the payments to household groups in Vietnam – could be built upon. Tenth, the issue of free, prior and informed consent of local communities and individual groups is critical. Eleventh, the design of the instrument should combine a market approach with regulatory approaches in order to work in the context of these countries. Indigenous people's groups have argued that they wish to bypass the nation states to access REDD resources directly. However, without the enabling conditions that a national system would potentially provide, efficiency, effectiveness and equity might not be achieved. Twelfth, it may be necessary to think in terms of regional approaches to avoid regional leakage.

At the same time, the institutional design of REDD is still evolving at the international level. It is not yet clear what scope the mechanism will have and how safeguards will eventually be protected, or what will be paid for compensating those who protect the forests and who precisely they will be. Nor is it absolutely clear how reference levels will eventually be designed – MRV may require very close scrutiny by external entities and this may be seen as a form of neo-imperialism.[48] What is clear is that REDD addresses only some proximate drivers and that the underlying drivers remain unaddressed.

11.4.2 Buying time or a REDD herring

Does REDD help to buy time or is it more of a 'REDD herring'? There are many indications that there is only a short and relatively small window of opportunity for making REDD work.

First, forest-related emissions as a fraction of total global greenhouse gas emissions have been decreasing over time. At Noordwijk (see Chapter 1), it was assumed that deforestation and degradation contribute substantially to climate change;[49] the IPCC 2007 stated that it was 17 per cent,[50] the G8 L'Aquilla Declaration that it was 20 per cent[51] and now the Global Carbon Project states that it is about 10 per cent.[52] Similarly, the contribution of greenhouse gas emissions from overall land-use change (LUC) to total emissions keeps falling in relative terms because of the continuing rapid rise of fossil-fuel-based emissions: in 1960 it was around 38 per cent of the total, in 1990 it was around 20 per cent and in 2010 it had dropped to 9 per cent.[53] This does not diminish the need to deal with deforestation as a GHG source, but suggests that dealing with deforestation will not necessarily buy the time needed to address other greenhouse gas emissions.

Second, there are no post-Kyoto targets as yet. The earliest legally binding targets are expected in 2020. This implies that for the next eight years, it is unlikely that offsetting can take place except in the context of banking (as was done in the past in relation to the Clean Development Mechanism).

Third, while, theoretically speaking, offsetting can raise resources quite easily for financing REDD, offsetting itself is problematic because it does not necessarily create a dent in Northern emissions. However, offsetting increases the costs of production in Northern industries resulting from the need to purchase credits, which will ultimately be passed on to consumers, so that they have less disposable incomes to spend on activities or goods with a high carbon footprint. REDD+ may, therefore, result in indirect reductions in emissions from the North. A second problem with offsetting is that lessons from the CDM reveal that less than 1 per cent of all projects and all certified emission reductions come from afforestation and reforestation projects because of the complex methods, temporary credits and exclusion from the EU's emission trading system; this may indicate that the practicalities may make the system less popular than initially envisaged (see 4.2.2).

Fourth, although US$ 4 billion was pledged for 2010–12, the source of this money is not always clear. Developing countries had asked for new and additional resources over and above official development assistance; but the case studies reveal that some of this money is coming from development assistance. Furthermore, there is a gap between what is pledged and what is paid (see Figure 4.1) and this fits into trends that show that the proliferation of funds does not imply that these funds are financed.[54] If REDD is to be financed through aid money, the news once more is not positive. Financial resources for official development assistance have been declining as a consequence of the recession since, first, as it is a percentage of national income, if national income decreases, the total amount decreases; and, second, in times of recession, there is less enthusiasm for spending on development cooperation. Furthermore, the apparent increase of resources for REDD+ from aid comes (a) at the cost of other fields of development cooperation (e.g. moving from other forest goals to REDD or from gender to REDD); (b) such funds are pledges and there has always been a gap between pledges and actual delivery of resources; and (c) the money will eventually only be paid if the reductions can be monitored and verified, which may create new challenges.

Norwegian pledges on REDD+ come from ODA (see Chapter 6); but there is apparently growing scepticism in Indonesian government circles about whether the benefits gained from participating in REDD will outweigh the costs of the MRV process, whether any benefits generated can be effectively distributed, and whether Norwegian resources for REDD+ projects will eventually be forthcoming.[55] Finally, even the pledged resources fall short of what the literature argues as necessary for dealing with avoided deforestation (see 4.4.3).

Fifth, the so-called cost efficiency of REDD has not taken into account the (a) need to include co-benefits and safeguards which considerably raise the price of REDD but may enhance its effectiveness, (b) the huge transaction costs involved in creating or building the institutions for implementing REDD which some perceptions suggest may be as high as 80–90 per cent of the total costs,[56] (c) and the costs of MRV which creates a new community of professionals who will cream off a large percentage of the rents (unless communities are also actively engaged in this). In other words, REDD may turn out to be significantly more expensive than initially anticipated. While the literature on REDD has taken the issue of opportunity costs into account, these costs are very different for subsistence farmers and for commercial farmers.

Sixth, some see REDD as merely shifting the focus from timber content (as many instruments in the past did) to the carbon content of forests, and that all the discussions on co-benefits and safeguards may not be enough to guarantee equitable protection of the forests at ground level. However, this argument is flawed in that timber and carbon are fundamentally different sides of the same coin – the timber value is only realized through extraction, whereas the carbon value is only realized if the forests remain intact.

Seventh, in 1998, deforestation in Brazil was 27,000 square km. In 2011, it was at 6,200 square km. President Lula stated at COP-15: 'By 2020, Brazil will reduce deforestation by 80 per cent relative to 2005.'[57] If Brazil implements this and afforests 80 per cent of illegal large farms by 2020, an area of 200,000 square kilometres will be afforested. This amounts to a sink of about 10 Gt Carbon in the period 2015–20. If successful, this could almost compensate the anticipated rise in greenhouse gas emissions of about 12 Gt of carbon by 2020. This amount is double the worldwide market of REDD at present and could bring down the price of REDD to levels that make the carbon market ineffective.[58] But this may not happen.

All this implies that there may be a very short and small window of opportunity to deal with forest-related carbon emissions as a way to address the greenhouse gas problem in a manner to keep it below 'dangerous' limits. This does not negate the significance of preserving forests for benefits from other ecosystem services. However, if the resources are not quickly generated, the interest generated for dealing with forests will not be rewarded with success. Let us recall here the issue-attention cycle theory of Downs.[59] An issue stays on the political agenda for a limited period. In the first pre-problem phase, the problem exists but is not the subject of public discussion. In the second phase, the public is alarmed and constructively aims to deal with the problem, thinking that this is possible 'without any fundamental reordering of society itself'.[60] During the third phase, social actors become aware of the costs of dealing with the problem. In the last phase public attention may decline as the problem is seen as too complex or expensive to address, or requiring major structural overhaul of societies. If policy measures are taken in the second phase, there is hope that the problem may be addressed. Applying Down's theory to REDD, the second phase ended in Copenhagen where no binding agreement was adopted. The global community has now entered the third phase, and although there is still progress being made on REDD policymaking, the magnitude of the challenges to be addressed in designing a workable REDD+ mechanism that takes all ecosystem services into account and is equitable is now being recognized. The question is whether these challenges can be overcome to develop lasting solutions where the benefits outweigh the costs, so that the global community can avoid entering Downs' fourth phase of disillusionment.

REDD has been framed as a win–win solution since it is a cost-effective solution to the problem of climate change,[61] that can also compensate those who actually look after forests. But there are risks that it evolves into a lose–lose situation (see 9.4), if (a) the commodification of forests[62] leads to an intensification of political struggles regarding who is to be the owner of forest land and how the bundle of rights is to be divided between social actors; these struggles are being resolved through the formal institutionalized regulatory instruments of nationalization, spatial planning, land tenure and concessions, and the criminalization of hitherto customary access,[63] which may hide the impacts on local farmers, communities and indigenous groups; (b) if subsequent to commodification and privatization, forests become subject to ownership and alienation rules, and are eventually sold to the highest bidder (who may not be an environmentalist), thus, inter alia, leading to 'land grabbing' (see Box 9.1), (c) if the economic instruments of funding, subsidies, tax returns, debt-for-nature swaps, offset schemes, REDD, and PES schemes are distributed in a way that they reward local stakeholders according to their individual opportunity costs thereby exacerbating local inequities between large landowners and small subsistence farmers; (d) if it is unable to raise the resources needed from the developed world; (e) if it is unable to effectively spend the resources especially in states with weak governance;[64] and (f) if it diverts global attention from putting pressure on the US, Canada and Russia to take greenhouse gas mitigation action and thus, in effect, allows consumption patterns in the North to continue unabated while asking for sacrifices from the South[65] compounded by intrusive one-way MRV processes.

11.4.3 A North–South analysis

This section examines the issue using a North–South Gap analysis (see 1.5)[66] which builds on the Third World Approaches to International Law (TWAIL) school of thought. This school believes that much of western international politics and law scholarship is biased towards dominant western discourses, theories and approaches.[67] This bias can only be addressed by explicitly looking at Southern interests. One way to do so is to examine Northern Goals (G), Arguments (A), Patterns (Pa) and compare these to Southern counter goals (CG), counter arguments (CA), counter patterns (CPa).[68] If the North–South patterns are identical, the TWAIL hypothesis is negated. If the patterns are dissimilar then such an explicit focus on the interests and arguments of developing countries may provide a richer analysis of the governance process.

One could argue that forests per se are not a North–South issue. Forest-rich countries have changed their positions over time (see Figure 11.2). With REDD, everyone seems to see forests currently as a global issue. However, climate change is a classic North–South issue as the differences in average emissions and average vulnerability to the impacts of climate change are considerable, and if climate change is to be addressed rapidly, the budget of emissions may expire as early as 2032 leaving little room for further growth (see 1.2.3). Within the context of climate change, some see the forest discussions as expanding the negotiating pie, as giving new impetus to the forests discussion. And yet, forests divert attention from the current inaction on climate change in the large Northern countries, distract attention from the unwillingness of the US, Canada and Russia to meaningfully participate in the climate change regime and, worse, REDD, if designed as an offset mechanism, will allow Northern countries to continue with their own lifestyles while the 'right to develop'[69] of the developing countries and its peoples may be compromised. However, the costs of offsetting will have to be passed on to Northern consumers and may lead to a change in consumption patterns, while the flow of REDD finance to Southern countries should help to provide revenue for their development rather than their obtaining it from deforesting.

The research reveals that the dominant 'Western' goal of the REDD instrument has been to create win–win options for all.[70] The underlying argument is the neo-liberal, ecological modernization vision that market instruments can be designed to address global problems. This is part of a longer trend of creating patterns of market-based instruments to address global and national 'commons' problems and within the climate change regime originated in the development of Activities Implemented Jointly and its follow-up instruments – the Clean Development Mechanism, Joint Implementation and Emissions Trading. When this issue is examined from the 'Southern' perspective, Southern countries have either in the past not wished to globalize the issue of forests and/or have been sceptical about REDD. Some small Southern countries with large forests have long asked to be compensated for the 'avoided emissions' that result from forest protection. However, many of these smaller countries have also continually put pressure on the developed countries to reduce their own emissions and were not very happy with the development of Activities Implemented Jointly[71] and the Clean Development Mechanism as an offset mechanism in 1997.[72] Furthermore, past experiences with these instruments have revealed that they do not often serve the multiple interests they are intended to do. For example, the CDM has delivered in terms of credit generation (which is the key goal of Northern investors) but its contribution to technology transfer and sustainable development has been questioned in the literature. If this is anything to go by, there are expectations that, if REDD is eventually implemented, the issue of co-benefits and safeguards may be ignored in actual implementation processes and projects will be inequitably distributed among countries.[73]

11.4.4 Going beyond REDD: the challenge of mainstreaming forests

This book has shown that most instruments in the forest sector deal only with proximate drivers, and often in an equitable manner (see 9.4 and 11.5). Very few deal with underlying drivers, and to the extent that they do so their effect is limited (e.g. debt-for-nature swaps, certification). This has been explored in Chapter 2 and in the case study countries. However, the very nature of the underlying drivers (e.g. demographic, economic, technological, political or cultural trends) is that they are often slow processes operating at national or global levels resulting from the aggregated behaviour of many regional, national, subnational and individual entities (in some cases), referred to in the Panarchy literature as slow variables.[74] Effecting change in these slow variables is either difficult due to their inertia, or unpredictable due to chance interactions with faster changing lower scale variables, and may therefore be beyond the power of any one of these entities to address. Collective action at the global level is clearly required, but there are often conflicting national interests (usually economic) that weaken the international resolve to find solutions to global environmental problems.

One way, however, to address underlying drivers of deforestation at the national level may be to mainstream forest protection into development paths. Forest policy needs to be integrated into sustainable development. This may not be easy in early stages of development or in early stages of the forest transition curve – as the motivation to develop often comes at the cost of resource extractions from forests and forest land. The implication of this is that without substantial financial and institutional support from external actors, mainstreaming forests into forest governance in countries where a large percentage of land is under forests is unlikely.

Regardless of the degree of deforestation present in a country, for lasting solutions to be developed, it is essential to see that forests are components of larger systems of land use, which also include arable agriculture, grasslands, wetlands, and human settlements. Deforestation is only one of several major problems that humans need to grapple with in the next century, together with concomitant increases in demand for food, water and energy against a backdrop of climate change, urbanization, and limited land resources; this has been referred to as the 'perfect storm'.[75] Dealing with any one land use component (such as forests) in isolation is likely to result in partial solutions at best as the law of unintended consequences starts to operate. Providing alternative employment opportunities to reduce dependence of people on forests for their livelihoods, for example, may result in increased GHG emissions from the industrial sector.

Although REDD began as an initiative to reduce deforestation, it is becoming apparent that if the global community is serious about reducing net GHG emissions, a systems approach is needed. This systems approach should take into consideration the many and complex interactions between forestry, agriculture and urban sectors, so as to optimize the fluxes of carbon and other nutrients, as well as social and economic flows, between different parts of the landscape. In this context, some[76] even ask whether REDD funding might be effectively used for funding agricultural research to increase production on existing agricultural land, reduce losses in the food chain, and return nutrients from urban to rural areas, thereby reducing the need to clear further forests. Investing in agricultural research is a mitigation strategy in its own right – it has been estimated that the avoided emissions through agricultural intensification since the 1960s have only cost around $4/tCO$_2$e saved, cheaper than many other abatement measures.[77]

11.5 Conclusions

Has the political situation changed so much over the last 20 years that the seeming complexity of global forest governance 20 years ago has evaporated over time; can a stable forest policy for the coming 20 years be expected?

Using the politics of scale approach, this book has argued that countries have reasons to scale up (see Table 11.1) or scale down (See Table 11.2) forest governance issues. The potential promise of REDD helped to generate global consensus on the need to deal with forests at a global level (see Figure 11.1). However, if the resources are not forthcoming, this may lead countries to revert to their original position of seeing forests primarily as a subject of national sovereignty and a source of income. This book, however, makes the case for 'glocal' forest governance – a process by which local through to global issues, trends, drivers and instruments are given due attention and an iterative multi-level governance framework is developed for sustainable long-term policy that goes beyond REDD. A general principle in designing such a governance system should be in devolving decision-making down to the lowest relevant level for the issue being addressed, and developing conflict resolution mechanisms for when these decision-making processes don't align.

To the extent that REDD can provide temporary relief to the climate change problem, this book has proposed 12 generic measures that can help to embed REDD within the domestic forest institutions (see 11.4.1) and has shown how these generic measures have contextual implications in Vietnam, Indonesia, Cameroon and Peru (see 5.6, 6.6, 7.6, 8.6). Countries are gradually investing in becoming ready for REDD projects in the hope that it will develop into a win–win situation. However, there are indications that if

REDD is poorly designed and/or implemented, it may turn into a lose–lose situation (see 11.4.2). But knowledge is power, and awareness of the risks may help countries and social actors mobilize themselves to prevent such risks. Furthermore, there is a possibility that the awareness of the actual costs of implementing forest conservation equitably may lead many to see this as less than cost-effective, and thus reduce the incentive to invest in it. REDD faces considerable challenges; if these challenges are not rapidly addressed, REDD may disappear from the global agenda in its current form as suggested by the issue-attention cycle.

Its enduring legacy, however, will be that it has mobilized global attention and social actors on the need to understand forests and human–forest interactions. There is far greater understanding of the drivers of deforestation in different countries and the limits of instruments in dealing with these forests. There is a realization that there is need to take a systemic view and see if mainstreaming forests in national development, agricultural, energy and mining policy leads to understanding how best to deal with forests. The next step thus is to understand how to mainstream forests and see if countries can reconcile their need and right to develop with the need to protect forests not just for themselves, but for the myriad local ecosystem services that they provide as well as the global ecosystem services. This knowledge and the mobilization of a large number of actors may be in itself critical for creating a more comprehensive, equitable, legitimate, effective, efficient and enduring system for dealing with forests. As in the myth of Sisyphus, it will remain an uphill struggle.

Notes

1 Cock, A.R. (2008) Tropical forests in the global states system, *International Affairs* 84 (2), 315.
2 Dimitrov, R.S. (2005: 3) Hostage to norms: states, institutions and global forest politics, *Global Environmental Politics* 5 (4), 1–24.
3 Dimitrov (2005) op. cit.
4 Dimitrov (2005: 3) op. cit.
5 Gupta, J. (2012) Negotiating challenges and climate change, *Climate Policy* [in press].
6 Cock (2008) op. cit.; Brown, K. (2001) Cut and run? Evolving institutions for global forest governance, *Journal of International Development* 13 (7), 893–905.
7 Gupta, J. (2008) Global Change: Analysing scale and scaling in environmental governance, in Young, O.R., Schroeder, H. and King, L.A., *Institutions and environmental change: principal findings, applications, and research frontiers* (pp. 225–58), MIT Press.
8 Bonan, G.B. (2008) Forests and Climate Change: Forcings, Feedbacks, and the Climate Benefits of Forest, *Science* 320, 1444–9; Anderson, R.G., Canadell, J.G., Randerson, J.T., Jackson, R.B., Hungate, B.A., Baldocci, D.D., Ban-Weiss, G.A., Bonan, G.B., Caldeira, K., Cao, L., Diffenbaugh, N.S., Gurney, K.R., Kueppers, L.M., Law, B.E., Luyssaert, S. and O'Halloran, T.L. (2012) Biophysical

considerations in forestry for climate protection, *Frontiers in Ecology and the Environment* 9, 174–82.
9 FAO (2010) *Global forest resources assessment 2010*, FAO Forestry Paper 163, Rome: Food and Agriculture Organisation of the United Nations, Table 2.21, p. 45.
10 Lambin, E.F. and Meyfroidt, P. (2011) Global land use change, economic globalization, and the looming land scarcity, *Proceedings of the National Academy of Sciences* 108 (9), 3465–72.
11 Lambin, E.F. and Geist, H.J. (2006) *Land-use and land-cover change. Local processes and global impacts*, Berlin: Springer-Verlag.
12 FAO (2011) *Forest Resources Assessment Working Paper 177. Assessing forest degradation: Towards the development of globally applicable guidelines*, Rome.
13 FAO (2010) *Global Forest Land-Use Change from 1990 to 2005. Initial Results from a Global Remote Sensing Survey*.
14 Lambin and Meyfroidt (2011) op. cit.
15 Shearman, P., Bryan, J. and Laurance, W.F. (2012) Are we approaching 'peak timber' in the tropics?, *Biological Conservation*.
16 Bäckstrand, K. and Lövbrand, E. (2006) Planting trees to mitigate climate change: contested discourses of ecological modernization, green governmentality and civic environmentalism, *Global Environmental Politics* 6 (1), 50–75.
17 For example, the EU: Dimitrov, R.S. (2006: 106) *Science and International Environmental Policy: Regimes and Nonregimes in Global Governance*, Lanham, USA: Rowman and Littlefield; Reischl, G. (2009: 96) *The European Union and the international forest negotiations: An analysis of influence*, doctoral thesis, Uppsala: Swedish University of Agricultural Sciences.
18 Rosendal, G.K. (1994: 95) *The Convention on Biological Diversity and Developing Countries*, Dordrecht: Kluwer Academic Publishers.
19 McCaffrey, S.C. (2001) *The Law of International Watercourses: Non-navigational uses*, Oxford, UK: Oxford University Press.
20 Schrijver, N. (1995) *Sovereignty over natural resources: Balancing rights and duties in an interdependent world*, thesis, Groningen: Rijksuniversiteit Groningen.
21 Stockholm Declaration. Report of the UN Conference on the Human Environment, Stockholm, 5–16 June 1972; UN doc.A/CONF.48/14/Rev.1; Rio Declaration and Agenda 21. Report on the UN Conference on Environment and Development, Rio de Janeiro, 3–14 June 1992, UN doc. A/CONF.151/26/Rev.1 (Vols. I–III).
22 Dimitrov (2006: 102–3) op. cit.; Humphreys, D. (1996) *Forest politics: The evolution of international cooperation*, London: Earthscan.
23 Humphreys (1996: 95) op. cit.
24 Australia, Brazil, Canada, China, Democratic Republic of Congo, the United States, India, Indonesia, Peru and Russia.
25 Gupta (2008) op. cit.
26 Humphreys, D. (2006) *Logjam. Deforestation and the crisis of global governance*, Earthscan forestry library, London: Earthscan.
27 Lipschutz, R.D. (2001: 159) 'Why Is There No International Forestry Law? An Examination of International Forestry Regulation, Both Public and Private', *UCLA Journal of Environmental Law and Policy* 19 (1).
28 Convention on Biological Diversity (Rio de Janeiro), 5 June 1992, in force 29 December 1993, 31 ILM 1992; United Nations Framework Convention on Climate Change (New York), 9 May 1992, in force 24 March 1994; 31 ILM 1992, 822.

29 Forest principles 1992; NLBI (2007) *Non-legally binding instrument on all types of forests*, New York: adopted by the seventh session of UNFF and by the United Nations General Assembly in 2007 (A/RES/62/98), http://www.un.org/esa/forests/nlbi-GA.html
30 Nagtzaam, G.J. (2008) *The International Tropical Timber Organization and Conservationist Forestry Norms: A Bridge Too Far*, ExpressO, available at: http://works.bepress.com/gerry_nagtzaam/1; Smouts, M.C. (2008) The issue of an International Forest Regime, *International Forestry Review* 10 (3), 429–32.
31 Humphreys (2006: 13) op. cit.
32 Chaytor, B. (2001) *The development of global forest policy: overview of legal and institutional frameworks*, London, UK: International Institute for Environment and Development (IIED) and the World Business Council for Sustainable Development (WCBSD); Humphreys (2006: 213) op. cit.; McDermott, C.L., O'Carroll, A. and Wood, P. (2007) *International forest policy – the instruments, agreements and processes that shape it*, New York: Department of Economic and Social Affairs.
33 Rosenau, J.N. (2004) Strong demand, huge supply: governance in an emerging epoch, in Bache, I. and Flinders, M., *Multi-level Governance*, 31–48, Oxford: Oxford University Press; Rosenau, J.N. (2002) Governance in a new global order, in Held, D. and McGrew, A., *Governing Globalization: Power, Authority and Global Governance*, 70–86, Malden: Polity Press and Blackwell Publishing Ltd.
34 Rosenau, J.N. (2003) *Dynamic proximities: dynamics beyond globalization*, 396–7, Princeton: Princeton University Press.
35 Cashore, B., Galloway, G., Cubbage, F., Humphreys, D., Katila, P., Levin, K., Maryudi, A., McDermott, C. and McGinley, K. (2010) Ability of institutions to address new challenges, in Mery, G., Katila, P., Galloway, G., Alfaro, R.I., Kanninen, M., Lobovikov, M. and Varjo, J., *Forests and Society – Responding to Global Drivers of Change*, 441–85, Tampere: International Union of Forest Research Organizations.
36 Kingsbury, B., Krisch, N. and Stewart, R.B. (2005) The emergence of global administrative law, *Law and Contemporary Problems* 68 (15).
37 FAO (2010) *Global Forest Resources Assessment 2010*, FAO Forestry Paper 163, Rome: Food and Agriculture Organisation of the United Nations.
38 Forest-related definitions adopted by the UNCBD are available online at: http://www.cbd.int/forest/definitions.shtml
39 Sasaki, N. and Putz, F.E. (2009) Critical need for new definitions of 'forest' and 'forest degradation' in global climate change agreements, *Conservation Letters* 2 (5), 226–32.
40 van Noordwijk, M. and Minang, P.A. (2009) If we cannot define it, we cannot save it, in van Bodegom, A.J., Savenije, H. and Wit, M., *Forests and climate change: adaptation and mitigation* (pp. 5–10), Wageningen: Tropenbos International.
41 United Nations (2001) *Framework Convention on Climate Change. Report of the conference of the parties on its seventh session*, held at Marrakesh from 29 October to 10 November 2001: Addendum (at 11/CP.7 Annex.).
42 ASB Partnership for the Tropical Forest Margins (2009) If we cannot define it, we cannot save it, Policy Briefs 14.
43 Mather, A.S. (1992) The forest transition, *Area* 24 (4), 367–79.
44 Meyfroidt, P. and Lambin, E.F. (2011) Global forest transition: Prospects for an end to deforestation, *Annual Review of Environment and Resources* 36, 343–71.
45 Ongoing research work of Meine van Noordwijk and his colleagues.

46 Obidzinski, K. and Chaudhury, M. (2009) Transition to timber plantation based forestry in Indonesia: towards a feasible new policy, in *International Forestry Review* 11 (1), 79–87; Meyfroidt, P., Rudel, T.K. and Lambin, E.F. (2010) Forest transitions, trade, and the global displacement of land use, *PNAS* 107 (49), 20917–22
47 ASB Partnership for Tropical Forest Margins (2009) op. cit.
48 Clements, T. (2010) Reduced expectations: The political and institutional challenges of REDD+, *Oryx* 44 (03), 309–10.
49 Noordwijk Declaration on Climate Change, adopted by the Ministerial Conference on Atmospheric Pollution and Climate Change, held at Noordwijk in the Netherlands on 6–7 November 1989, Leidschendam : Climate Conference Secretariat.
50 IPCC (2007) *Climate Change 2007: The Physical Science Basis. Contribution of Working Group 1 to the Fourth Assessment Report of the Intergovernmental Panel on Climate Change*, Cambridge: Cambridge University Press.
51 Declaration of the Major Economies Forum on Energy and Climate, adopted at the G8 meeting held at L'Aquila in Italy on 8–10 July 2009.
52 Peters, G.P., Marland, G., Le Quéré, C., Boden, T., Canadell, J.G. and Raupach, M.R. (2012) Rapid growth in CO_2 emissions after the 2008–2009 global financial crisis, *Nature Climate Change* 2, 2–4.
53 Le Quéré, C., Raupach, M.R., Canadell, J.G., Marland, G. and et al. (2009) Trends in the sources and sinks of carbon dioxide/Global Carbon project, *Nature Geoscience*.
54 Michaelowa, A. (2012) Manouvering climate finance around the pitfalls – Finding the right policy, in Michaelowa, A. (ed.) *Carbon markets or climate finance? Low carbon and adaptation investment choices for the developing world*, Abingdon, UK: Routledge.
55 Le Quéré *et al.* (2009) op. cit.
56 Suyanto, M. E. and van Noordwijk, M. (2009) *ASB Policy Brief No. 8*: Fair and efficient? How stakeholders view investments to avoid deforestation in Indonesia, ASB Partnership for the Tropical Forest Margins, Kenya: Nairobi.
57 President Lula's address at COP-15, Copenhagen, Denmark, 17 December 2009. http://www.brasil.gov.br/para/press/speeches/president-lula-at-cop15/br_model1?set_language=en (accessed 6 June 6 2012).
58 Câmara, G. (2012) National Institute for Space Research, Brazil, presentation at Planet Under Pressure Conference, March, London.
59 Downs, A. (1972) Up and down with ecology – the issue-attention cycle, *The Public Interest* 28, 38–50.
60 Downs (1972: 39) op. cit.
61 Stern, N. (2006) *Stern review: the economics of climate change*, London: HM Treasury; Eliasch, J. (2008) *Climate change: financing global forests*, London: Office of Climate Change.
62 McDermott, C.L., Levin, K. and Cashore, B. (2011) Building the forest–climate bandwagon: REDD+ and the logic of problem amelioration, *Global Environmental Politics* 11 (3), 85–103; Clements (2010) op. cit.
63 McElwee, P. (2004) You say illegal, I say legal: the relationship between 'illegal' logging and land tenure, poverty, and forest use rights in Vietnam, *Journal of Sustainable Forestry* 19 (1–3), 97–135; Curran, B.E. (2009) Are Central Africa's protected areas displacing hundreds of thousands of rural poor?, *Conservation and Society* 7 (1), 30–45.

64 Karsenty, A. and Ongolo, S. (2012) Can 'fragile states' decide to reduce their deforestation? The inappropriate use of the theory of incentives with respect to the REDD mechanism, *Forest Policy and Economics* 18, 38–45.
65 Maya, R.S. and Gupta, J. (1996) *Joint implementation: carbon colonies or business opportunities? Weighing the odds in an information vacuum*, Harare, ZW: Southern Centre for Energy and Environment; van Asselt, H. and Gupta, J. (2009) Stretching too far – Developing countries and the role of flexibility mechanisms beyond Kyoto, *Stanford Environmental Law Journal* 28, 311.
66 Gupta, J. (2011) Climate Change: A GAP analysis based on Third World approaches to international law, *German Yearbook of International Law* 53, 341–70.
67 Chimni, B.S. (2006) Third World Approaches to International Law: A Manifesto, *International Community Law Review* 8 (1), 3–27; Khosla, M. (2007) The TWAIL discourse: The emergence of a new phase, *International Community Law Review* 9 (3), 291–304.
68 Gupta (2011) op. cit.
69 UNGA Right to Development, A/RES/41/128, 4 December 1986.
70 Stern (2006) op. cit.; Eliasch (2008) op. cit.; McKinsey & Company (2009) *Pathways to a Low-Carbon Economy: Version 2 of the Global Greenhouse Gas Abatement Cost Curve*, McKinsey & Company.
71 Maya and Gupta (1996) op. cit.
72 Gupta, J. (1997) *The Climate Change Convention and Developing Countries – From conflict to consensus?*, Dordrecht: Kluwer Academic Publishers.
73 ASB Partnership for the Tropical Forest Margins (2009) Global Survey of REDD projects: What Implications for global climate objectives? Policy Brief 12.
74 Gunderson, L.H. and Holling, C.S. (2001) *Panarchy*, Washington DC: Island Press.
75 Beddington, J. (2009) *Food, energy, water and the climate: a perfect storm of global events?*, London: Government Office for Science.
76 Matthews, R. and De Pinto, A. (2012) Should REDD+ fund 'sustainable intensification' as a means of reducing tropical deforestation?, *Carbon Management* 3 (2), 117–20.
77 Burney, J.A., Davis, S.J. and Lobell, D.B. (2010) Greenhouse gas mitigation by agricultural intensification, *Proceedings of the National Academy of Science USA* 107 (26), 12052–7.

INDEX

access issues 194–7
additionality 87, 245–6
African Timber Organization (ATO) 62
agriculture 26–33, 100–102, 121–4, 144–6, 164–70, 185–92, 201, 207–14, 230–1, 243
allocation issues 194–6
Amazon Cooperation Treaty Organization (ACTO) 62–3
Asia Forest Partnership (AFP) 62

'BASIC' group of countries 6
benefit-sharing 199–200, 247
binding forest rules 36, 42, 66, 127, 136, 243
biodiversity 7, 202, 245; *see also* Convention on Biological Diversity
biofuels 63, 211–12

Cameroon 17, 143–57, 210; community-based management 153–4; decentralization 148–50, 157; deforestation and forest degradation 144–5; forest certification 153; forest policy 147, 157; forest taxes 152–3; influence of international treaties and bodies 147–8; land rights 152; law enforcement 152; logging concessions 150–2; organizational framework 146–7; REDD 157
Cancun Agreement 4, 88
carbon: commodification 91; leakage 221, 223, 225; offset funds 38, 108, 193–4; sequestration 217, 237, 245; sinks 60, 239

carbon leakage 87, 219–25, 241, 245–7; people-based 223–4; simulations of 221–5
Central African Forest Commission (COMIFAC) 62
Central American Forest Convention (CAFC) 6
certification *see* forest certification
Clean Development Mechanism (CDM) 1, 14–15, 38, 59, 78–9, 91, 229, 240, 248, 251–2
climate change: and forests 1, 13, 229, 232, 251; governance process 3–4; physical problem of 3; political challenges 4–6
Collaborative Partnership on Forests (CPF) 56
community-based management (CBM) 41, 153–4, 194, 243
concessions 37, 122–5, 128–36, 144–7, 150–1, 155–6, 165–78, 186–201, 243–4
Congo Basin Forest Partnership (CBFP) 62
Convention on Biological Diversity (CBD) 11, 14, 58, 68–9, 238, 240, 246
Convention on International Trade in Endangered Species (CITES) 9–10, 58, 65–7
Copenhagen Accord (2009) 4, 6, 250
corporate social responsibility (CSR) 39, 43, 66, 109, 111, 131, 136, 153, 156, 175, 178

debt-for-nature swaps 38–9, 131
decentralization 105–6, 127, 148–50, 157, 202, 244

deforestation 2, 9, 71, 199–200, 207–14, 218–19, 225; baseline scenario 213–14; cost of avoiding emissions 218–19; definitions 15–16, 210; drivers 29–34, 185–7, 232, 244, 252, 254; future prospects 210–14
displacement effects from forest conservation 220–1
downscaling of forest issues 233–8, 253
DPSIR (Drivers-Pressures-State-Impacts-Responses) methodology 17
drivers: global 44, 232, 235, 238, 243–4; local 2, 188, 192, 232, 244; national 234, 236; proximate 29–31, 44, 102, 107, 111–12, 123–4, 136, 145, 156–7, 164–5, 171, 178, 186, 199, 202, 244, 247, 252; underlying 18, 29–32, 44, 101–2, 107, 111–13, 123–4, 136–7, 145, 156–7, 165, 171, 178, 187, 200–12, 247, 252
Dynamic GTAP model 209

Earth Systems Governance Project 17
ecological modernization 197–8, 251
economic instruments *see* instruments
ecosystem approach to forest governance 65, 239
ecosystem services of forests 7–8, 13, 60–1, 239; *see also* payment for ecosystem services
ecotourism 39
education 135, 153, 194
enforcement 34–7, 88, 122, 126–30, 135, 155–6, 166, 172, 174, 190–3; *see also* law enforcement
Environmental Kuznets Curve 4–5, 25, 27–8
extraction 29–32, 42–3, 100, 102, 123–5, 129, 137, 145–8, 165–6, 171–6, 186, 190, 194, 201, 214, 230, 243–4, 249, 252
extraterritorial impacts 70, 111, 125, 136, 156, 178, 235

food: demand 211; global market 215–16
Food and Agriculture Organization (FAO) 12, 57, 69, 239
Forest Carbon Partnership Facility (FCPF) 82–3

forest certification 40, 56, 68, 109, 131–2, 153, 175–6, 194, 197
forest concessions 129, 150–2, 172–4, 190–2
forest conservation 245; costs 225; displacement effects 220–1; economic effects 216; induced by REDD 214–19
forest cover 9, 25–32, 61, 69, 82, 85, 99, 106–7, 111, 121, 123, 129, 143–4, 163, 185, 189, 221, 233, 239, 243–5
forest definitions 239–40
forest degradation 9; definition 210; drivers 185–7, 232, 244
forest governance 9–11; global 237–8; 'glocal' 231–8, 253; institutions 52–71; national 241–5; phases 77–80, 230–1; regional 62, 167–70
Forest Law Enforcement and Governance (FLEG) initiative 62–3, 246
forest management: policy instruments 33–5, 41–4; *see also* community-based management
Forest Principles 64–5
Forest Stewardship Council (FSC) 56, 239
forest transition: concept and theory 17, 25–9, 44, 241–3; curve 27–8, 33–4, 85, 199, 241–3, 252
forests: and climate change 1, 13, 229, 232, 251; current status 7–10; 'mainstreaming' 252–4; political challenges 12–13

gender-related issues 195–6
Global Carbon Project 248
Global Environment Facility (GEF) 38, 58

Harmon Doctrine 234

income distribution 211
indigenous peoples: policies 59–60; REDD 89–90
Indonesia 17, 121–37, 210; community-based management 132; debt-for-nature swaps 131; decentralization 127; deforestation and forest degradation 122–4, 135–7;

Index   261

forest certification 131–2; forest policy 125, 127–8, 135–7; influence of international policies and bodies 125–6; land rights 129–30; logging concessions 129; NGO-based management 133; organizational framework 124–5; REDD 133–5; reporting, monitoring and enforcement arrangements 130–1; spatial planning 128–9; trade restrictions 126–7
infrastructure 26, 30–2, 42, 67, 100, 102, 123–4, 128, 136, 145, 149, 157, 164–6, 173, 176, 186, 208, 243
institutional design 17, 247
Institutional Dimensions of Global Environmental Change (IDGEC) 16, 209
instruments: economic 69, 136–7, 201, 239, 244, 250; regulatory 34–8, 64–7, 244; suasive 40, 68–9, 202, 244
Intergovernmental Panel on Climate Change (IPCC) 3–4, 13, 248
International funds 38, 43, 67, 244
International Human Dimensions Programme 16
International Labour Organization (ILO) 59, 67
International Monetary Fund (IMF) 57
International Tropical Timber Agreement (ITTA) 10, 54
International Tropical Timber Organization (ITTO) 54, 56, 68–9, 210, 214, 230, 239

Johannesburg Summit on Sustainable Development (2002) 62
joint management 43

Kyoto Protocol 1, 4, 6, 11, 14, 58–9, 69, 78–9, 229

Lacey Act (US, 1900) 35, 63, 106, 111, 188–9, 191, 234
land, future demand for and supply of 212–13
land grabbing 195, 234, 250
land rights 129–30, 152, 174, 192, 202

land tenure 40, 91, 103, 106, 111–12, 129–36, 152, 174, 178, 192, 194, 202, 244, 250
land use, land-use change and forestry (LULUCF), emissions from 78–9
land-use changes, baseline scenario for 213–14
law enforcement 42, 53, 57, 62–3, 106, 111, 126, 133, 136, 145–7, 152, 188, 197, 243–4
leakage from emissions control *see* carbon leakage
logging 10–11, 15, 29–33, 40, 42, 54–5, 62–6, 71, 101–7, 111–12, 122, 126, 129–36, 144–57, 164–6, 169, 172–8, 185–94, 209, 221, 223, 238

management measures 40
Marrakesh Agreement (2001) 1, 14, 79
micro-credit 38, 197
Millennium Development Goals 40
Millennium Ecosystem Assessment 7
mining 29–33, 67, 89, 122–4, 134–6, 144–5, 150–1, 155, 157, 165, 168, 171, 186–7, 193, 201–2, 223–4, 230, 244–5, 254
monitoring, reporting and verification (MRV) systems 2, 86, 88, 245–9

national forest programmes (NFPs) 67
national protected areas *see* protected areas
nationally appropriate migration actions (NAMAs) 4, 6, 81, 245
NGO-based management 43, 133, 136, 176, 178
Non-Legally Binding Instrument on All Types of Forests (NLBI) 55, 68–9
Noordwijk Declaration (1989) 13, 78
North–South divide 5, 250–2

official development assistance (ODA) 248
opportunity costs of forest conservation 214–15

payment for ecosystem services (PES) 37–8, 107–8, 175, 197, 203, 244
permanence concerns 86–7, 245–6

262   Index

Peru 163–80, 210; deforestation and forest degradation 164–6; forest certification 175–6; forest policy 168–72, 179; influence of international treaties and bodies 169; land rights 174; logging concessions 172–4; NGO-based management 176; organizational framework 167–8; payment for ecosystem services 175; REDD 176–80
Programme for the Endorsement of Forest Certification 56
property rights 29–31, 37, 59, 106, 170, 176
protected areas (PAs) 8, 13, 32, 36–7, 40, 42, 61, 66–7, 70, 88, 100, 134, 136, 149–51, 154–7, 167–72, 176–8, 190–1, 196, 198, 235, 238, 244–6

Ramsar Convention on Wetlands 10, 58, 67, 238
REDD (Reducing Emissions from Deforestation and Forest Degradation) 1–2, 6, 13–15, 20, 44–5, 52, 56, 60, 68, 229–30, 237; basic premise 77; challenges 84–8, 92, 254; climate change 250; emergence 80–4; financing 85–6, 245, 248, 253; impact 88, 214–19, 224–5; leakage 220; practical options for implementation 245–7; readiness 253–4; safeguards 87–8, 245–6
reference levels 84–5, 245, 247
reforestation 9–15, 27, 36, 42, 44, 59, 66, 78–9, 82, 104–7, 112–13, 146, 149, 168, 172, 186, 191–4, 197, 199, 213, 229, 241–4, 248
regulatory instruments *see* instruments
rents 207–8
Rio Declaration on Environment and Development 14, 64, 66

scale, politics of 231–9, 245, 253
soft law 40, 70, 239
sovereignty 12–15, 20, 55, 64, 67–70, 234–7, 253; over natural resources 234, 237
spatial planning 36, 42, 66, 128–9, 148, 170, 188–91, 196–7, 202, 244, 250

suasive instruments *see* instruments
subsidies 32, 38, 43–4, 66, 104–7, 111, 113, 145, 193, 197, 243–4, 250
sustainable forest management (SFM) 10–13, 36, 43, 54–5, 65, 69–71, 106, 109, 113, 126, 148, 150, 153, 157, 169, 189–92, 198, 202, 238–9

'take-off' point in economic development 25
tariffs 38, 43, 66, 68
taxes 38, 43, 57, 66, 145, 152–3, 156, 169, 190–3, 208, 243–4
third world approaches to international law (TWAIL) 17, 250–1
trade liberalization 68, 193, 230
trade restrictions 34, 42, 65–6, 126–7, 188, 191, 244
Tropical Forest Action Plan (1987) 10–11

United Nations Conference on Environment and Development (UNCED) (1992) 11, 52–5, 64; *see also* Rio Declaration on Environment and Development
United Nations Convention to Combat Desertification (UNCCD) 59
United Nations Declaration on the Rights of Indigenous Peoples (2007) 90
United Nations Development Programme (UNDP) 57
United Nations Environment Programme (UNEP) 57–8
United Nations Forum on Forests (UNFF) 11, 54–5, 69
United Nations Framework Convention on Climate Change (1992) (UNFCCC) 1, 4–6, 11, 13, 78, 240
United Nations Permanent Forum on Indigenous Issues (UNFII) 59
upscaling of forest issues 233–8, 253
Uruguay Round 68

Vietnam 17, 99–113; carbon offset funds 108; decentralization 105–6;

deforestation and forest degradation 100, 112–13; forest certification 109; forest policy 103–5, 112–13; land rights 106; law enforcement 106; organizational framework 101–3; payment for ecosystem services 107–8; REDD 109–13; subsidies 106–7

Von Thünen land allocation model 207–10, 218, 225

Western Hemisphere Convention 63

World Bank 57, 67, 238–9; carbon financing 82–3

World Heritage Convention (1975) (WHC) 10, 58, 67

World Trade Organization (WTO) 38, 59, 68, 193, 238